Systems Modelling for Energy Policy

Systems Modelling for Energy Policy

Edited by
Derek W. Bunn
London Business School
and
Erik R. Larsen
University of Bologna

JOHN WILEY & SONS
Chichester · New York · Weinheim · Brisbane · Singapore · Toronto

Baffins Lane, Chichester,
West Sussex PO19 1UD, England

National 01243 779777
International (+44) 1243 779777
e-mail (for orders and customer service enquiries): cs-books@wiley.co.uk
Visit our Home Page on http://www.wiley.co.uk
or http://www.wiley.com

Other Wiley Editorial Offices

John Wiley & Sons, Inc., 605 Third Avenue,
New York, NY 10158-0012, USA

VCH Verlagsgesellschaft mbH
Pappelallee 3
D-69469 Weinheim
Germany

Jacaranda Wiley Ltd, 33 Park Road, Milton,
Queensland 4064, Australia

John Wiley & Sons (Canada) Ltd, 22 Worcester Road,
Rexdale, Ontario M9W 1L1, Canada

John Wiley & Sons (Asia) Pte Ltd, 2 Clementi Loop #02-01,
Jin Xing Distripark, Singapore 129809

Library of Congress Cataloging-in-Publication Data

Bunn, Derek W.
Systems modelling for energy policy / Derek W. Bunn and Erik R. Larsen.
p. cm.
Includes bibliographical references and index.
ISBN 0-471-95794-1 (hardcover)
1. Energy policy—Mathematical models. 2. System analysis.
I. Larsen, Erik R.
HD9502.A2B86 1997
333.79'01'5118—dc21 96–47600
CIP

British Library Cataloguing in Publication Data

A catalogue record for this book is available from the British Library

ISBN 0-471-95794-1

Typeset in 10/12 pt Times by Mathematical Composition Setters Ltd., Salisbury, Wiltshire.
Printed and bound in Great Britain by Bookcraft (Bath), Midsomer Norton, Somerset.
This book is printed on acid-free paper responsibly manufactured from sustainable forestation, for which at least two trees are planted for each one used for paper production.

Contents

List of Contributors		vii
Chapter 1	**Systems Modelling for Energy Policy** *Derek W. Bunn and Erik R. Larsen*	1
Chapter 2	**The National Energy Modelling System: Policy Analysis and Forecasting at the US Department of Energy** *Andy S. Kydes and Susan H. Shaw*	9
Chapter 3	**The IDEAS Model and Its Use in Developing the US Climate Change Action Plan** *Frances P. Wood and Jay C. Geinzer*	31
Chapter 4	**MIS: A Model-based Macroeconomic Information System for Energy Analysis in Germany** *Claudia Kemfert and W. Kuckshinrichs*	47
Chapter 5	**Benefits from Electricity Trade in Northern Europe under CO_2 Constraints** *Tomas Larsson*	67
Chapter 6	**Energy Planning and Multi-level Optimization: The Australian Energy Planning System Optimization Modelling Study** *Sardar M.N. Islam*	79
Chapter 7	**Application and Limitations of Annual Models for Electricity Capacity Development** *Poul Erik Grohnheit*	89

Chapter 8	**Energy–Economy–Environmental Interactions in a General Equilibrium Framework: The Case of Turkey** *Y. Arikan, Ç. Güven and G. Kumbaroglu*	117
Chapter 9	**The Economics of the CO_2 Problem: What About the Supply Side?** *Jürgen E. Blank and Wolfgang J. Ströbele*	141
Chapter 10	**Exploring Oil Market Dynamics: A System Dynamics Model and Microworld of the Oil Producers** *John D.W. Morecroft and Brian Marsh*	167
Chapter 11	**Diary of an Oil Market Model: How a System Dynamics Modelling Process Was Used with Managers to Resolve Conflict and to Generate Insight** *David C. Lane*	205
Chapter 12	**The Changing Role of Simulation Models: The Case of the Pacific Northwest Electric System** *Andrew Ford*	241
Chapter 13	**A Systems Simulation Platform to Support Energy Policy in Colombia** *Isaac Dyner and Derek W. Bunn*	259
Chapter 14	**Preparing for a Competitive Environment: Developing Strategies for America's Electric Utilities** *James M. Lyneis*	273
Chapter 15	**Complementary Modelling Approaches for Analysing Several Effects of Privatization on Electricity Investment** *Derek W. Bunn, Erik R. Larsen and Kiriakos Vlahos*	303
Index		327

Contributors

Y. ARIKAN
Middle East Technical University, Industrial Engineering Department, Ankara, Turkey

JÜRGEN E. BLANK
Westf. Wilhelms-Universität Münster, Lehrstuhl für Volkwirtschaftstheorie, Universitätsstr. 14–16, D-48143, Müster, Germany

DEREK W. BUNN
London Business School, Sussex Place, Regent's Park, London NW1 4SA, UK

ISAAC DYNER
Facultad de Minas, Universidad National de Colombia, AA 1027 Medellin, Colombia

ANDREW FORD
Program in Environmental Science and Regional Planning, Washington State University, Pullman, WA 99164-4430, USA

JAY C. GEINZER
AES Consulting, Inc., 1001 North 19th Street, Arlington, VA 22209, USA

POUL ERIK GROHNHEIT
Risø National Laboratory, Systems Analysis Department, Postbox 49, DK-4000 Roskilde, Denmark

Ç. GÜVEN
Middle East Technical University, Industrial Engineering Department, Ankara 06531, Turkey

SARDAR M.N. ISLAM
Sustainable Economic Growth Program, Centre for Strategic Economic Studies,

Victoria University of Technology, PO Box 14428 MMC, Melbourne, Victoria 8001, Australia

CLAUDIA KEMFERT
Working Group Energy and System Modelling, Oldenburg University, PO Box 2503, D-26111 Oldenburg, Germany

W. KUCKSHINRICHS
Programme Group Systems Analysis and Technology Evaluation, Research Centre Jülich, D-52425 Jülich, Germany

G. KUMBAROGLU
Middle East Technical University, Industrial Engineering Department, Ankara 06531, Turkey

ANDY S. KYDES
Energy Information Administration, US Department of Energy, 1000 Independence Avenue, SW, Washington, DC 20585, USA

DAVID C. LANE
The London School of Economics and Political Science, Houghton Street, London WC2A 2AE, UK

ERIK R. LARSEN
Dipartimento di Discipline Economico-Aziendali, University of Bologna, Piazza Scaravilli, 2, 40126 Bologna, Italy

TOMAS LARSSON
Energy Systems Technology, Chalmers University of Technology, S-412 96 Göteborg, Sweden

JAMES M. LYNEIS
Pugh-Roberts Associates, 41 William Linskey Way, Cambridge, MA 02142, USA

BRIAN MARSH
St Andrews Management Institute, 3 St Mary's Place, St Andrews, Fife KY16 9UY, UK

JOHN D.W. MORECROFT
London Business School, Sussex Place, Regent's Park, London NW1 4SA, UK

SUSAN H. SHAW
Energy Information Administration, US Department of Energy, 1000 Independence Avenue, SW, Washington, DC 20585, USA

WOLFGANG J. STRÖBELE
Westf. Wilhelms-Universität Münster, Lehrstuhl für Volkwirtschaftstheorie, Universitätsstr. 14–16, D-48143, Müster, Germany

KIRIAKOS VLAHOS
London Business School, Sussex Place, Regent's Park, London NW1 4SA, UK

FRANCES P. WOOD
On Location, Inc., 8100 Oak St, Dunn Loring, VA 22027, USA

Chapter 1

Systems Modelling for Energy Policy

Derek W. Bunn and Erik R. Larsen

INTRODUCTION

The use of systems models has become increasingly popular in the energy sector to investigate energy–economy interaction, regulatory policies, conservation, environmental issues, strategic competitive behaviour and, more recently, the impact of privatization. The energy sector is one where there has always been a serious modelling tradition, and, in particular, the concerns about global warming that have emerged in the 1990s have certainly stimulated an intense and methodologically diverse level of analytical activity in this respect, aimed ultimately at greater international coordination in policy and planning. At the same time we are also witnessing a worldwide trend away from planned national energy policies to new market structures for the utilities, such that the risks, uncertainties and competitive aspects that need to be incorporated into energy market analyses pose modelling challenges of a different sort than before, in order to evaluate imperfect competition within new regulatory regimes. Consequently, with these two strong, and apparently countervailing, themes dominating energy policy in the current decade, it is therefore quite opportune to take stock of recent practice in the development

Systems Modelling for Energy Policy. Edited by D.W. Bunn and E.R. Larsen

and application of systems methods in this wider context. That is the motivation behind this collected volume.

Traditionally, energy models have developed as large econometric or optimization-based systems of modules to serve the needs of evaluating the impact of different policies in areas such as pricing, economic growth, environment impact and national resources. The need to take a broad system-wide perspective in the analysis arises from the long time scales involved, the broad economy-wide and environmental impacts, the inherited capital and technological stock providing sub-optimal infrastructures, and the extensive potential for substitution on both the supply and demand sides. For example, in looking at alternatives for a new electricity generating plant, a comparison of alternative costs cannot be made without some assumptions for each plant's pattern of operation (load factor), which in turn can only be properly evaluated in the context of a full, multi-period, plant-by-plant model of the electricity generating and transmission system, which might involve competitive power pools effectively scheduling the generation. Optimization models such as EGEAS (EPRI, 1982) and ECAP (Bunn & Vlahos, 1989) were developed for electricity planning with this in mind, and, more generally, models such as MARKAL and its derivatives have been widely used to optimize across several energy sectors (Zhang, 1992).

Although such large-scale optimization models are rich in detail, at a level that is often necessary for investigating such issues as conservation measures, system security and regional disparities, they generally take as exogenous assumptions a number of variables that in a broader perspective should be evaluated as part of the model. Demand–price interactions and feedback, for example are difficult to represent in these models, and, indeed, a whole macroeconomic set of assumptions generally surrounds their specification. In a sense, while they are system-wide in detail, they can be relatively unsystematic in scope. It is therefore not surprising that recent research has been as much directed to linking with macroeconomic models as in developing new techniques for handling larger and more detailed specifications. The MARKAL–MACRO link (Manne & Wene, 1992), for example, uses a simplified economic model (MACRO) in order to give the MARKAL analysis a longer term framework and mutual feedback between demand, prices and substitution. Clearly, as in all modelling applications, the appropriateness of different techniques, with different balances of detail and scope, depends very much upon the focus of analysis under consideration, and in this volume, a number of useful MARKAL and similar optimization-based applications are described quite effectively.

On the other hand, econometric methods using large sets of macroeconomic equations can provide the broader scope of energy–economy interaction, and determine demand and prices more endogenously. As macroeconometric modelling has evolved to provide the basis of most governments' policy simulations, it is inevitable that the energy sector is becoming more completely represented within them (see, for example, Italianer, 1986). However, the energy representations are still typically highly aggregated and, because the methodology relies upon time-series estimation, it is not robust to periods of rapid structural changes.

As far as representing more detail is concerned, input–output analysis is an alternative, long-standing, econometric technique that can handle more subsectors and explicitly focus upon sectorial interactions. It has been the basis of energy modelling for many years, despite some limitations (see Pearson, 1989; Proops, Faber & Wagenhals, 1993). Fixed parameters with respect to substitution and returns to scale, as well as exogenously determined demand, limit its longer term application, however.

For the longer term, the use of Computable General Equilibrium (CGE) approaches has become quite popular among econometricians. Set up as a system of supply–demand equations for all of the markets involved in the specification, the model is solved endogenously for prices, demand and substitution. As such, it is useful for long-term analyses of future socioeconomic–environmental scenarios, but is not without its criticisms (see Borges, 1986). Based upon traditional microeconomic principles of market clearing, it is more difficult to incorporate imperfect competition and disequilibrium than in more conventional econometric methods. It also relies upon calibration, rather than estimation, for its validity. Most of the assumptions regarding elasticities etc. have to be determined outside the model and used as inputs, rather than being estimated as parameters in a statistical model. This is not a serious problem, but does set it aside from more traditional econometric practice.

This issue of validation through calibration, rather than estimation, is also shared with the use of system dynamics modelling in this context. Here, systems of behavioural equations are specified for all of the decision making agents described in the model, together with feedback links between variables over time. Scope, rather than detail, has generally been seen as their strength, and although they have been used for energy policy for many years (see Naill, 1992), their ability to take a strategic perspective to imperfect competition has made them particularly attractive for applications to restructured and privatized energy markets (Bunn & Larsen, 1992). These are sometimes just small models which provide strategic insight and understanding on a particular aspect of policy, or which have been used for training and scenario building, without necessarily seeking to model the energy sector in any comprehensive way. Indeed, the issue of facilitating the usefulness of system dynamics modelling, its behavioural scope and flexibility, with the scale, detail and authenticity of more conventional energy planning models is still an active research theme (see Ninios, Vlahos & Bunn, 1995). Nevertheless, because of its rising applicability to energy policy, where strategic competition is an important new element in the analysis, several different examples of its use are included in this volume.

PREVIEW

The fourteen contributions to this volume exemplify most of the approaches and issues indicated above. Most were prepared specially for this volume and some were presentations made at the Energy Modelling Symposium, held at the London Business School in July 1995. Practical modelling issues can be compared across the broad range of approaches discussed, together with the scope of policy issues they support.

In Chapter 2, "The National Energy Modeling System: Policy Analysis and Forecasting at the US Department of Energy", by Andy S. Kydes and Susan H. Shaw, we have the background for the development of the National Energy Modeling System (NEMS) by the Energy Information Administration within the US Department of Energy. This is a very large, modular system of sectorial energy models. The authors describe the motivation for a new national energy model, the structure of NEMS and the requirements for modularity and full documentation to disseminate the model within broader energy and academic circles. The chapter also describes the main modules of the NEMS model and concludes with a discussion of the use of the model for energy policy.

National energy modelling in the US continues as the theme in Chapter 3 with "The IDEAS Model and its Use in Developing the US Climate Change Action Plan", by Frances Wood and Jay Geinzer, but, for contrast, quite a different modelling platform is described. Here the use of system dynamics, rather than optimization, provides the underlying methodology. The Integrated Dynamic Energy Analysis Simulation (IDEAS) model has had a long history of providing policy support for long-term investigations of supply–demand balances, with a particular strength in the area of conservation. It is large scale, and the authors discuss some aspects of its development over 15 years. The recent application to strategies for the reduction of greenhouse gases is then described.

In Chapter 4, "MIS: A Model-Based Macroeconomic Information System for Energy Analysis in Germany", by C. Kemfert and W. Kuckshinrichs, we have a further modelling variation applied to the greenhouse issue. They describe a dynamic input–output model, an interesting feature of which is that it allows the user to check for economic consistency in traditional bottom-up models, like MARKAL and IKARUS-LP, within economic growth scenarios. The authors use the MIS model to look at a scenario for the introduction of a CO_2 tax in Germany. This scenario is then compared to a "business-as-usual" scenario to analyse the consequences of such a tax. The conclusion is that high reduction targets can be reached but at various macroeconomic prices, which the multisectorial nature of the input–output framework is able to evaluate in some detail.

Closely related to this is the application presented in Chapter 5, "Benefits from Electricity Trade in Northern Europe under CO_2 Constraints", by Tomas Larsson. He uses an extended version of the IEA-MARKAL model to analyse the implications for CO_2 reductions of electricity trade between the Scandinavian countries. The author starts by setting up a model of trade between the countries and discusses what results one can obtain from the use of IEA-MARKAL, which is a multi-period linear programming model of the full energy system from production to end use. The results are discussed in detail using Norway as an example, and it is shown how trade can be used to reduce the cost of reductions in CO_2 emissions in Europe.

The link between optimization and macroeconomic modelling continues as a methodological theme in Chapter 6 with a presentation of the "Energy Planning and Multi-Level Optimization: The Australian Energy Planning System Optimization Modelling Study", by Sardar Islam. The multi-level optimization framework links government policy with the behaviour of economic agents. Some aspects of its solution technique are indicated, as are various Australian policy issues which have been illuminated by its application.

In Chapter 7, "Application and Limitations of Annual Models for Electricity Capacity Development", Poul Grohnheit describes the structure, use and methodological issues associated with the European Commission's Energy Flow Optimization Model (EFOM). The author uses Denmark as an example to highlight the assumptions in the largely optimization-based framework to show a typical set of policy recommendations from the model. The second part of the chapter shows how the EFOM model can be used to analyse emergent competition in the electricity sector, again using Denmark as an example. First, there is a discussion of a case with only two competitors in the market. This is then extended to a case where there is a market with several competitors, and the whole issue of extending this approach to more restructured markets is finally raised.

The use of an ETA-MACRO type of approach is used in Chapter 8, "Energy–Economy–Environmental Interactions in a General Equilibrium Framework: The case of Turkey", by Y. Arikan, Ç. Güven and G. Kumbaroglu. This develops an integrated model of the economy, energy policy and environmental factors, and the model allows for feedback between the different sectors. They argue that a model needs to integrate at this level, when the consequences of economic policy for energy, and energy policy for the environment and economics have to be evaluated. Applying the framework to Turkey, they conclude that Turkey can reduce emissions considerably without slowing economic growth, if the government intervention is done carefully.

A critical evaluation of the CGE class of models continues in Chapter 9, "The Economics of the CO_2 Problem: What about the Supply Side?" The authors, J.E. Blank and W.J. Ströbele, compare three CGE models of energy supply and demand, and their application to scenarios over the next 100 years, in order to evaluate the CO_2 problem. In considering the introduction of a carbon tax, the conclusion they reach is that the modelling of the demand side generally is at a good level, but that the modelling of the supply side, especially strategic matters, e.g. a possible reaction from the oil and coal producing countries, has several shortcomings within all three models. They conclude that these shortcomings need to be addressed to get a better understanding of the impact of carbon taxes.

Chapter 10, "Exploring Oil Market Dynamics: A System Dynamics Model and Microworld of the Oil Producers", by John Morecroft and Brian Marsh, picks up this issue directly by presenting a system dynamics feedback model of the oil market. The authors describe how the model was developed with one of the major oil companies focusing upon the feedback within and between the different sectors. They then continue to show the development and use of the model as a "management flight simulator" to facilitate strategic thinking. The chapter ends with a discussion of how useful the model/game was within the company.

David Lane provides a complementary paper to this in Chapter 11, "Diary of an Oil Market Model: How a System Dynamics Modelling Process Was Used with Managers to Resolve Conflict and to Generate Insight". Again in the context of the oil industry, and focusing upon strategic issues, this chapter describes a process of using system dynamics modelling with a management team. The application investigates the relationship between output and cash generation in a case where managers involved in refining and trading had conflicting views. The author uses a diary format to recreate the process of how the model evolved through meetings with the managers. The last part of the paper is a reflection on the longer term impact of the work within the organization.

Chapter 12, "The Changing Role of Simulation Models: The Case of the Pacific Northwest Electricity System", by Andrew Ford, continues many of these system dynamic modelling issues and also, rather like Chapter 2, reflects upon an approach developed over many years to support long-term policy making. The chapter describe the development and use of the Conservation Policy Analysis Model (CPAM) for the Bonneville Company in the Pacific Northwest. The models are based on feedback modelling and the author describes how the CPAM models are designed to help the end users to evaluate the effect of conservation measures. As well as the main parts and the structure of the model, there is a discussion of how better value can be gained from simulation models through turning them into "management games".

One of the issues associated with strategic modelling is whether to develop a small behavioural model focusing upon just one well-defined policy problem, or to develop a capability for strategic modelling upon a more general model of the energy system under consideration. The latter situation then raises the task of managing a larger, less focused model. In Chapter 13, "A Systems Simulation Platform to Support Energy Policy in Colombia", by Isaac Dyner and Derek Bunn, the use of a system dynamics platform is considered from this modelling perspective, and applied to the current dilemma in Colombia of pursuing market liberalization initiatives at the same time as maintaining a commitment to integrated energy planning.

Chapter 14, "Preparing for a Competitive Environment: Developing Strategies for America's Electric Utilities", by J.M. Lyneis, also describes a generic feedback model which can be used by utilities to understand the consequences for their future finances of major structural changes in their markets. These include the regulatory process as well as new competition, and have implications for corporate downsizing and mergers. The model has been used with a number of utilities over the past decade with apparently good results.

The final chapter, "Complementary Modelling Approaches for Analysing Several Effects of Privatization on Electricity Investment", by Derek Bunn, Erik Larsen and Kiriakos Vlahos, comes back to one of the main themes that has run through the volume. This is the issue of using different modelling techniques in a coordinated way, which has been referred to several times, whether at an economy-wide level (e.g. system optimization and macro-economic modelling) or, as in this case, linking the traditional power of optimization to understand multi-period investment technology choice in some detail, with the scope of system dynamics to model strategic behaviour in new market regimes. The application area relates to industry restructuring, with corporate and regulatory behaviour being important independent elements to consider.

Overall, the volume provides a representative cross-section of many of the modelling challenges exercising analysts in the late 1990s. Not only are there technical challenges in handling large models using different techniques in order to become more comprehensive, but there are strategic challenges in maintaining focus on key industry restructuring issues, for which past data is not available, and to which smaller, policy-specific behavioural models may give greater insight. These issues are likely to be with us for some time.

ACKNOWLEDGEMENTS

The authors are grateful for the support of the London Business School, the Danish Social Science Research Council, the International Federation of Operational Research Societies, for helping to facilitate a symposium where some of the contributions originated, and the Operational Research Society of the UK for permission to republish the paper upon which Chapter 15 is based.

REFERENCES

Borges, A.M. (1986) Applied general equilibrium models: an assessment of their usefulness for policy analysis, *OECD Economic Studies* no. 7, 7–43.

Bunn, D.W. and Larsen, K. (1992) Sensitivity of reserve margin to factors influencing investment in the electricity market of England and Wales, *Energy Policy* **29**, 420–429.

Bunn, D.W. and Vlahos, K. (1989) Evaluation of the long-term effects on UK electricity prices following privatisation, *Fiscal Studies* **10**, 104–116.

EPRI (1982) EGEAS, report EL-2561, Electric Power Research Institute, 3412 Hillview Avenue, Palo Alto, CA 94304, USA.

Italianer, A. (1986) The HERMES model: complete specification and first estimation results, Commission of the European Communities, Brussels.

Manne, A.S. and Wene, C.-O. (1992) MARKAL–MACRO: a linked model for energy-economy analysis, BNL-47161, Brookhaven National Laboratory, Upton, New York.

Naill, R. (1992) A system dynamics model for national energy policy and planning, *System Dynamics Review* **8**, 1–19.

Ninios, P., Vlahos, K. and Bunn, D. (1995) Industry simulation: system modelling with an object-orientated/DEVS technology, *European Journal of Operational Research* **81**, 521–534.

Pearson, P.J.G. (1989) Proactive energy–environment policy strategies: a role for input–output analysis? *Environment and Planning A* **21**, 1329–1348.

Proops, J.L.R., Faber, M. and Wagenhals, G. (1993) *Reducing CO_2 Emissions: A Comparative Input–Output Study for Germany and the UK*, Springer-Verlag, Berlin.

Zhang, Zhong Xiang (1992) The Markal model, Department of General Economics, Landbouwuniversiteit, Wageningen, The Netherlands.

Chapter 2

The National Energy Modelling System: Policy Analysis and Forecasting at the US Department of Energy*

Andy S. Kydes and Susan H. Shaw

INTRODUCTION

The Energy Information Administration (EIA) of the US Department of Energy (DOE) has responsibility for the collection, analysis, and publication of data related to energy and energy's relationship to the economy. Part of the mission of EIA, as stated in the legislation creating DOE in 1977, includes the

*This product was prepared as an account of work sponsored by an agency of the United States Government. Neither the United States Government nor any agency thereof, nor any of their employees, makes any warranty, express or implied, or assumes any legal liability or responsibility for the accuracy, completeness, or usefulness of any information, apparatus, product, or process disclosed, or represents that its use would not infringe on privately owned rights. Reference herein to any specific commercial product, process, or service by trade name, trademark, manufacturer, or otherwise does not necessarily constitute or imply its endorsement, recommendation, or favoring by the United States Government or any agency thereof. The views and opinions of the authors expressed herein do not necessarily state or reflect those of the United States Government or any agency thereof.

Systems Modelling for Energy Policy. Edited by D.W. Bunn and E.R. Larsen
Published 1997 John Wiley & Sons Ltd.

mandate to produce an annual report of long-term energy trends. Since 1974, EIA and its predecessor organization, the Federal Energy Administration, have fulfilled this mandate with a series of computer modeling systems representing domestic energy–economy markets and the projected trends.

These modeling systems have been used by EIA on an annual basis to produce the baseline projections of domestic energy markets published in a series of reports, currently titled the *Annual Energy Outlook*. The models are also used to analyze the potential impacts of proposed government policies and of alternative energy market assumptions in Service Reports and special studies prepared in response to requests by policy analysts and decision makers, including the DOE Office of Policy, other DOE offices, the US Congress and other government agencies.

In 1974, the Federal Energy Administration developed the Project Independence Evaluation System (PIES), the first large-scale integrated model of domestic energy markets (Hogan, 1975). PIES was first used to produce projections of energy production, consumption, imports, and prices for the *Project Independence Blueprint*, addressing the specific issue of the feasibility of the United States obtaining energy independence by 1980. For the series of annual reports that was subsequently initiated, PIES was later modified and updated to incorporate more complex market and regulatory features than was possible at first. Later renamed the Midterm Energy Forecasting System, PIES was used for a number of studies of energy policies, including proposed tax structures, production incentives, and regulations, particularly as applied to natural gas and crude oil resources.

With interest in energy policy issues evolving to issues requiring greater temporal definition, greater macroeconomic feedback, and greater modularity, PIES was replaced in 1982 by a newly developed modeling system, the Intermediate Future Forecasting System (IFFS) (Murphy et al., 1988). Typical of the emerging issues that PIES could not satisfactorily address were transitional costs of natural gas deregulation and other regulatory provisions. In addition, by solving for snapshots of energy markets in discrete future years, PIES could not represent other annual transitions, such as reserve accounting of oil and gas or financial accounts. Less cumbersome than PIES, IFFS was planned with a more flexible, modular design and operated by solving in a year-by-year mode. IFFS was used for the *Annual Energy Outlook* through 1993.

From 1989 to 1991, DOE prepared the National Energy Strategy (NES), which contained long-term policy recommendations and analyses. The DOE Office of Policy, Planning, and Analysis used a collection of existing models to support the development of the NES, including IFFS. However, IFFS alone could not support the analysis of many energy issues of the 1990s. Typical among these were issues of new technology penetration, including "learning-by-doing" and "technological optimism", inability to insure consistency between the international oil markets and the US oil market, the impact of integrated resource planning and demand-side management programs, and the role of non-utility generation in the emerging electricity market. Within the Department, the need was identified for a new comprehensive modeling system. This effort began in 1990 with a task to the National Academy of Sciences to review existing energy models and provide guidance on the development of a National Energy Modeling System (NEMS).

DEVELOPMENT OF NEMS

The National Academy of Sciences Recommendations

Within the context of the current and future efforts to produce a comprehensive National Energy Strategy, the DOE Office of Policy, Planning, and Analysis tasked the National Research Council of the National Academy of Sciences to provide recommendations to the Department on the design, implementation, and use of a new energy modeling system. The committee formed by the National Research Council reviewed current energy models, particularly focusing on the existing EIA models, and analyzed the potential uses of energy models in the formation of national policy and decisions on research and development priorities.

The Committee concluded that the current EIA models provided a good starting point for further model development and strongly recommended that EIA itself develop, implement, and maintain the modeling system (National Research Council, 1992). While recommending the EIA system as a starting point, they acknowledged the need to update the modeling system to focus on the energy issues foreseeable in the next decade. An example of such updating, although not specifically mentioned in the Committee's report, is the treatment of natural gas markets. IFFS, designed as it was in the early 1980s, was heavily oriented to the rigid regulatory structure of the Natural Gas Policy Act of 1978. Throughout the 1980s, the gas markets evolved toward a less-regulated, market-oriented structure which clearly highlighted the need for a redesign of the natural gas component of the system.

The Committee recommended the following features for NEMS:

- Use of an iterative, problem-directed approach which requires an awareness of the policy initiatives that are expected to be analyzed. Such an approach will determine the priority design and implementation decisions for NEMS.
- Use of a modular structure to facilitate and enable the model builders to work with particular aspects of the model independently. In principle, implementation of this feature allows the development and use of less, or more, detailed modules, depending on the policy issues being analyzed.
- Integration of engineering and economic approaches to represent actual producer and consumer behavior.

- Output measures of concern to decision makers, including key economic, environmental, and national security measures relevant to policy decisions, such as income and regional distribution characterizations, economic measures, environmental measures, energy security measures, regional and international measures, and income distribution measures.
- Initial focus of NEMS development to be on the midterm time period.
- Use of widely available computers to maximize the likelihood that NEMS would be used by many different parties with interest in the national energy debates.
- Outreach to the broader energy analysis community, encouraging interface with existing or future models outside of NEMS. Modularity of design and good documentation will encourage such outreach.
- Interaction with outside peer groups to encourage critical reviews through participation in workshops, oversight committees, and other forms of interaction in the development.
- Transparency—ease of understanding—of results to facilitate analysis.
- Fast turnaround capability for policy analysis. Rapid response should be facilitated by reduced form modeling that represents a simplification of the more complex model.
- High levels of quality control in all phases of NEMS development, including code and methodological documentation, and documentation to confirm validity, as appropriate.

Decisions on the Development of NEMS

Model development decisions on NEMS were guided both by EIA's own long experience with energy policy modeling and by the recommendations of the National Academy of Sciences and other policy analysts within the Department. The size, content, and level of detail of the model were necessarily constrained by financial and computer resources available for the initial development of the system and for its continuing maintenance, update, and use. In many cases, data availability served as an additional constraint on EIA's ability to provide further meaningful and detailed representations within the model. NEMS was completed in August 1993 and was first used to produce the *Annual Energy Outlook 1994* (Energy Information Administration, 1994). A complete list of the NEMS documentation is provided in the Appendix.

Content

The purpose of NEMS is to project the energy, economic, environmental, and security consequences to the United States of alternative energy policies or assumptions concerning energy markets. The focus of NEMS is on domestic energy markets, including those aspects of the international markets that significantly impact domestic production, supply, and prices. The intended use for analysis determined the level of detail incorporated into NEMS. Where policy initiatives or analyses were deemed likely, detailed representations and options were designed and implemented. For example, unlike the predecessor model, NEMS includes sufficient detail in the refining and transportation demand sectors to analyze the cost and penetration of alternative or reformulated fuels and transportation technologies.

NEMS has been developed and implemented to provide annual forecasts to the year 2015. This defines a time period in which technology, demographics, economic conditions, and other factors can be sufficiently characterized to represent energy markets in considerable detail. Although the *Annual Energy Outlook 1994* and *Annual Energy Outlook 1995* published projections to 2010, projections to 2015 were first produced in the *Annual Energy Outlook 1996*, published in January 1996. Projections to 2020 are scheduled for the *Annual Energy Outlook 1998*.

The results produced by NEMS are projections of the energy future, given current laws and policies, and do not formulate a plan or strategy for achieving a particular goal. The forecasts are not considered to be absolute predictions of the future but are conditional forecasts, contingent upon the specifications of key assumptions concerning the US energy system and international energy markets. EIA's view of these key assumptions is formally reviewed and adjusted periodically to reflect improved information or fundamental market changes. For example, key assumptions include the behavior of the Organization of Petroleum Exporting Countries (OPEC), the size of the economically recoverable resource base for fossil fuels, characteristics of the world supply and demand for energy, the rate of development and penetration of new energy technologies, and existing or intended government actions or policies. NEMS provides a consistent framework for representing the complex interactions of the US energy system and its response to a wide variety of alternative assumptions and policies, without attempting to determine which of many plausible energy futures is most likely to occur.

Modularity

NEMS is designed as a modular system, as shown in Figure 2.1. The modules of NEMS represent each of the fuel supply markets, conversion sectors, and end-use consumption sectors. NEMS also incorporates macroeconomic activity and international energy modules. The primary flows between these modules are the delivered prices of energy and the quantities consumed by product, region, and sector. The information flows are not, however, limited to prices and quantities, and include other information such as economic activity, capital expenditures, and the impacts of demand-side management programs. The delivered prices of fuels encompass all of the activities necessary to produce, import, and transport fuels to the end user.

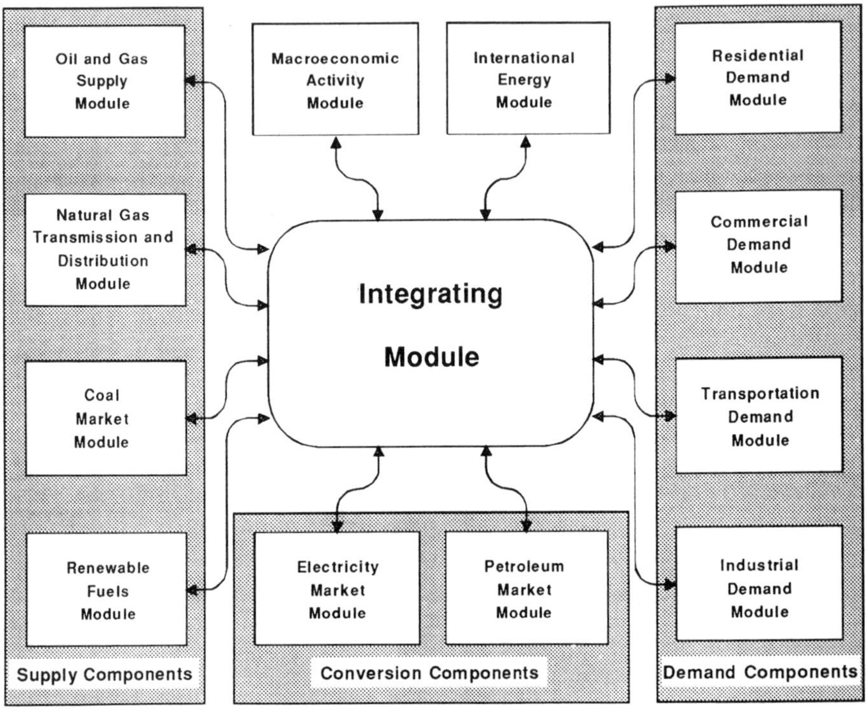

Figure 2.1 National Energy Modeling System

The integrating module controls the independent execution of the component modules. The modules are executed from the integrating module, and, to facilitate modularity, the components do not pass information to each other directly but communicate through a central data file. This modular design provides the capability to execute modules individually or to substitute alternative modules, thus allowing decentralized development of the system and independent analysis and testing of individual modules. Furthermore, this modularity allows the flexibility to use the methodology and level of detail that is most appropriate to represent each energy sector.

Solution is achieved by equilibrating on the delivered prices of energy and quantities demanded, thus assuring an economic equilibrium of supply and demand in the consuming sectors. Each fuel supply, conversion, or end-use demand module is called in sequence by the integrating module and solved, assuming all other variables in the energy markets are fixed. For each forecast year, the modules are called iteratively until the end-use prices and quantities remain constant within a specified tolerance, a state defined as convergence. This equilibration is achieved annually through the midterm period to 2015. The algorithm also checks for convergence on variables that represent petroleum product imports, crude oil imports, and several macroeconomic indicators.

Regionality

Because of the diversity of energy supply, consumption, and conversion in the United States, NEMS is designed to support regional modeling and analysis in order to portray transportation flows, represent the regional differences in energy markets, such as proximity to supplies or differences in the infrastructure or demographics, and provide estimates of policy impacts at the regional level. The level of regional detail differs for each sector given data availability, energy issues to be addressed, and commonly-used regions in each sector. In addition to data limitations, the desire to maintain a manageable model system precluded finer regional detail.

The four end-use demand sectors produce projections at the level of the nine Census divisions, although, due to data limitations, the industrial demand module represents consumption for the four Census regions and disaggregates the demand to the nine Census divisions. Each of the supply and conversion modules uses its own regional structure:

- Oil and gas supply: twelve supply regions, including three offshore and three Alaskan regions.
- Natural gas transmission and distribution: twelve regions, based on the nine Census divisions with further subdivisions to represent key transportation issues.
- Coal: eleven supply regions, representing the major coal geologic formations.
- Electricity: thirteen North American Electric Reliability Council regions and subregions.
- Petroleum market: five Petroleum Administration for Defense Districts (PADDs).

Transformations between these regional structures and the nine Census division structure are accomplished by explicit mapping and sharing algorithms or by representations of aggregate transportation networks. Overall convergence is determined at the nine Census division level.

Design reports and the independent expert review program

To assure that NEMS was state-of-the-art in terms of model design and use of best available data, the EIA developed and implemented a new rigorous model development and review process. The US energy–economy system was divided into approximately forty market segments, called "components". For example, electricity generation capacity expansion was defined as a component of the electricity market module of NEMS. Each component was assigned to a specific group of modelers with industry expertise in that area.

Prior to any component implementation, each group was required to develop a proposal for model design which included: (i) a review of the relevant model literature for that component, (ii) a description of the strengths and failings of existing models of that component with respect to the policy issues that the new model would have to address, (iii) a rationale, including data requirements and support, for the new modeling approach proposed, and (iv) a detailed mathematical description of the proposed component. Each of these proposals was reviewed and revised as appropriate through the normal EIA management structure. The proposals were finally reviewed and commented on by the Senior Technical Advisor, other Office Directors at EIA, the Deputy Administrator, and the Administrator. Model implementation could only begin after all of EIA's reviewers had approved the proposal.

Following the Administrator's approval, the proposals were forwarded to outside industry experts, other DOE technical staff, and university faculty, for further review and comment on the proposals. The degree of open examination of a proposed product prior to implementation was unprecedented in EIA history. In addition, another EIA office responsible for quality assurance, the Office of Statistical Standards, utilized their Independent Expert Review (IER) Program under which acknowledged specialists and experts in each model area were formally directed to review, analyze, and comment on the proposals, and follow up with recommendations for change as needed. The EIA modelers and the IER experts met and exchanged information through the component development process. In addition to these forums for vetting the NEMS model components, EIA held annual NEMS conferences and invited a broad spectrum of government and private sector participants to discuss the NEMS modules at intermediate and more mature states of development. The net result is a state-of-the-art energy–economy model of the United States with wider acceptance and review than any predecessor model.

Documentation

In accordance with established EIA standards, all of the NEMS modules are thoroughly documented. The required documentation provides complete theoretical discussions and equation listings. Although the component proposal and review process invited open review and comment prior to initial implementation, the documentation process allows for complete model dissemination and review by the broader energy community on an ongoing basis.

Each module or component was initially documented following the first use of NEMS for the *Annual Energy Outlook 1994*. As required, each documentation report included the following items: summary; discussion of the purpose and scope, including the relationship to the other components of NEMS;

theoretical approach and rationale; structure; inventory of input data, parameters, and outputs; detailed mathematical description; bibliography; abstract; and discussion of the data quality and estimation procedures. Following each update of NEMS, usually in accordance with the recurring *Annual Energy Outlook* cycle, the documentation is updated with replacement pages or chapters, as needed, depending on the scope of the updates. This ensures that the documentation of NEMS will always be complete and current.

Fast turnaround capability

A model with the scope, complexity, and level of disaggregation of NEMS also becomes computationally troublesome. In its initial implementation, NEMS required as many as 36 hours to execute through the entire forecast horizon. Fortunately, the modular structure of NEMS allowed model development and debugging to be accomplished with individual modules of NEMS that solved much more rapidly than the fully integrated system. However, the goal of NEMS to be a tool for policy analysis required a much more rapid turnaround capability. Policy questions frequently require a response within a few hours, not a few days.

Since the initial model use, a number of efforts have significantly reduced the execution time of NEMS to one-third the time initially required without compromising the content of the model. Extensive efforts to examine and correct instances of nonconvergence and slow convergence have reduced the number of iterations required. In several instances, a restructuring of data files for more rapid input–output has yielded significant time savings. In addition, as a recommendation from the National Academy of Sciences committee, an effort for reduced-form models was introduced. Additional work that moved NEMS to a faster platform is described in the next section.

Due to the modular nature of NEMS, the substitution of alternative modules is straightforward as long as input and output variables for the module are defined to be compatible. A reduced-form module could be a restructured or aggregated version of the larger module or an econometrically-estimated, response surface version of the module. In either case, the reduced-form module could then be substituted for the larger module when that sector is not of particular interest for the policy question being analyzed, thus yielding faster turnaround. Initial efforts for reduced-form modeling targeted the electricity market module and the petroleum market module as the most time-consuming sectors of NEMS. The sensitivity analysis with the electricity market module proved that the number of slices in the load curve could be reduced significantly with no impact on the results. Thus, the reduced-form version became the default version in NEMS and resulted in large time savings. The efforts with the petroleum market module produced a two-region reduced-form version that can be used to reduce turnaround time.

Alternative platforms and outreach

The initial version of NEMS was developed on an IBM 3090 series mainframe computer and required 24–36 hours to achieve equilibrium in all years. The EIA 3090 was purchased in the early 1980s and represents a somewhat dated technology. To make the model more usable and accessible to EIA customers, two new platform developments were pursued.

First, individual modules of the 1994 and, later, the 1995 versions of NEMS were ported to a Windows-based environment on the personal computer. A user interface was developed, using the C++ language, to allow a user to run the modules and view results. In addition, interactive spreadsheets allowed the user to view and modify key input assumptions for the modules, including input normally received from other modules. This allowed the modules to be disseminated for wide review and use in analyzing specific sectors. Although budget limitations forced the curtailment of this effort in 1996, the code and data files are still available for users to install on their own.

Second, after evaluating the state of the computing technology, EIA decided to implement NEMS on a Reduced Instruction Set Computer (RISC) system, operating under a UNIX operating system. This shift was accomplished for the development of the forecasts for the *Annual Energy Outlook 1996*. Implementing NEMS on the RISC reduced execution time of the full model to approximately 5 hours, enabling ongoing model and forecast development to proceed in a far more expeditious manner and allowing EIA to answer policy requests more rapidly. In addition, the port to the RISC, as well as the availability of the computer code, allows for wide accessibility of NEMS on key computing platforms.

Beyond documentation and model accessibility, further outreach activities continue with the annual NEMS conference, workshops on special topics, and frequent presentations at professional and industry meetings. The first two NEMS conferences focused on the model as it was being developed and on the results following the publication of the *Annual Energy Outlook 1994* with plans for future development. Subsequent conferences were structured around topics in the developing energy markets. Industry experts and policy analysts are invited to share opinions on the new developments and the incorporation of the changes within NEMS.

MODULES OF NEMS

The modules of NEMS represent individual sectors of domestic energy markets, an international energy module, and a macroeconomic activity module, directly interacting with one another through the integrating module. In general, the modules interact through values representing the prices of energy delivered to consuming sectors and quantities of end-use energy consumption. The international and macroeconomic modules provide information on international energy market activities and macroeconomic conditions that impact domestic energy. Impacts of recent legislation and policy initiatives are included throughout, including the Clean Air Act Amendments of 1990, the Energy Policy Act of 1992, and the Climate Change Action Plan. All modules represent the impact and cost of environmental and other regulations that affect the sector and report the levels of key emissions produced by energy combustion and related activities; carbon or BTU taxes can be imposed globally on the system. The key features of each module used in the *Annual Energy Outlook 1997* are summarized below.

The International Energy Module

The international energy module calculates the average price of imported crude oil and provides supply curves for five grades of crude oil for import into the United States. It also provides supply curves for petroleum products imports and for oxygenates (methyl tertiary butyl ether (MTBE) and methanol).

World oil prices are computed to provide a balance of supply and demand for crude oil in the international market as a function of OPEC production decisions, supply of non-OPEC oil, and international demand for oil, which depends upon worldwide economic growth. Import supply curves for five qualities of crude oil for each of the five PADDs are generated as a function of the world oil price, production sources, transportation rates, and assumptions about the utilization and/or expansion of worldwide refining capacity. Supply curves for ten petroleum products imported to each of the five PADDs reflect international refinery capacity, transportation costs, and the effects of US demand on world markets. The petroleum market module uses the supply curves to determine the quantity of crude oil and products to be imported.

Supply curves for MTBE and methanol are developed from current price and production capacity data and assumptions about new production capacity. It is assumed that ethanol, ethyl tertiary butyl ether (ETBE), tertiary amyl methyl ether (TAME), and tertiary butyl alcohol (TBA) are all supplied from domestic sources.

The Macroeconomic Activity Module

The macroeconomic activity module forecasts the impacts on the aggregate economy of changes in energy market conditions. Forecasts of economic variables are provided to the energy modules within NEMS. A wide range of economic variables drives other parts of NEMS, including interest rates, final demands for goods and services, disposable income, and industrial output.

The national submodule is a response surface representation of the Data

Resources, Inc. (DRI) US Quarterly Model. The national submodule simulates the reaction of the economy to changing energy prices and to exogenously specified information on labor force growth and productivity. The inter-industry submodule is a response surface version of the DRI input–output model. It provides industrial output projections and calculates inter-industry impacts associated with deviations from a baseline inter-industry projection. Initially, a set of equations that share national economic and industrial output data to the regions, derived from the DRI Regional Macroeconomic Model (RMM), provides regional forecasts of economic variables.

Supply Modules

The oil and gas supply module

The oil and gas supply module projects the availability of domestic crude oil production and dry natural gas wellhead production from onshore, offshore, and Alaskan reservoirs; imported pipeline-quality gas from Mexico and Canada; and imported liquefied natural gas. It is assumed that domestic oil and gas exploration and development activities are undertaken if the discounted present value of the recovered resources covers the present value of taxes and the cost of capital, exploration, development, and production, subject to a budget constraint. International gas trade is determined in part by scenario-dependent, non-economic factors. Crude oil is transported to refineries, modeled in the petroleum market module, for conversion and blending into refined petroleum products.

Lower 48 supplies consist of oil and gas produced by conventional recovery methods in onshore and offshore regions, onshore enhanced oil recovery, and onshore unconventional gas recovery produced from low-permeability or tight sandstones, Devonian shale, and coalbed methane. Capital expenditures in the lower 48 are developed by well class region and fuel type as a function of expected profitability for a representative well. Expenditures divided by drilling cost for each fuel, well class and region yield the number of wells drilled. Regional equations forecast new field discoveries and reserve revisions from development drilling. Enhanced oil recovery was modeled explicitly in NEMS for the first time for the *Annual Energy Outlook 1995* projections. Actual production levels for oil and gas are calculated within the market equilibration process. Natural gas production and prices are determined within the natural gas transmission and distribution module. Domestic oil production and oil imports are determined by the petroleum market module, based on fuel demands, reserves, productive capacity, and flow from new wells.

Alaskan oil and gas supplies are modeled as discrete projects. Transportation costs to lower 48 facilities are used with the market prices of oil or gas to calculate the net price received at the wellhead. A discounted cash flow method is used to determine the economic viability of each project at the netback price. Production is determined based on assumed drilling schedules and production profiles for new fields and developmental projects, historical production patterns, and announced plans for currently producing fields.

Net Canadian natural gas supplies to the United States at six border crossings are computed over a range of gas prices. Canadian drilling activity, supply, and domestic Canadian demand are modeled to determine gas available for export. Gas supply is allocated to regional Canadian/US border crossing points, accounting for pipeline capacities and the price responsiveness of supplies. Mexican gas trade is influenced by a range of non-economic factors, which are so uncertain that both the volume and direction of the flow of gas between the United States and Mexico are speculative. Thus, Mexican gas trade is determined on a scenario basis, which is fixed and not responsive to energy price changes. The volume of liquefied natural gas available for import into the United States is computed at various price levels, based upon available gasification capacity, schedules for constructing facilities, tanker availability, expected utilization rates, worldwide liquefaction capacity, and lead times for restarting operations at inactive terminals.

The natural gas transmission and distribution module

The natural gas transmission and distribution module models the natural gas transmission and distribution network in the lower 48 States, linking domestic and foreign suppliers and consumers. The module determines regional market-clearing prices for natural gas supplies, border prices, domestic production, imports, tariffs, pipeline and storage capacity expansion, and interregional flows. The module uses a simplified representation of the transmission and distribution system, structured as a network of nodes and arcs, with twelve internal nodes and nine border trans-shipment nodes. End-use prices are derived for both firm and interruptible markets. The decision making process for expanding pipeline and/or regional underground storage capacity is simulated to determine the amount of capacity to be added between or within regions and to establish maximum annual utilization rates for each of the interregional pipeline routes.

The pipeline tariffs are computed for transportation and storage services provided by interstate pipeline companies with an accounting system that tracks costs and computes rates under cost of service (straight fixed variable) rate design for firm and interruptible services. Distributor tariffs include both the distributor markups charged by local distribution companies for distribution from the city gate to the end user and the markups charged by intrastate pipeline companies for intrastate transportation services. A volumetric charge is computed that covers the cost of providing distribution or transportation services from the city gate to the end user. Distributor margins typically decline in real terms.

The coal market module

The coal market module represents the production, transportation, and distribution of coal on a cost-minimizing basis to meet the domestic and international demand for coal. Thirteen different coal types are differentiated with respect to thermal grade, sulfur content, and underground or surface mining. Production and distribution of coal are forecast for 11 supply and 13 demand regions, and the world trade is modeled by 16 export and 20 import regions.

Annual coal supply curves, relating annual production to marginal costs, are produced for each mine type (surface or underground), coal type, and supply region. The factors accounted for in constructing the supply curves are reserve depletion, capacity utilization, lead-time constraints on the opening of new mines, labor productivity, and real labor and fuel costs. A linear program determines the least-cost distribution of coal for a set of demands by sector and region. Existing supply contracts constrain the solution. Transportation costs are derived from reported data and adjusted by mode-specific trends in factor input costs, such as labor productivity and fuel cost charges.

World coal trade from the US for each coal export region and coal type is determined within the linear program, by minimizing overall costs for delivered coal, subject to US supply costs, including mining, preparation, and transportation costs. Limits on sulfur dioxide emissions and concerns about diversity of coal sources are also considered.

The renewable fuels module

The renewable fuels module represents the various types of renewable energy. Information on grid-connected electricity applications is passed to the electricity market module so that the technologies can be built and dispatched in competition with other generating technologies on the basis of capital costs, capacity factors, and fixed and variable operation and maintenance costs. Data on dispersed applications are input exogenously.

A hydroelectric file in the electricity market module provides currently available and planned regional hydropower capacity and capacity factors. Unplanned capacity changes are incorporated by offline estimates based on hydropower relicensing, water availability, land use and environmental regulations, and electricity demand growth. The wind-electric submodule projects the availability of wind resources and the cost and performance of wind turbines, expressed in the form of energy supply curves, which provide maximum turbine capacity, given the available land area, wind speed, and capacity factor.

The geothermal-electric submodule provides forecasts of new geothermal capacity that can be built, along with related cost and performance data, in the form of a supply curve from each geothermal site based on installed capacity and potential capacity. Only hydrothermal (hot water and steam) resources are currently considered. The solar-electric submodule represents solar-photovoltaic and thermal-electric installations. The intermittent nature and regional availability of solar energy are dealt with by the regional load shapes in the electricity market module with different time slices to represent intermittency.

Biomass is used to produce liquid fuels such as ethyl alcohol (ethanol). Demand for ethanol as a gasoline blending component is determined by the petroleum market module. Supply curves for ethanol are computed by PADD for ethanol produced from corn, accounting for feedstock costs, conversion costs, and energy prices. Wood fuel prices and technology characterization information are provided to the electricity market module, represented by a supply curve based on the accessibility of wood resources to the electricity generation sector. Dispersed consumption of wood energy is modeled in the industrial, commercial, and residential demand modules. Projections of energy produced from the incineration of municipal solid waste are calculated using the quantity of waste produced, derived from an econometric equation using gross domestic product as the principal driver, the heating value, and shares combusted for energy recovery.

Conversion Modules

The electricity market module

The electricity market module represents the generation, transmission, and pricing of electricity, with a cost-minimizing mix of facilities and fuels, subject to environmental constraints. Non-utility supply and electricity trade are included in the build and operation decisions; however, co-generators are represented in the demand and fuel supply modules. Demand is represented by load curves, which vary by region and season and by the penetration of demand-side management programs. In conjunction with the coal module, facility development, retrofitting, and dispatch are constrained to comply with the pollution constraints of the Clean Air Act Amendments of 1990 and other pollution constraints, including the system of trading sulfur emissions allowances.

The electricity capacity planning submodule projects the construction of new utility and non-utility plants, the penetration of demand-side management programs, the level of firm power trades, and the addition of scrubbers for environmental compliance to meet expected growth in electricity demand, given available resources, expected load curves, expected demands and fuel prices, environmental constraints, and costs for utility and non-utility technologies. When new capacity is the best alternative for meeting future increases in demand, then the timing of the demand increase, the expected utilization for base, intermediate, or peak load, and the construction and operating costs of various technologies primarily determine what technology is chosen. The ownership of the facility—utility or non-utility—influences the decision, because the capital costs differ according to the ownership and corresponding capital structure of the facility. The submodule looks up to six periods into the future, performing a multi-period optimization. After performing its levelized cost analysis, a market penetration algorithm shares capacity additions among existing and new technologies of similar types to address uncertainty about the costs and performance of generating technologies. Two important new features incorporated in the 1995 version of NEMS for the new technologies in the electricity market module are a correction for technological optimism and "learning-by-doing" to reflect reduced costs due to manufacturing experience.

Given available capacity, interregional firm purchased-power agreements, fuel prices, and load curves, the electricity fuel dispatch submodule solves for generation facility utilization and economy power exchanges to satisfy a demand that varies by time and region. Utility, independent power producer, and small power producer plants are dispatched until demand is met, in a sequence based on minimizing fuel plus operating costs. Limits on emissions of sulfur dioxide from generating units and the engineering characteristics of units serve as constraints. The operation is simulated over six seasons-time-of-day periods to reflect the seasonal variation in electricity demand. Load following, capacity derating, and interregional economy trade are also represented. International economy trade is determined exogenously.

The costs of building capacity, instituting demand-side management programs, buying power, and generating electricity are totalled, and the cost-of-service method often used by State regulators is simulated to determine the price of electricity. Revenue requirements shared over sales by customer class yield the price of electricity for each class. Utility-sponsored demand-side management programs, which attempt to alter the load curves by reducing or shifting demand, are evaluated to determine which programs are competitive. These are factored into the capacity expansion decisions and are passed to the demand modules so that their impacts are reflected in the future demands for electricity.

The petroleum market module

The petroleum market module is a linear programming representation of petroleum product pricing, crude oil and product import activity, and domestic refinery operations and fuel consumption. Refining operations are represented for each of the five PADDs, with the most detailed representation for PADD

III (Texas and nearby states), where more than 30 distinct refinery processes are modeled. Gasoline production is modeled within NEMS to account for the requirements of the Clean Air Act Amendments of 1990, with stipulations on oxygenated and reformulated gasolines, and for the more stringent California gasoline standards.

A mix of refinery operations is chosen that minimizes the costs of meeting demand. Crude oil and petroleum product import quantities are chosen as part of the process of minimizing costs, using supply curves for imported crude oil and products developed by the international energy module. Both domestic and imported crude oil are aggregated into five types, defined by a range of gravity and sulfur categories, allowing for changes in crude oil composition over time. The outputs of natural gas processing plants—ethane, propane, butane, isobutane, etc.—are modeled and either meet demand or are inputs to the refinery. The linear program provides prices of products sold in the refinery regions (refinery gate) and in the demand regions (wholesale). Regional end-use markups are added to produce retail prices.

Demand Modules

The residential demand module

The residential demand module projects fuel consumption by region for three housing types—single-family, multi-family, and mobile—seven major end uses, and three minor end uses. Housing stock, by type, is tracked from base year data by subtracting retirements based on assumed rates of decay and adding new housing starts, which are provided from the macroeconomic activity module. Appliance stocks are also calculated by fuel, end use, and technology by retiring appliances from retired houses, retiring equipment from surviving housing stock using a retirement function, and adding new appliances to either existing or new houses. The integrity of housing shells is also modeled. Shell efficiencies for existing and new housing improve in response to increases in real energy prices and on recent trends which are independent of price.

Fuel-specific equipment choices are made by comparing relative initial capital costs and first-year operating costs. Equipment type and fuel combinations are projected based on technology shares from recent data on new construction. Each equipment choice is selected from the range of efficiencies projected to be available in the year the choice is made. Energy consumption for each end use is based on energy intensity estimates from recent data. These base energy intensities are modified for increases in efficiency, changes in shell integrity, changes in real fuel prices, and changes in weather relative to the base year.

The commercial demand module

The commercial demand module projects the use of energy by region for eleven building types, non-building uses, nine building end uses, and more than one thousand technology types. Commercial floorspace is tracked by type of building by estimating retirements from existing stock and construction based upon demographic and economic variables from the macroeconomic activity module.

Demand for energy services is projected for each service based on the number, size, type, and location of the buildings, adjusted by short-term elasticities in response to changes in real energy prices. Given the level of service demanded, new equipment is purchased to meet demands in new construction, to meet demands created by the retirement of equipment at the end of its useful life, and to meet demands created by the retirement of equipment at the end of its economic life. Three consumer behavior rules are represented: minimum life-cycle cost, minimum cost constrained by the use of the same fuel as in existing or retired equipment, and retaining the same technology as the existing or retired equipment. Shares of consumers are assigned to each rule. The technologies available are characterized by commercial availability, capital costs, operating and maintenance costs, efficiencies, and lifetime.

The industrial demand module

The industrial demand module forecasts energy consumption for each of three major industry categories: energy-intensive manufacturing, non-energy-intensive manufacturing, and non-manufacturing. For energy-intensive industries, consumption is projected for heat and power and for feedstocks and raw materials at a process or end use level. The most detailed industries include food, paper, chemicals, petroleum refining—modeled in the petroleum market module—stone/clay/glass, and primary metals. The influence of energy prices on consumption is modeled in terms of the efficiency of both new and existing capital stock and the mix of fuels used, given the capital stock.

The module uses an implicit vintaged capital stock accounting framework that represents energy use in the existing stock and in new additions for each process step. Capacity is implicitly retired at a fixed rate, leading to intensity improvement in each process step. New production capacity is added when the capacity of the existing stock cannot produce the output forecast by the macroeconomic activity module. Each industry is modeled, where appropriate, as three components: buildings, boilers/steam/co-generation, and process/assembly. Use of energy by buildings is relatively small and is represented as growing at the same rate as employment in the industry. The boilers/steam/co-generation component produces steam to meet the requirements of the other two components and for co-generation. The process/assembly component accounts for the largest share of energy used for heat and power. The energy-intensive/large energy-consuming industries are modeled by explicitly describing the major process flows or stages of production in the industry. Aggregate production and cost functions are used to model factor shares, primarily energy, at each stage. Technology penetration for each of the major processes is based on technology possibility curves that represent the declining energy intensity of the capital stock. Additional conservation results from changes in absolute and relative energy prices. The module also includes some fuel switching in existing boiler applications, recycling, and by-product consumption.

The transportation demand module

The transportation demand module projects the use of energy by travel mode based upon fuel prices and a variety of macroeconomic variables. Light-duty vehicles are represented by ten vintages and six passenger car and six light truck size classes. Fifteen alternative-fuel technologies are also included.

New light-duty vehicle fuel efficiency is estimated by size class as a function of fuel prices and income variables. These efficiencies are combined with estimates of new car and truck sales from the macroeconomic activity module to yield projections of regional sales and characteristics of light-duty vehicles. Shares of sales of alternative-fuel vehicles are estimated as a function of time, technology characteristics, costs, and fuel prices in a three-stage decision process. The stock of light-duty vehicles is represented, accounting for the new sales and retirements. Finally, vehicle-miles traveled for light-duty vehicles are estimated based on the fuel cost, per capita disposable personal income, and demographics, from which fuel consumption is computed. Travel demand, efficiency, and consumption for fleet vehicles are estimated separately.

Air travel is estimated for business and personal travel and for dedicated freight and air freight based on the costs of air travel and economic growth. Average aircraft fuel efficiency is estimated by size class as a function of fuel prices and economic growth, which affects the age distribution of the aircraft. Estimates of industrial production are used to project ton-miles traveled for truck, rail, and ship freight travel. Efficiency projections are based on energy price impacts, technology trends and stock turnover. Miscellaneous travel consumption—military, mass transit, recreational boating, etc.—is based upon a variety of efficiency and macroeconomic variables.

USES OF NEMS

At this time, NEMS has been used to produce the baseline projections of energy markets for the *Annual Energy Outlook 1994* through *Annual Energy*

Outlook 1997, published in December 1996. These publications provide a reference projection plus four additional cases that incorporate higher or lower assumptions of world oil prices or domestic economic growth. The more recent *Outlooks* exploit many of the capabilities of the NEMS modules by presenting additional sensitivity cases that use single modules or small groupings of the modules, as opposed to fully integrated model runs. These sensitivities generally explore the potential impacts of higher and lower rates of technology penetration, more and less rapid nuclear retirement, and higher electricity demand, as examples.

NEMS has also been used to support a variety of Departmental studies, frequently with some modifications to the baseline assumptions at the request of the office conducting the study. The results have been used to analyze the impacts of lower electricity prices due to industry restructuring, more rapid efficiency improvements of consumption technologies due to research and development, and lower capital costs of renewable generation technologies. These analyses are typically conducted for the Office of Policy or the Office of Energy Efficiency and Renewable Energy. In a study conducted for the Office of Naval Petroleum Reserves, the petroleum market module was modified to represent Los Angeles and San Francisco area refineries in order to determine the value to refiners of crude oil from the Naval Petroleum Reserve, exploiting the capability to analyze crude oil and product specifications, product slates, and refining units. Recent studies include *The Impacts on US Energy Markets and the Economy of Reducing Oil Imports* for the General Accounting Office (Energy Information Administration, September 1996) and *An Analysis of Carbon Mitigation Cases* for the Environmental Protection Agency (Energy Information Administration, June 1996).

NEMS results are widely used and disseminated electronically, and the model itself is also available. Individual modules are being utilized on the personal computer, and the entire model has been installed on the workstation at the National Renewable Energy Laboratory and the Lawrence Berkeley Laboratory.

CONCLUSIONS

NEMS is an effective analytical tool for a wide variety of detailed policy issues. Critical energy and economic issues on the horizon that NEMS can assist in analyzing include energy security, further efficiency improvements in the energy production, conversion, and consumption technologies, continued regulatory reform in the natural gas and electricity sectors, and the increased use of renewable sources of energy. In addition, the International Energy Agency has proposed using NEMS to develop the US component of the *World*

Energy Outlook. Through the use of models like NEMS, the energy debate in the United States has been and should continue to be enhanced. The public availability of NEMS expands the debate to an even wider group of participants.

APPENDIX: THE NEMS DOCUMENTATION

Energy Information Administration, *The National Energy Modeling System: An Overview*, DOE/EIA-0581 (96) (Washington, DC, March 1996).

Energy Information Administration, *National Energy Modeling System Integrating Module Documentation Report*, DOE/EIA-M057 (97) (Washington, DC, forthcoming).

Energy Information Administration, *Model Documentation Report: Macroeconomic Activity Module of the National Energy Modeling System*, DOE/EIA-M065 (97) (Washington, DC, forthcoming).

Energy Information Administration, *Documentation of the DRI Model of the US Economy*, DOE/EIA-M061 (Washington, DC, December 1993).

Energy Information Administration, *NEMS International Energy Module, Model Documentation Report*, DOE/EIA-M071 (Washington, DC, April 1994).

Energy Information Administration, *EIA Model Documentation: World Oil Refining Logistics Demand Model, "World" Reference Manual*, DOE/EIA-M058 (Washington, DC, March 1994).

Energy Information Administration, *Model Documentation Report: Residential Sector Demand Module of the National Energy Modeling System*, DOE/EIA-M067 (97) (Washington, DC, forthcoming).

Energy Information Administration, *Model Documentation Report: Commercial Sector Demand Module of the National Energy Modeling System*, DOE/EIA-M066 (97) (Washington, DC, forthcoming).

Energy Information Administration, *Model Documentation Report: Industrial Sector Demand Module of the National Energy Modeling System*, DOE/EIA-M064 (97) (Washington, DC, forthcoming).

Energy Information Administration, *Model Documentation Report: Transportation Sector Model of the National Energy Modeling System*, DOE/EIA-M070 (97) (Washington, DC, forthcoming).

Energy Information Administration, *Model Documentation Report: Electricity Market Module, Load and Demand-Side Management Submodule*. Vol. 1: *Model Description*, DOE/EIA-M068-A/1 (Washington, DC, March 1994).

Energy Information Administration, *Model Documentation Report: Electricity Market Module, Load and Demand-Side Management Submodule*. Vol. 2: *Model Code Listing*, DOE/EIA-M068-A/2 (Washington, DC, March 1994).

Energy Information Administration, *Model Documentation Report: Electricity Market Module, Electricity Capacity Planning Submodule*, DOE/EIA-M068-B (Washington, DC, March 1994).

Energy Information Administration, *Model Documentation Report: Electricity Market Module, Electricity Finance and Pricing Submodule*, DOE/EIA-M068-C (Washington, DC, March 1994).

Energy Information Administration, *Model Documentation Report: Electricity Market Module, Electricity Fuel Dispatch Submodule*, DOE/EIA-M068-D (97) (Washington, DC, forthcoming).

Energy Information Administration, *Documentation of the Oil and Gas Supply Module*, DOE/EIA-M063 (97) (Washington, DC, forthcoming).

Energy Information Administration, *Model Documentation: Natural Gas Transmission and Distribution Model of the National Energy Modeling System*, DOE/EIA-M062 (96) (Washington, DC, December 1996).

Energy Information Administration, *EIA Model Documentation: Petroleum Market Model of the National Energy Modeling System*, DOE/EIA-M059 (97) (Washington, DC, forthcoming).

Energy Information Administration, *Model Documentation: Coal Market Module of the National Energy Modeling System*, DOE/EIA-M060 (97) (Washington, DC, forthcoming).

Energy Information Administration, *Model Documentation Report: Model Documentation Renewable Fuels Module of the National Energy Modeling System*, DOE/EIA-M069 (97) (Washington, DC, forthcoming).

REFERENCES

Energy Information Administration (January 1994) *Annual Energy Outlook 1994*, DOE/EIA–0383(94), Washington, DC.

Energy Information Administration (June 1996) *An Analysis of Carbon Mitigation Cases*, SR/01AF (96–01), Washington, DC.

Energy Information Administration (September 1996) *The Impacts on US Energy Markets and the Economy of Reducing Oil Imports*, SR/01AF (96–04), Washington, DC.

Hogan, W.W. (1975) Energy policy models for project independence, *Computers and Operations Research* **2**, 251–271.

Murphy, F.H., Conti, J., Sanders, R. and Shaw, S. (1988) Modeling and forecasting energy markets with the Intermediate Future Forecasting System, *Operations Research* **36**(3), 877–902.

National Research Council (1992) *The National Energy Modeling System*, National Academy Press, Washington, DC.

Chapter 3

The IDEAS Model and Its Use in Developing the US Climate Change Action Plan

Frances P. Wood and Jay C. Geinzer

INTRODUCTION

In 1992 world leaders came together at the Earth Summit in Rio de Janeiro, Brazil, to discuss the global ecological crisis. Scientists believe that a wide range of human activities are increasing the atmospheric concentrations of greenhouse gases, such as carbon dioxide, methane, and nitrous oxides. Increasing concentrations of these gases could lead to large, unpredictable, and relatively fast changes in the climate, on both local and global levels. The speed, unpredictability, and magnitude of these changes are likely to have an adverse effect on the ecology, food production, and human health. It is equally likely that any effort to reverse these trends will require tens if not hundreds of years to take effect.

At the Earth Summit, the United States and other countries signed the Framework Convention on Climate Change. The Convention calls on nations to work to stabilize the concentrations of greenhouse gases so as to:

> ... allow ecosystems to adapt naturally, to ensure that food production is not threatened, and to enable economic development to proceed in a sustainable manner.

Systems Modelling for Energy Policy. Edited by D.W. Bunn and E.R. Larsen

As part of its commitment to live up to the goals of the Convention, the White House put together an inter-agency task force to formulate a Climate Change Action Plan that would reduce US emissions of greenhouse gases to 1990 levels by the year 2000. This task force included participants primarily from the White House, the State Department, the Department of Energy (DOE), the Environmental Protection Agency (EPA), the Department of Transportation, the Department of Agriculture, and the Office of Management and Budget. Within six months this task force accomplished its goal, and in the process:

- Established a baseline projection of greenhouse gas emissions that would likely occur over the remainder of the decade with current policies in place.
- Identified (over 200) new or expanded initiatives that would reduce greenhouse gas emissions.
- Estimated the savings and costs for implementing these initiatives.
- Selected a package of fifty initiatives that would achieve the goal in a fiscally responsible manner.

It was recognized at the outset that the proposed initiatives would affect all parts of the energy system, and that a system-level model would be needed to perform the final integrating analysis when the initiatives were "bundled". The Integrated Dynamic Energy Analysis Simulation (IDEAS) model, a long-term simulation model of US energy supply and demand, was selected for this role. The IDEAS model has been used by the Policy Office at the DOE for over 15 years to assess the impacts of national energy policy, and was well suited to this task.

THE HISTORY OF THE IDEAS MODEL

In April of 1977, Carter Administration staff produced a document called "The National Energy Plan" (NEP I 1977). By that time, the United States had faced its first energy crisis (caused by a selective Arab oil embargo), and there was a realization that supply interruptions and shortages were symptoms of an underlying imbalance of energy supply and demand. NEP I proposed a number of major policy initiatives to reduce US dependence on oil and gas. These included policies to:

- Increase energy conservation and fuel efficiency.
- Raise the price of oil and natural gas.
- Encourage the conversion to coal.
- Promote limited development of nuclear power.
- Stimulate the long-term development of renewable resources.

Included in the first national energy plan were many policy initiatives that later became law as a part of the National Energy Act of 1978 (for example, fuel efficiency standards and a "gas guzzler" tax for automobiles; residential conservation tax credits; the Public Utilities Regulatory Policy Act, or PURPA; a windfall profits tax on oil; and expansion of the strategic petroleum reserve) as well as some proposals that never became law (a standby gasoline tax of up to 50 cents per gallon).

Immediately after publication of this plan, the House Subcommittee on Energy and Power asked the Dartmouth System Dynamics Group to evaluate the plan using the FOSSIL1 model. A report entitled "FOSSIL1: Evaluation of the National Energy Plan" was prepared in June 1977 and used as the basis for testimony before the subcommittee in hearings about the plan that summer. The evaluation concluded that the plan was primarily an energy conservation strategy. It projected that the plan would likely meet its goal of reducing energy demand but would fall short on its other goals, particularly its goal of reducing net imports.

Soon after this analysis was published, the FOSSIL1 model was implemented and modified extensively at the Department of Energy. The new version of the model, called FOSSIL2, was used in the second National Energy Plan (NEP II 1979) and in subsequent DOE analyses of new energy policy initiatives. For the second National Energy Plan, the FOSSIL2 model was used to analyze the net effect of a combination of supply-side policy initiatives on oil imports. These policies included a proposal to decontrol oil prices (coupled with a windfall profits tax on oil revenues), a tax credit for shale oil, and additional solar energy tax credits. The FOSSIL2 model was also used at the DOE for other energy policy analyses, including the Oil Vulnerability Study, undertaken in 1980 to respond to a heightened concern for oil supply disruptions due to a doubling of world oil prices and also to political developments in the Middle East. Projections prepared for the Synfuels Task Force and other special studies prepared at the request of the Secretary of Energy also made use of FOSSIL2.

In the mid-1980s there was a growing interest in understanding the consumer side of the energy system. In 1985/86 new demand sectors for FOSSIL2 were built to analyze buildings, industry, and transportation uses of energy. The sectors were constructed around the principle that consumers act to minimize their cost for energy services (i.e. heat, light, mobility), not fuel expenditures. During the 1980s, the model continued to be used for the biennial National Energy Plan projections and for analyzing various policy proposals.

From 1990 to 1994 the model was used by the Department of Energy's Policy Office for numerous analyses concerning the National Energy Strategy (NES 1991) and NES-related legislation, which became the National Energy Policy Act of 1992. Other energy policy initiatives in both the House and Senate, including the Cooper–Synar CO_2 Offsets bill, and various tax initiatives (carbon, BTU, gasoline, oil import fees, etc.) were analyzed by DOE using the IDEAS model. Measures used to evaluate the effectiveness of these policies include economic measures (fuel prices), energy security measures (oil imports, domestic fuel supply and production capacity, fuel switching capabilities for electric generators in the utility and industrial sectors, and conservation potential), and environmental measures (SO_2, NO_x, and greenhouse gas emissions reductions). Extensive improvements have been made to the transportation and electric utility sectors, and the model has been renamed the IDEAS model.

For a model to be useful over a long period of time, it must be modified continuously to stay current with policy issues and the energy system. Some updates are relatively minor in nature, such as adding carbon tax policy levers, while others require the rebuilding of whole sectors of the model. An example of the latter is the new demand sectors in the mid-1980s and the new transportation sector in the early 1990s. In both cases the sectors were constructed from scratch when new policies, such as appliance standards and alternate-fueled vehicle promotion, were being proposed and the existing structure had no way of addressing them. The new transportation sector now has a detailed representation of consumer choice decisions for new vehicle purchases, including alternate fueled vehicles. Other structures may atrophy over time and be removed, such as synthetic fuels development. In most cases there is a trend toward adding detail and making the model more complex. This tendency must always be balanced with the intended use of the model and the desired level of transparency. Much of the strength of IDEAS stems from its quick turnaround nature and its ability to integrate the various parts of the energy system. If too much detail is added, the model becomes too cumbersome to run and the results become hard to understand.

IDEAS is also relatively data intensive compared to many system dynamics models. The base case assumptions, including technology characteristics, are reviewed annually. For the past several years, we have calibrated the model to the Energy Information Administration's (EIA) annual forecasts in order to create an apolitical baseline as a starting point for any policy analysis.

THE STRUCTURE OF THE IDEAS MODEL

The IDEAS model has mechanisms that simulate the behavior of energy markets: they adjust energy prices to ensure that supply equals demand for certain energy fuels. Only rarely does a situation occur in which demand is unmet (such as during an oil embargo). The model simulates price-setting and market-clearing mechanisms for the four major energy commodities: oil, natural gas, coal, and electricity. All other prices used in the model (e.g. the price of new capital) are set exogenously by the user.

The demand side of the model is represented by residential, commercial, industrial and transportation sectors (Figure 3.1). The demand sectors assume

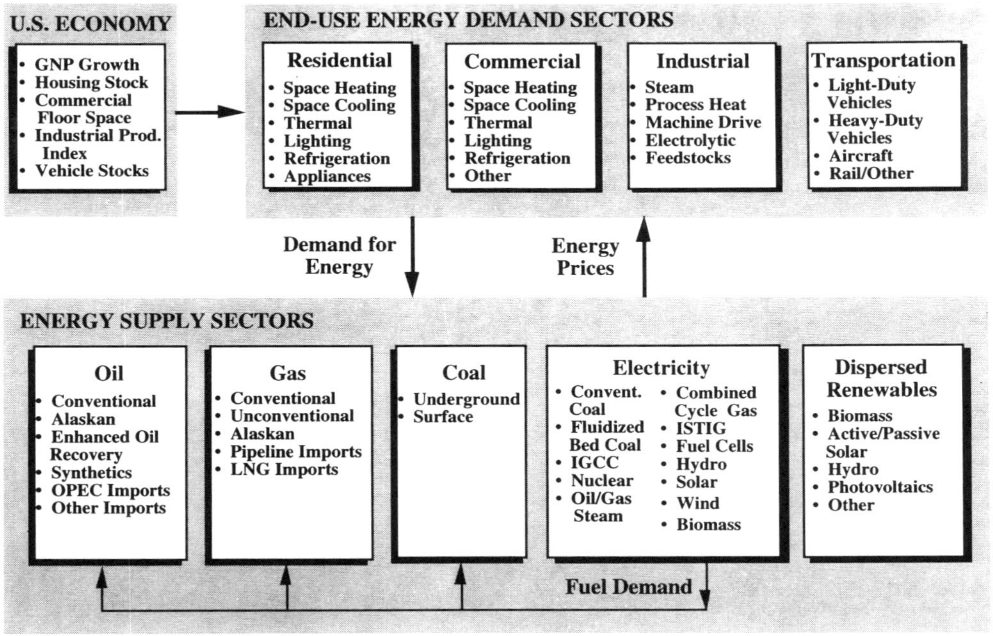

Figure 3.1 The IDEAS model

that consumers act to minimize their energy service costs, using financial variables consistent with historical behavior. Following this approach, the model first projects the demand for energy services by end use and then calculates the share of end use markets captured by various technologies, including, for example, conservation, co-generation, or conventional technologies. Conservation technologies are represented in conservation cost and savings curves that allow multiple technology options to be represented in an aggregate form.

Each of the four supply sectors in IDEAS represents the behavior of an industry whose purpose is to satisfy the demand for one of the four major energy sources in the US economy (oil, gas, electricity, and coal). Although the energy products of (and government restrictions on) each industry are different, the four industries share certain common structural properties. Each industry must finance new capital investment, allocate the investment to alternative energy sources, and convert the capital, labor, and resource inputs it purchases into a finished product. Conventional and new resources are included in each sector.

The following discussion shows how the model's generic market-clearing structure is consistent with economic theory. It should be noted, however, that two of the four model sectors determine price in a non-generic manner: crude oil prices are set exogenously in this model (because oil prices are determined in a world oil market) and electricity prices are regulated (that is, prices are based on costs plus a return on "rate base").

Like most other models of US energy supply and demand (for example, EIA's energy models, the DRI model, GEMINI, or the GRI model), IDEAS might be described as a partial equilibrium model, because it models the market-equilibrating processes of only a part of the total economy (the energy sector part).[1] Market-clearing in the IDEAS model occurs over two different time frames, consistent with economic theory: short-run and long-run (Figure 3.2; see Mansfield, 1970, pp. 222ff.). The explicit representation of short- and long-run market-clearing in IDEAS is an innovative way to translate economic theory into practical application in a descriptive simulation model

Over the long run, the IDEAS model seeks to balance demand and production by investing (or not investing) in new energy supply and demand technologies. An imbalance between demand and production capacity causes price to adjust upwards or downwards relative to costs. For example, below-normal production capacity (as measured by an industry's capacity factor compared to the long-run industry norm) would cause prices to increase. Higher prices would raise profits, resulting in more investment in new energy production capacity on the supply side. Higher prices also would stimulate more investments in energy-conserving technologies on the demand side. These responses tend to balance energy supply and demand over the long run—after some time lag, the investments add to supply and reduce demand, offsetting

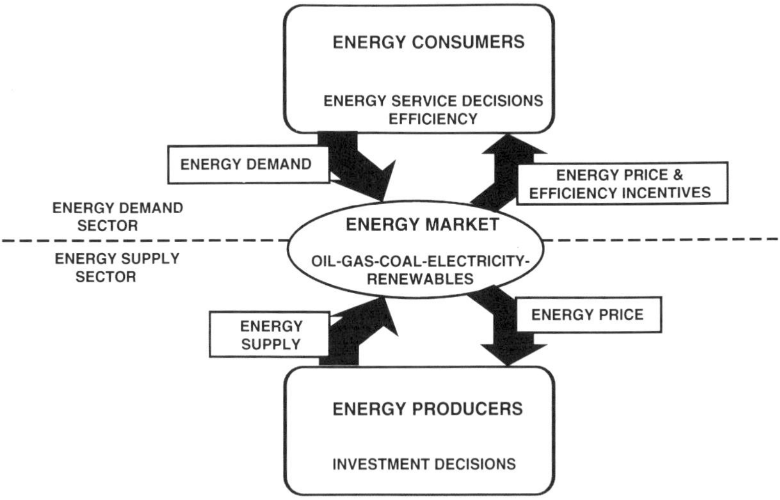

Figure 3.2 Market-clearing in the IDEAS model

the initial price movement (which in this case was a price increase). This response of energy capital investments to prices with the goal of balancing supply and demand over the long term is how the model represents the economic concept of long-run market equilibrium. The behavior of energy supply and demand in pursuit of the goal of long-run equilibrium is what in large part explains the model's energy forecasts.

In addition, the IDEAS model contains structures that represent short-term market-clearing mechanisms. These mechanisms adjust a commodity's supply, demand, and price over a shorter time period (less than one year) to insure that a short-run equilibrium is maintained. In the short run, energy-using capital stocks and production capacity are fixed. Demand and supply can only be adjusted by changing the intensity of use of these capital stocks. On the demand side, an example of such an adjustment included in the model would be adjusting the thermostat setting for heating and cooling. For supply, the primary short-term adjustment would be changes in *capacity utilization*, which adjusts upward or downward to match production with short-term demand. Price is set as a function of the balance between demand and production capacity (supply). If supply and demand are balanced, prices are set at a level that will result in a normal return on investment for that industry. This price function (price as a function of supply and demand) is derived from historical behavior in each energy sector.

Economic theory also recognizes that "disequilibrium, rather than equilibrium, has to be the usual state of most real-life markets" (Mansfield, 1970, p. 240). The IDEAS model is a *dynamic* model of energy supply and demand.

Dynamic models indicate the movement over a period of time of economic variables and the way such variables move from one equilibrium (or disequilibrium) to another (Mansfield, 1970, p. 241). For example, the cobweb theorem in economic theory is a simple dynamic model that explains how a commodity's price and output might adjust over time toward equilibrium. Technically, the IDEAS model is a very complex cobweb model.[2] It is a nonlinear dynamic simulation model whose structure is designed to explicitly represent the dynamic behavior of US energy supply and demand.

PREVIOUS USE OF IDEAS FOR GLOBAL WARMING POLICY ANALYSIS

The US Congress in late 1988 directed DOE to conduct a study of energy technology and policy options to mitigate greenhouse gas emissions.[3] The study mandate was quite broad, allowing DOE to evaluate a comprehensive set of policies aimed at reducing emissions by 20% (from 1989 levels) by the year 2000 and 50% by 2010. The focus of the study was energy technologies and policy. The energy sector is a major producer of greenhouse emissions, accounting for about 57% of US emissions in the 1980s (US Environmental Protection Agency, 1989).

In this study the IDEAS model was used to assess the emissions reduction potential and costs of a broad set of technology policies, as well as several individual policies and combinations of policies. This study found that reforestation offsets compared favorably to other energy-related policy initiatives, because they could potentially achieve the goal of 20% reduction of CO_2 from then-current levels at a relatively low cost (compared to, for example, a carbon tax). There were also considerable opportunities for conservation to limit the growth of energy use and carbon emissions, and possible cost savings (reductions in energy service costs) were associated with at least one case (the high conservation standards case). In the electricity sector, the need to find an inexpensive, reliable, and environmentally benign fuel source to replace coal for electricity generation was underscored.

The natural feedback interaction of energy supply, demand, and price illuminated another important insight regarding policy design. By having policies address only one sector and ignoring their ramifications on other sectors, unintended results occurred. For example, in the policy promoting coal efficiency, a penalty was assessed against inefficient coal plants. This reduction in demand brought down coal prices and made coal more attractive to industrial users. As a result of this insight, policies were redesigned to affect all sectors. Although insights like these seldom make it into official policy documents, they become part of the intellectual capital for policy makers and policy design evolves down a different path once these truths are understood.

THE CLIMATE CHANGE ACTION PLAN

The Climate Change Action Plan (CCAP; for a further description of the modeling, see DOE, 1994) represented a next logical step in the evolution of a coherent policy that would address global climate change. This study occurred four years after the previous study, and in the intervening time a wide array of promising initiatives had been identified by researchers around the country that would help mitigate global warming. Literally hundreds of ideas were brought to the table and discussed, such as agro-forestry techniques, alcohol production, automobile efficiency improvements, land-use planning changes, building and appliance performance standards, and more efficient or renewable electricity generation technologies. Most of these initiatives had some other benefit to offer as well.

The process was kicked off with a White House-sponsored Conference on Global Climate Change, which was held on June 10th and 11th, 1993 in Washington, DC. Over 800 participants attended this conference, including researchers, industry and trade association representatives, and policy makers. Most of the discussion took place in a series of parallel workshops that were aimed at energy-using sectors or discrete issues:

- Residential energy demand
- Commercial energy demand
- Industrial energy demand
- Auto/light truck energy demand
- Commercial transportation energy demand
- Urban infrastructure
- Energy supply
- Methane and other gases
- Joint implementation
- Sinks

The purpose of the workshops was to identify issues and initiatives that needed to be addressed in the upcoming analysis. After the conference the analysis team assembled a master list of all initiatives and produced a document that included a one-page description of each initiative, the steps required for implementation, costs and/or savings associated with implementation, a list of stakeholders, and the expected impact on greenhouse gas emissions.

A base case projection of carbon emissions was also produced in conjunction with the list of initiatives. Some initiatives involved expanding current policies or programs, and it was important to make sure that the base case reflected those programs at their current level of implementation. The production of the base case was a surprisingly difficult part of the process, but it also illustrated some of the contributions a model can make. At the beginning of the discussions for the base case transportation sector assumptions, for example, it became clear that there were three different views on how the light duty vehicle sector would evolve, and each one had some merit. Analysts were quickly (overnight, in fact) able to produce cases that reflected the assumptions inherent in each point of view using the IDEAS light duty vehicles sector. Once the participants in that working group were able to see the consequences of each viewpoint, the discussion quickly focussed on the assumptions themselves rather than on the validity of each party's point of view, and a consensus outlook was quickly achieved. In this case the model served an accounting function, making sure that everything was being counted the same way, and this function was sufficient to avoid an unpleasant logjam that could have unnecessarily delayed the process.

After the base case was produced the initiatives were tested individually to see whether the model results agreed with the estimated carbon savings of each initiative. Discrepancies generally fell into two categories: those where the modelers did not understand the full impacts of the policies (which were corrected), and those where the analysts who estimated the impacts used different base case assumptions (which were not corrected). This part of the process served as a validation exercise since the intention was to test bundles of policies, not individual ones.

The real insights began to emerge as the initiatives were tested in combination with each other. In some cases the combined effect of two initiatives was different from the effect of each initiative in isolation. The most interesting cases were those where two initiatives tended to offset each other. In some cases we found that the two initiatives included one that increased the efficiency of an existing technology and another that increased the market share of a new technology. In the electric utility sector we found that aggressive conservation improvement programs reduced the need for new generation in general, which reduced the effectiveness of programs designed to increase the penetration of renewable technologies. In the light duty vehicles sector we found that increasing Corporate Average Fuel Economy (CAFE) standards for gasoline vehicles decreases the relative attractiveness of alternate-fueled vehicles. The validity of this simple insight was affirmed by the participants in the working groups and by participants in an ongoing Energy Modeling Forum study on energy conservation. This insight did not get reported in the final document, but it helped inform policy makers during the process.

The final plan included 44 policy actions, including 31 in the energy sector. The individual actions span the range of economic sectors, although they are concentrated in energy efficiency improvements in the demand sectors. Many of the actions are voluntary or industry/government partnership programs, such as EPA's Green Lights Program and DOE's Motor Challenge Program. In both of these programs, firms are asked to commit to making cost-effective improvements in their lighting or motor efficiencies. In return, firms receive technical assistance and public recognition from the agencies. On the supply side, actions included those aimed at reducing the carbon intensity of fuels used to generate electricity, as well as those to increase efficiency. Examples of fuel switching actions included the formation of a renewable energy market mobilization collaborative and seasonal use of natural gas for NO_x control. Improved efficiency was targeted through standards and voluntary adoption programs for high-efficiency transformers. Only a few programs required legislative action. These were reforming federal tax subsidies for employer-provided parking and improving hydroelectric generation at existing dams.

In addition, two foundation actions, Climate Challenge and Climate-Wise, were included in the plan, although no carbon savings were quantified for them. These are voluntary programs where electric utilities and industrial firms commit to lowering their emissions through whatever methods they choose. They were not quantified because of the concern of potentially double-counting these savings because they might overlap with other actions in the plan. For example, an electric utility might encourage its commercial customers to participate in EPA's Green Lights program in conjunction with utility efficiency rebates. However, these commitments increase the likelihood that the projected emission reductions in the plan will be achieved.

The net result of the plan is to reduce projected carbon emissions from the energy sector by 66 million metric tons carbon equivalent (MMTCE), and total emissions by 109 MMTCE. Energy-related emissions increase from 1990 levels, but emissions from other gases, such as methane and N_2O, decline below 1990 levels.

Total primary energy consumption is projected to be reduced by 2.7 quads in 2000 or 3% by the plan. This is less than the 5% reduction in energy-related carbon emissions because of fuel switching to less carbon-intensive fuels. Oil and coal consumption, the fuels with the highest carbon intensity, decline, while natural gas and renewable energy use increase slightly by 2000 (Figure 3.3).

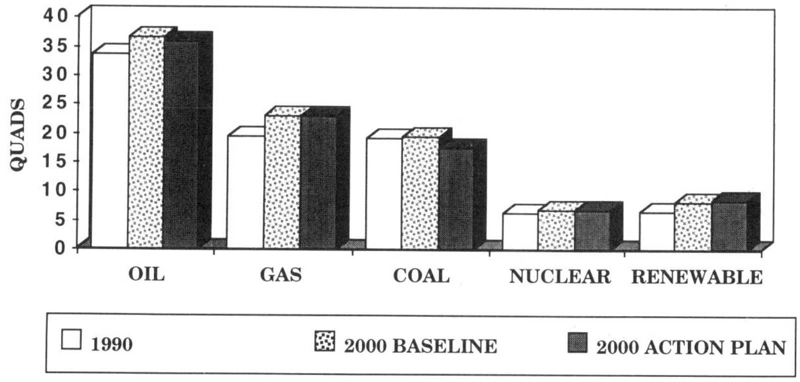

Figure 3.3 Primary energy consumption

Coal reductions stem from reduced demand for electricity and increased reliance on natural gas and renewable sources for electricity generation.

From an end user's perspective, the largest carbon reductions are associated with electricity (Figure 3.4). These reductions result from energy efficiency improvements that reduce the demand for electricity and from actions that reduce the carbon-intensity of generation. Many of the actions specifically target end uses that use electricity, such as lighting and motors through the Green Lights and Motor Challenge programs. These and other actions aimed more broadly at improving energy efficiency in buildings and industry are projected to reduce electricity demand growth by 0.5% per year from 1990 to 2000.

On the supply side, increasing generating capacity at existing hydro facilities and encouraging natural gas co-firing in existing coal plants would lead to lower carbon emissions to meet current demand. The renewable energy consortium and cost-sharing for high efficiency gas technologies are projected to reduce the carbon-intensity of capacity additions. These latter policies do not lead to significant savings by 2000, but were included in the plan due to their significant potential to reduce emissions over the longer term.

One of the roles that an integrated model, such as IDEAS, can play in this type of policy analysis is to analyze the interactions of electricity demand reductions and supply mix changes. The carbon saved from demand-side efficiency improvements is related to the carbon-intensity of electricity generation at the marginal rather than the average intensity. The marginal carbon-intensity will change through time as a result of many factors: the base case level of demand, the size of the demand reduction, and whether new capacity is being constructed. This is illustrated in Figure 3.5, which shows the electricity sales reductions and associated carbon reductions resulting from the demand-side actions of the Plan. In the near term, utility carbon emission reductions are greater than sales reductions because the marginal intensity is greater than the average. The average intensity reflects a mix of coal, oil, gas,

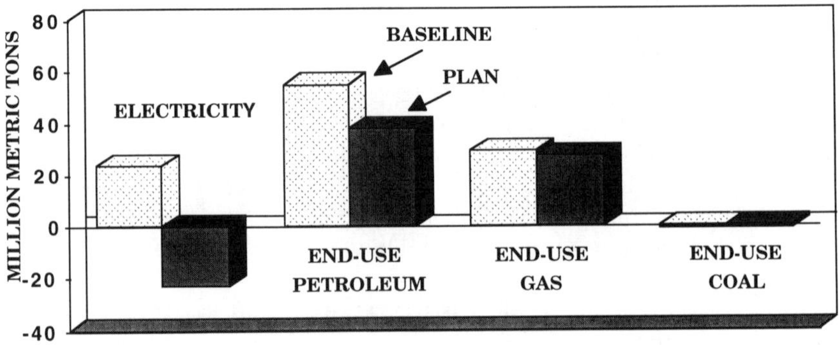

Figure 3.4 Year 2000 carbon emissions relative to 1990

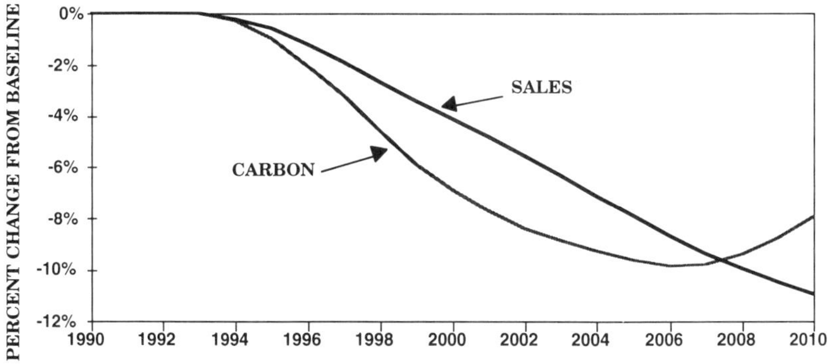

Figure 3.5 Annual utility sales and carbon emissions (demand reductions from baseline)

nuclear, and renewable (mostly hydro) facilities. When demand is reduced in the near term, utilities will back out fossil fuel generation from existing plants because nuclear and renewable plants have low operating costs. In addition, in the CCAP greater reductions come from coal plants because the base case electricity growth rate is already relatively low and much of the oil and gas capacity has already been reduced. Over the longer term, lower electricity demand leads to less construction of new facilities. The new facilities include a large proportion of gas and renewables. By 2008 the marginal carbon-intensity is below average intensity, and demand reductions lead to proportionally lower reductions in carbon.

When supply-side policy actions are added to the plan, they change the mix of new resources. The success of programs to stimulate the addition of less carbon-intensive new generation capacity is dependent on growth in the electricity demand and the resulting need for new resources. In a way, they "compete" with the energy efficiency programs for the same opportunity to reduce carbon emissions. If energy efficiency programs are very successful, there will be less opportunity for new capacity construction. At the same time, the construction of low carbon-intensive generation sources leads to smaller carbon reductions associated with electricity savings. As a result, the combination of the two types of programs will lead to lower savings than the simple sum of savings. This overlap of savings was insignificant in the CCAP by 2000, but increases by the year 2010.

Other types of interactions and feedback occur in any policy analysis, whether including only one policy or multiple policies, that integrated modeling can help quantify and explain. These interactive effects include impacts among individual programs within an area, price changes resulting from changes in demand levels, and potential changes in the demand for energy services.

In the CCAP, many of the interactions among programs in the same sector were handled by modeling them as a group. For example, for modeling purposes the residential building standards, residential energy efficient mortgages programs, and a program to create energy efficient housing centers to make builders more aware of energy efficient building options were combined. A judgment by DOE program managers was made of what type of savings would result from these programs and how many buildings would be affected by 2000.

Price feedbacks usually dampen the impact of programs. For example, programs that increase energy efficiency will lead to lower natural gas demand, which in turn lowers the projected price of natural gas. With lower prices the program will usually have a smaller impact. An exception to this is electricity prices. Because regulated electricity prices are based on embedded cost, demand reductions can lead to price increases in the near term. The fixed costs of electricity generation must be covered by fewer kilowatt-hour sales. In the long run, however, reduced demand leads to less construction and prices will be lower.

Feedback to the level of energy services can occur through changes in economic activity (more buildings to be heated and lit or more widgets to be produced) or through changes in the level of service demanded due to changing comfort levels. For the CCAP, the macroeconomic feedback was purposely excluded from the analysis. In the IDEAS model, the "behavioral" changes in service levels are assumed to occur when fuel bills change as a percentage of income. Behavioral impact can occur either because of sudden changes in energy prices or as take-back effects after conservation investments are made. This latter effect tends to dampen the impact of energy efficiency programs, because consumers may "take-back" some of the energy savings in increased comfort. In addition, lower energy prices resulting from a variety of programs could lead to higher service levels.

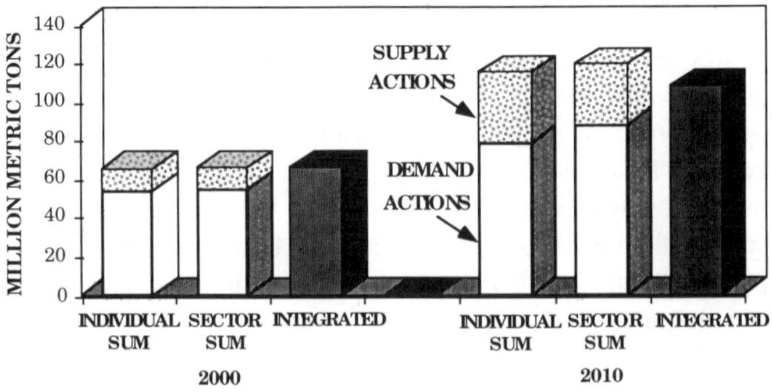

Figure 3.6 Annual carbon reductions

As illustrated in Figure 3.6, the combined interactive effects in the plan were negligible in 2000, and roughly 6% by 2010. The price impacts were offsetting and the electricity supply and demand overlap does not have any affect because there is little new construction of capacity by 2000 even in the base case. By 2010 the overlap among supply options and between supply and demand reductions is greater. The combined carbon reductions from all the supply options is 13% less than the sum of the individual supply actions. One of the biggest reasons for this combined effect is that high efficiency gas generation technologies compete with advanced renewables for expanded market share.

CONCLUSIONS AND REFLECTIONS

The Climate Change Action Plan provides a road map for the Federal government to stimulate reductions in greenhouse gases to attain the goal of reaching 1990 emission levels by the year 2000. A variety of actions were selected that span the energy-using sectors of the economy. The IDEAS model provided a framework for the analysis of the projected savings from the energy-related actions. The total emissions reductions from these actions was 66 MMTCE or 5% of projected emissions in 2000.

The process of creating this plan was a huge undertaking in a relatively short amount of time. Analysts from multiple agencies were involved, with DOE providing the coordinating role for the energy-related actions. DOE's use of the IDEAS model helped focus the analysis and make sure that all the analysts involved clearly stated their assumptions. Computer modeling in general adds rigor to any analysis; integrated modeling especially helps because all options go through a common point where they can be assessed on an equal basis.

Multiple programs can lead to overlapping savings, but may be necessary to help ensure overall program success. Because some individual programs may fall short, the inclusion of multiple policies acts as insurance. Therefore, for example, it makes sense to stimulate investment of high efficiency gas as well as renewable generation technologies, even though there may be some overlap in target markets. Integrated modeling can help policy makers to avoid overcounting the total impact of multiple policies and to understand interactions among policies.

The size of the interactive effects depends on the type and magnitude of the programs and the time horizon considered. Programs involving both the demand and supply of electricity are likely to have significant overlap, especially if the programs are large. Other cases of targeting the same market with multiple programs might be improving the efficiency of gasoline powered vehicles and at the same time promoting alternate-fueled vehicles. For the CCAP, the interactive effects were offsetting in 2000, and the total program impact was equivalent to the sum of the individual bundles of programs. Intra-sector interactions were mostly calculated offline and were taken into account through the inputs for the bundled actions. Over a longer time horizon, there were more interactions among programs, mostly related to new electricity capacity.

Even with single policies, it can be important to capture feedback effects, such as energy price responses. In most cases feedbacks dampen program impacts. For example, programs that are expected to lead to a significant drop in natural gas demand will also result in lower gas prices, which will offset some of the gas reductions. However, because electricity prices are regulated, the price may not move in the opposite direction to demand, especially in the near term.

NOTES

(1) However, because IDEAS captures some of the feedback relationships between the energy sector and the economy, it has some "general equilibrium" properties.

(2) The cobweb model does not always converge to equilibrium. Depending upon the slopes of the supply and demand curves the cobweb model can sometimes explode, leading to an ever-increasing gap between quantities demanded and supplied. Convergence to equilibrium has not been formally tested in IDEAS, but in fifteen years of use it has not been known to exhibit behavior typical of an exploding system. It is therefore felt that an equilibrium exists even though it has not yet been formally tested.

(3) This was mandated by the 1989 Energy and Water Authorization Bill.

REFERENCES

DOE (1994) *The Climate Change Action Plan: Technical Supplement*, DOE/PO-0011, US Department of Energy, Office of Policy, Planning, Program Evaluation, Washington, DC, March (available from the National Technical Information Service, US Department of Commerce, 5285 Port Royal Rd., Springfield, VA 22161, USA).

Mansfield, E. (1970) *Microeconomics: Theory and Applications*, W.W. Norton, New York.

US Environmental Protection Agency (1989) Policy options for stabilizing global climate, draft report to US Congress.

Chapter 4

MIS: A Model-based Macroeconomic Information System For Energy Analysis in Germany

Claudia Kemfert and W. Kuckshinrichs

INTRODUCTION

A large number of countries have decided to reduce energy-related CO_2 emissions drastically (see OECD/IEA, 1994 for an (incomplete) survey). In 1990, the German Federal Government pledged itself to make a 25–30% reduction in national CO_2 emissions by 2005 on the basis of 1987 CO_2 emissions. After the unification of the two German states, this target is valid now for the entire Federal Republic. Apart from that, the Study Commission on Preventive Measures to Protect the Earth's Atmosphere recommended a CO_2 reduction of 50% by 2020. Based on the work of the commission, the Federal Ministry of Education, Science, Research and Technology (BMBF) initiated the IKARUS (Instrumente für Klimagasreduktionsstrategien) project in October 1990. The aim of the project is to provide tools for developing strategies to reduce energy-related emissions of greenhouse gases in Germany. These tools comprise energy models and a database. To analyse economic aspects of CO_2 reduction strategies, the macroeconomic simulation model MIS (Macroeconomic Information System) has been developed by AGEP/

Systems Modelling for Energy Policy. Edited by D.W. Bunn and E.R. Larsen

Oldenburg University (Energy and Systems Modelling Working Group; Pfaffenberger & Ströbele, 1994) within the framework of cooperation with STE (Programme Group Systems Analysis and Technology Evaluation).

The concept and structure of MIS were chosen in order to reproduce in a macroeconomic context the energy flow and the technology structure of the German economy, which are determined by a bottom-up energy system optimizing model. Besides the production of base data for the energy system optimizing model, the main tasks of MIS are to evaluate within the sectoral structure of intermediate production the economic consistency and to show the macroeconomic effects of strategies calculated by the technology-orientated energy system optimization model. Because of that, the input–output (I/O) analysis seems to be the best concept to fulfil the claim. MIS has the advantage of disaggregating the economy into 30 sectors within an IO concept with special consideration of energy-intensive and emission-intensive sectors.

We use the MIS approach to study the economic effects of a specific CO_2 tax proposal. The driving force of the IO approach, the demand vector, is exogenously determined and therefore not price-dependent. Nevertheless, it is useful to determine the production adjustment due to a tax-induced energy price increase.

MIS MODEL

Overview

The central element of the IKARUS project is the technology-orientated optimization model (IKARUS-LP). Based on the method of linear programming, this model determines the cost-minimal technology structure and energy carrier mix that fulfils exogenously given energy demands regarding emission targets. In the context of the IKARUS project, the MIS model is developed to analyse strategies to reduce greenhouse gases within a macroeconomic context. In particular, the MIS model is developed to prove the macroeconomic consistency and to demonstrate the consequences of the given energy technology bundles within the structure of intermediate production. The technology bundles can be calculated by the IKARUS-LP. (The MIS model is a "top-down" model while the linear optimization model follows the "bottom-up" approach.)

The MIS model is an open, dynamic and demand driven IO model, based on the national input–output table of the German Federal Statistical Office.

IO models determine the impacts on macroeconomic sizes induced by exogenous changes of final demand. Final demand is exogenous in the MIS model, except for investment regarding the dynamic aspect of the IO approach (see the appendix). Exogenous final demand comprises consumption, government expenditure and exports. The model calculations occur after exogenous modifications for the forecasting periods 2005 and 2020.

The model determines structural effects resulting from specific exogenous premises: on the one hand, the Autonomous Energy Efficiency Improvement (AEEI) factors improve the energy efficiency; on the other hand, energy demand is influenced by energy prices. The substitution effect describes the consequence of changes in energy price relations: capital replaces energy. In the model this effect is described by the Constant Elasticity of Substitution (CES) production function. Changes of relative prices of the energy input factors induce further substitution effects between the individual energy factors. Furthermore, the structure of the energy quantities is influenced by energy imports. The user is also able to set assumptions about development in the electricity sector or simulate various demands of transport (modal split), which all change the structure of the interconnected IO table.

MIS iteratively computes the inevitable investment, capital stock and employment for a given demand. Modifications of scenario parameters are possible to describe the structure of demand,[1] the lifetime of buildings and equipment, the development of sectoral productivity, population and information about the target import/export structure.[2] A further submodel determines the demand for housing.

Macroeconomic changes, the development of energy and CO_2 emissions and transport flows will be calculated for the forecasting periods 2005 and 2020 (Figure 4.1).

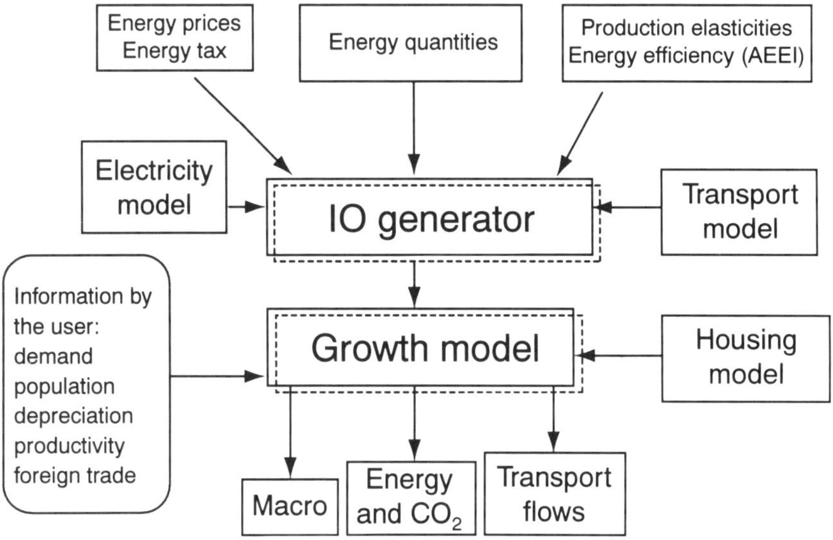

Figure 4.1 Overview of the MIS model and important dependencies

The IO Approach

As a rule, IO models are based on a national IO table of a specific base year. On the one hand, IO tables point out how the revenue of domestic production and import is distributed on individual goods; and on the other, how these goods are used in the economy and what income will arise in the individual production sectors (Leontief, 1986). The direct and indirect impacts on macroeconomic sizes due to exogenous changes of the final demand in the period of time are computed endogenously.

The MIS model is based on the 1989 IO table of the German Federal Statistical Office. To get better answers to energy policy questions and a better view of the energy flow and its interlinkages, the 58 sectors of the functionally orientated IO table of the national production statistics are aggregated to 8 energy and 22 non-energy sectors. The sectoral structure is shown in Figure 4.1.

Because of their particular significance from the aspect of CO_2 reduction, the space heat and motorized individual transport (MIT; private cars and motorcycles) sectors were introduced as so-called artificial production sectors. The space heat sector produces space heat and delivers it to all sectors and consumption; the MIT sector produces the artificial good mobility and delivers it to all sectors and consumption.

Primary inputs include depreciation, indirect taxes minus subsidies, wages, profits and imports. Final demand comprises consumption, government expenditure, and investment, subdivided into buildings and machinery/equipment, and export (Figure 4.2).

Investment

Generally, the dynamic IO approach includes time-dependent variables. In this approach investment is not determined autonomously, but endogenously. On the one hand, investment raises final demand; on the other, it alters the capital stock and therefore the production possibilities. Dynamic IO models are solved by considering the changes of capital stock due to sectoral investment. Therefore, MIS balances the stock and flow effect of sectoral investment.

MIS computes investment by a submodel differentiating net investment and replacement investment. Net investment is determined by the growth rate of value added and the respective capital coefficient. Increasing investment depends on the depreciation rate and the capital stock of the current period. Sectoral investment activity determines investment demand, which has to be satisfied by the investment producing sectors. The interconnected investment is obtained by the combination of investment activity and investment demand (see the appendix).

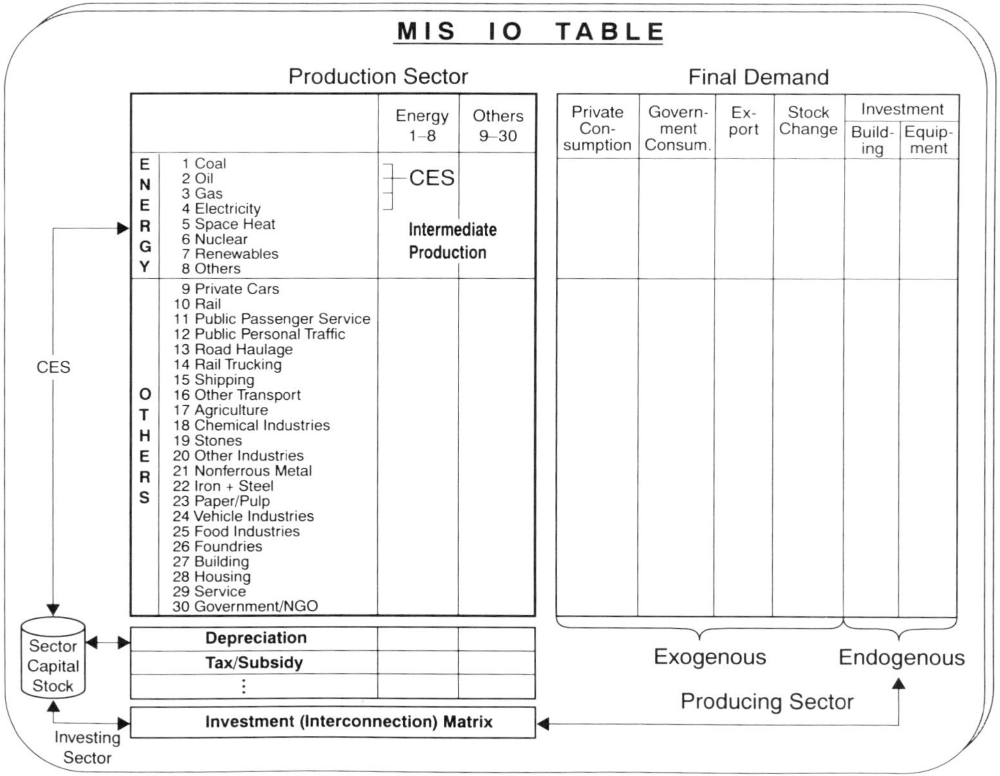

Figure 4.2 MIS IO structure

Population

Based on the data of the base year, development of population is computed by an exogenous growth rate for the period of time. It follows from this that the employment potential is received by an exogenous activity rate.

Employment

The model checks the total availability of the necessary employment for production. The necessary quantity of employment is obtained from the productivity function, which deduces the productivity growth from the growth rate of value added (see the appendix). After that the employment is computed for the target period by multiplying the quotient of value added growth rate and productivity growth with the employment of the last period. The model offers the advantage of modifying the labour productivity by varying the relevant parameters.

In the future it is planned to develop the model considering employment as an input factor of the production function. Therefore, substitution possibilities between labour, capital and energy will be reproducible.

Housing

Future housing requirement is determined in a separate submodel. The rental value of the demand for housing is computed based on forecasting requirements and living space. The age structure of buildings is differentiated: on the one hand, finished housings are registered, which will be reduced by a demolition quota; on the other, housing that will be finished within the period of time is included, therefore the quantity of housing is preserved. All stocks are characterized by the share of houses for one or several families, specific building costs and specific heat requirements for houses. Consequently, the total value of housing stock, rental value and the total heat requirement are received.

For several periods of time an annual modernization can be additionally simulated, describing the respective costs/qm and thus the achievable energy reductions. The necessary investment requirement per year, the growing building value, the resulting change of rent and the influence on the total heat requirement of the building will then be calculated.

Exogenous parameters comprise the persons per housing, share of new or old houses for one or more families, living space per house, building costs, heat requirement and modernization share.

Transport

Similarly to the housing sector, transport is described in a separate submodel. In this submodel the transport requirement is generated for the periods 2005 and 2020. A technological model preceding the MIS model determines technological changes and substitutions between individual transport sizes.

The user has the opportunity to give informations about the transport technology (for example, specific fuel consumption), the diversification of the transport demand in the base year, how the transport sectors are distributed and how transport requirement is disaggregated in transport purpose and transport distance.

Energy

The energy model is constructed to obtain detailed answers to specific energy economic questions, such as changes of energy structures in the macroeconomic context. Increasing energy prices induced by an energy/CO_2 tax have effects on the sectoral energy structure and total energy prices. In particular, the structure of the electricity sector can be modified by the user: the degree of effectiveness, share, capacity utilization and costs of generating stations are exogenous for the target periods 2005 and 2020.

Changes of the energy structure are described by the energy efficiency improvement factor (AEEI factor) and by a production function explaining substitution of energy carriers by capital and/or other energy carriers.

CES Production Function

The sectoral and altering energy prices are set exogenously by the user. In the MIS model the effects of increasing energy prices are described as follows: rising energy prices induce total sectoral price effects and therefore substitution effects between capital and energy and between energy carriers themselves. These substitution effects are calculated by the CES production function (Kemfert, 1994):

$$Y^{-\beta} = a \cdot K^{-\beta} + b \cdot \sum_{i=1}^{4} E_i^{-\beta_i} \tag{4.1}$$

where Y is production, K is capital, E is energy, a, b are distribution parameters and β is the substitution parameter $-1 \leqslant \beta \leqslant +\infty$. The elasticity of substitution $\sigma = 1/(1+\beta)$.

Because of relative prices changing, the structure of all sectoral prices will be calculated by a price model (see the appendix) and therefore sectoral substitution processes are induced (Figure 4.3).[3] Therefore, the production structure of the Leontief type will be replaced by a production structure with substitution possibilities (Arrow–Solow type).[4] Because of that, the coefficients of the IO table will be varied and the new IO structure will be determined by the so-called IO generator.

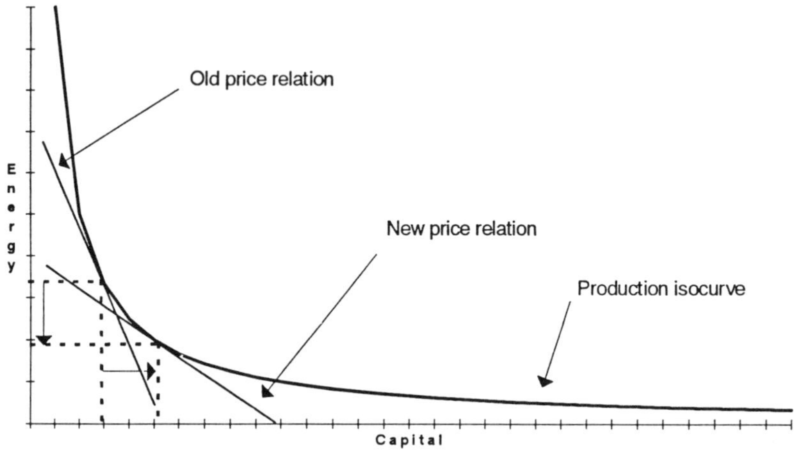

Figure 4.3 Substitution effect

AEEI Factor

The model includes sectoral autonomous energy efficiency improving technical progress (AEEI) to describe non-price induced energy efficiency gains freely available for the economy. This AEEI factor covers several aspects, such as technical improvements due to replacement-investing and organizational restructuring, structural change based on increasing share of less energy-intensive produced goods within a given sector and administrative measures, e.g. speed limits.

The CES production function including AEEI therefore changes to

$$Y^{-\beta} = a \cdot K^{-\beta} + b \cdot \sum_{i=1}^{4} (\gamma_i(t) \cdot E_i)^{-\beta_i} \tag{4.2}$$

where $\gamma_i(t)$ represents AEEI. $\gamma_i(t)$ represents energy-augmenting technical progress, with $\gamma_i(t) \cdot E$ measuring energy in efficiency units.

Energy-augmenting technical progress crumples the production isocurve. Therefore, assuming no price effect, the optimal factor input relation changes.

In the MIS model, sectoral AEEI rates must be given externally. Different specific energy needs result in different coefficients for the IO tables. MIS calculates the new structure of the IO table by an IO generator.

CO_2 Tax Calculations for Germany

In the context of the IKARUS project, the macroeconomic consistency of results calculated by the LP model is proved by MIS. Linking the two models is carried out by implementing several parameters, resulting from an optimization run of the IKARUS-LP, into the economic model. Beyond the energy optimization model, the economic model calculates not only the direct investments induced by the reduction of CO_2, but also the indirect effects through the capital goods producing sectors.

The LP model determines, inter alia, sectoral primary energy inputs, CO_2 emissions and costs for the target periods 2005 and 2020. These reduction costs are implemented as a CO_2 tax in MIS. In this case it is not the purpose to check the linear programmed results for macroeconomic consistency. Our intention is to show the macroeconomic consequences of changing energy policy in the IKARUS context.

The calculations concentrate on the years 1989 and 2005 for the region of the German Previous States.[5] To describe the effects of an energy-related CO_2 restriction for the structure of the energy supply and energy saving techniques, as well as for the overall economy, two cases were considered.

Business-as-usual

The first was a reference case for 2005 based on a "business as usual" (BaU) scenario. This means that no specific measures to reduce CO_2 emissions are taken up, but as a result of structural effects in the production sector and technical progress implemented via ordinary investment measures, the energy-related specific emissions can decrease.

Reduction case

A CO_2 reduction (Red) case for 2005 is based on the reference case (BaU) but includes a restriction of CO_2 to 75% of the 1989 value.

The base case according to 1989 statistics represents the starting point for the calculations. To figure out the effects of an emission reduction, the BaU and Red cases have to be compared.

Key assumptions

To run the MIS model, basic assumptions comprise economic and demographic developments, and energy-related political decisions. This includes the expected development of macroeconomic variables like GDP and prices of import energy, as well as political decisions and measures restricting the use of primary energy carriers and conversion techniques.

Based on the predicted economic development (e.g. by PROGNOS), detailed sectoral analysis, and own estimations, AGEP and STE expect GDP to increase in real terms from 2011 billion DM_{1989} in 1989 to 2895 billion DM_{1989} in 2005 (Figure 4.4). Population is expected to increase from 61.7 million in

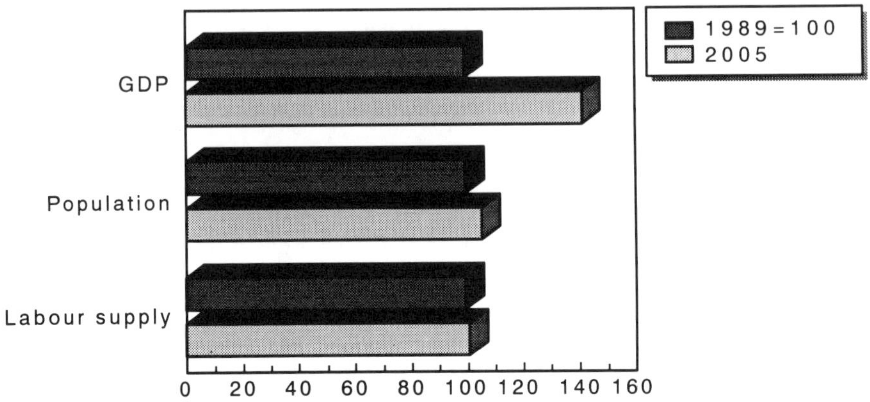

Figure 4.4 Basic assumptions

1989 to 65.0 million in 2005. Demographic effects result in a smooth increase of labour supply to about 30.3 million people.

Additional assumptions to run the MIS model comprise sectoral AEEI factors, sectoral elasticities of substitution and energy prices.

Sectoral AEEI

It is necessary to implement sectoral AEEI factors into MIS to model the energy efficiency improvement development that the optimization model calculations are implicitly based upon. Figure 4.5 describes the underlying parameters of those sectors with AEEI exceeding 1% per year. Particularly for

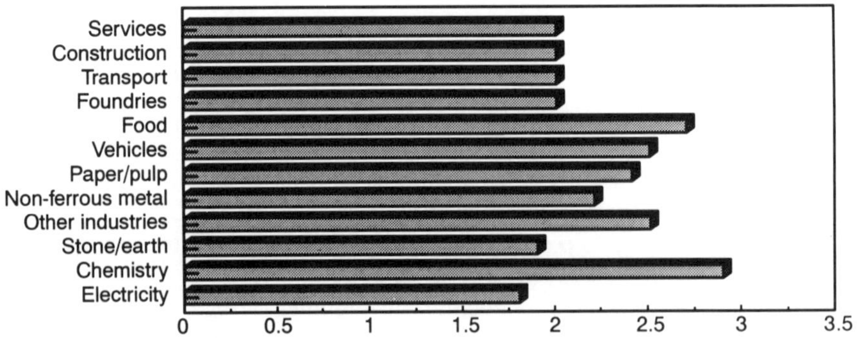

Figure 4.5 Sectoral AEEI in per cent p.a. (for the industry sectors, see Deutsches Institut für Wirtschaftsforschung, 1985)

the industry sectors, the parameters comprise structural effects depending on the change towards less energy-intensive goods within a sector and technological effects based on non-price induced technical progress, resulting in reduction of specific energy use in production.

CES parameter

Substitution possibilities within the CES structure of sectoral production are summarized in the elasticity of substitution (ESUB). This parameter describes the ability of production to revise factor input combinations as a reaction to changes of relative prices of these factor inputs.

Figure 4.6 demonstrates the linkage structure of the nested CES production functions and summarizes the estimated elasticities.

Energy prices

The prices of the energy carriers are expected to develop differently: for hardcoal the price is nearly stable; oil increases at a rate of 1.3% p.a. and gas at a rate of 1.4% p.a. until the target period 2005 (Tables 4.1 and 4.2).

Costs/CO_2 tax

The linear programming model IKARUS-LP calculates the marginal and average costs of reducing CO_2 emissions. Figure 4.7 shows the CO_2 shadow price curve.

Latest calculations determine marginal costs of CO_2 reduction of about 350 DM/t CO_2 whereas the average costs are about 150 DM/t CO_2.

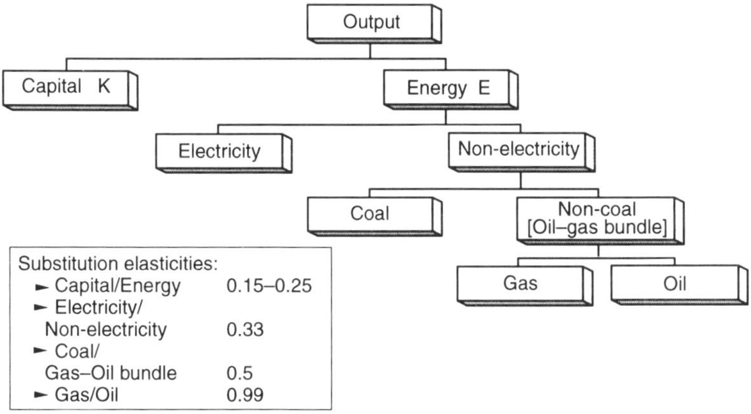

Figure 4.6 CES production structure

Table 4.1 Energy prices

Energy carriers	Base (DM/GJ)	Increase (% p.a.)	2005 (DM/GJ)
Coal*	7.64	0.36	8.13
Oil	4.83	1.3	6.02
Gas	3.92	1.4	4.97

*Coal is determined by a conglomerate of coal shares given externally by the user. According to LP assumptions, the shares differ as shown in Table 4.2.

Table 4.2 Coal shares

Coal shares	Base case (%)	2005 (%)
Brown coal (lignite)	27.53	24
German hardcoal	65.01	38
Imported coal	7.46	38

CO_2 factors

Table 4.3 shows CO_2 emission factors and corresponding tax rates for the fossil energy carriers, with tax rates based on the assumption of a CO_2 tax of 350 DM/t CO_2.

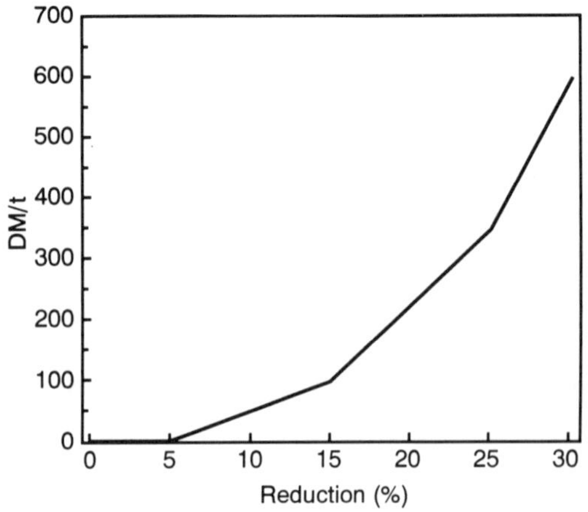

Figure 4.7 CO_2 shadow price

Table 4.3 CO_2 emission factors and tax rates

Energy carriers*	CO_2 emission factors (kg/GJ)	Tax rates (DM/GJ)
Coal	83.52	29.23
Oil	69.44	24.31
Gas	56.86	19.90

*The energy carriers and therefore the emission factors are aggregated according to the MIS energy structure.

CALCULATED MACROECONOMIC EFFECTS

Reference Scenario: Business-as-Usual

The reference case (BaU) is set as the standard to compare variations of energy policy. This scenario therefore demonstrates the purely structural effect without explicit energy policy implementations.[6] To give a short survey about the macroeconomic consequences, employment and emissions development are demonstrated as follows. As a measure to show the effects of a CO_2 tax implementation, employment effects are used.[7] Because of the significant AEEI factors, the reference case demonstrates considerable energy improvements and therefore CO_2 emission reductions within the target period until 2005 (Figure 4.8). A 25% CO_2 reduction can be received in the reduction scenario, where AEEI factors and the CO_2 tax are active.

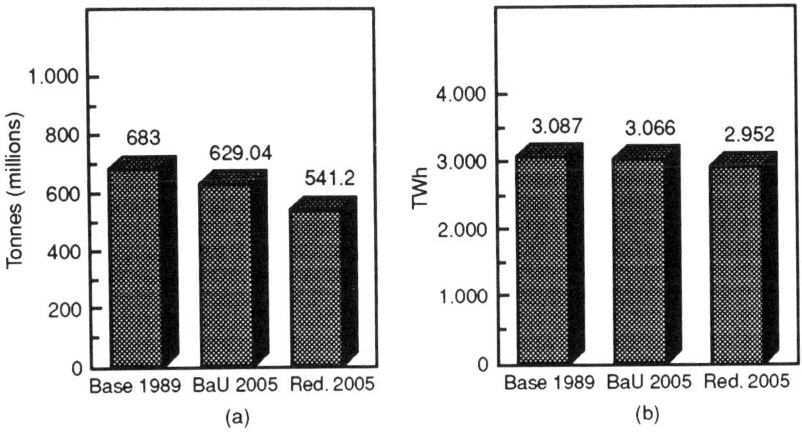

Figure 4.8 (a) CO_2 emissions and (b) primary energy

Figure 4.9 shows the development of sectoral employment due to the structural effect. It indicates the difference between sectoral employment in 1989 and 2005. The winners due to the structural effect are the service sector, construction, transport and other industries, whereas the main losers are agriculture and coal mining. But in total, the structural effect results in additional employment of 1.2 million. This additional employment concentrates in the service sector. This structure effect indicates the trend from energy intensive to labour intensive industries.

The strong effect in the coal sector reflects an assumption used to run the linear programming model. In that model an upper limit for the use of domestic hardcoal has been set to reflect political pressure in Germany to decrease the highly subsidized domestic coal mining industry. In conjunction with further efforts to increase labour productivity in this sector, this results in a strong drop of the German coal mining sector in the BaU case. The assumed demographic effects results in a smooth increase of labour supply to about 30.3 million people. Therefore, the net effect of 1.2 million additional employment reduces labour market disequilibrium. But forcing improved labour mobility needs additional labour market policy measures, which have to be discussed outside this model.

Reduction Scenario

According to the marginal reduction costs of 350 DM/t CO_2, calculated by IKARUS-LP, a CO_2 tax is implemented. Ideally, the implementation of a CO_2 tax encourages economic actors to install the optimal energy technology

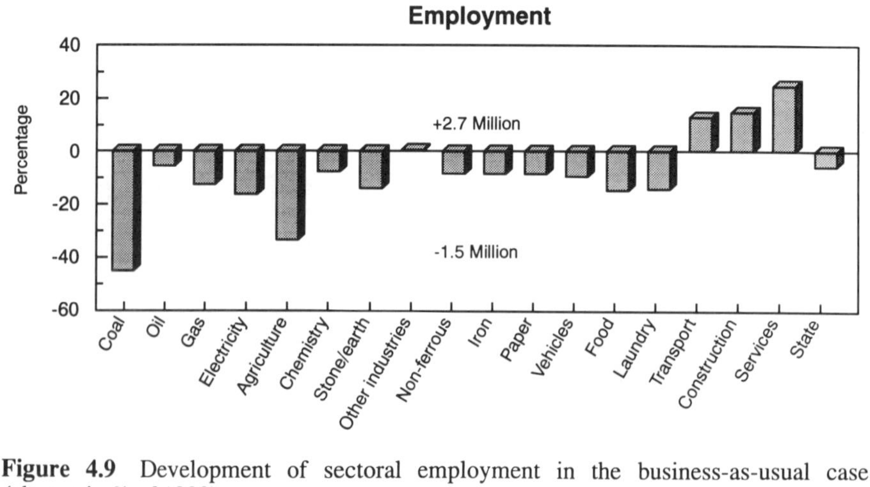

Figure 4.9 Development of sectoral employment in the business-as-usual case (change in % of 1989)

structure when CO_2 emissions are restricted. A CO_2 tax equal to the marginal reduction costs of 350 DM/t CO_2, raised for CO_2 emissions of 500 million t in 2005, amounts to a tax revenue of about 175 billion DM. Thus, high tax payments would influence the growth of the overall economy as well as structural change. In particular, the energy intensive producing sectors can get into trouble. The tax differently increases the cost of fossil energy carriers according to their carbon content, so that the relative prices of the energy carries change. Thus, based on sectoral assumptions of the elasticity of substitution, the CES production structure determines a revised combination of the input factors for sectoral production. Table 4.3 shows the tax rates for the fossil energy carriers, based on the assumption of a CO_2 tax of 350 DM/t CO_2.

To get a clear distinction between the price effect and the influence on the competitiveness of some energy-intensive producing sectors, the import/export assumptions of the BaU case remain unchanged in the reduction scenario. Nevertheless, it can be reasonable to adjust the parameters in a further step. This would result in an additional decrease of employment in some sectors.

Within the IO background we do not describe the effect of a revenue-neutral fiscal policy, where the CO_2 tax is transferred back, e.g. with decreased labour costs. Therefore, the model describes the worst case, where fiscal policy does not intend to minimize the cost burden of the economic actors.

Figure 4.10 shows several sectors outside the energy sectors. As Figure 4.10 demonstrates, energy intensive sectors like iron and transport have significant employment reductions compared with the BaU case. If these sectors are not subsidized, they will have to accept significant deficits.

Within the energy sectors severe reduction in the employment structure is

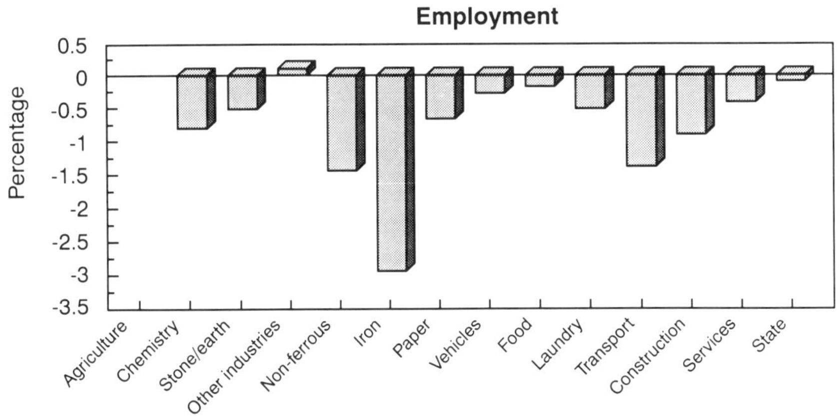

Figure 4.10 Employment in the reduction scenario: comparison with business-as-usual

recognizable. The oil industry and the coal industry will have the main employment reductions because of the high CO_2 tax. In the case of coal, employment decreases by nearly 65% compared to the BaU case. The oil and gas sectors show similar results. For coal, this effect is due in part to the increased share of coal imports in conjunction with decreasing coal use. Therefore, this result not only reflects the direct effect of a CO_2 tax, but comprises the employment effect of a more market-orientated German coal policy, reducing the use of highly subsidized domestic hardcoal in favour of cheap imported coal.

The total employment reduction is 178 000 people in comparison with the BaU case.

Whether this development is acceptable, is a political and social question that the MIS model cannot give an answer to. The model checks macroeconomic consistency and shows the macroeconomic effects of CO_2 reduction measures, whose political and social acceptability can only be discussed outside the model.

CONCLUSION

Within this chapter the status of the MIS model, the interdependent impacts and effectiveness by changing exogenous parameters have been represented precisely. MIS is developed within the framework of the German IKARUS project. Its main tasks in this context are to evaluate within the sectoral structure of the intermediate production the economic consistency and to show the macroeconomic effects of strategies calculated by a technology-orientated energy system optimization model. MIS is based on a dynamic IO approach, disaggregating the economy into 30 sectors.

Because of the autonomous effectiveness of the model, additional economic measures could be handled. This is demonstrated by the implementation of a national CO_2 tax according to special assumptions and externally calculated results.

As a measure to show the macroeconomic effects, employment figures are used. Changes in other macroeconomic sizes, especially information about the development of foreign trade, are neglected in this context.

Model calculations show that the high CO_2 reduction targets of the entire Federal Republic are achievable, but with significant economic effects based on policy interventions. On the one hand, the analysis is based on very high autonomous energy efficiency improvement assumptions; on the other, a significant CO_2 tax proposal is supposed.

Against the background of the assumptions, energy intensive sectors will get into trouble within the IO model concept. But this scenario represents the case with a non-neutral fiscal policy, e.g. CO_2 tax revenues are not transferred back to lower labour costs.

Finally, it must be said that the MIS model as an IO model has some weaknesses, e.g. of possibility to modelling feedback effects of macroeconomic changes. But with the model development and enlargement, the MIS model as a dynamic IO model represents a suitable macroeconomic analysis tool, especially to figure out the effects of a national CO_2 tax implementation. A great strength is showing the sectoral substitution effects between energy and capital and between energy carriers themselves.

APPENDIX: MODEL FORMULATION

Static Input–Output Analysis

The solution of an open, static IO system is given by:

$$\begin{aligned} X &= AX + Y \\ X[I - A] &= Y \\ X &= [I - A]^{-1}Y \end{aligned} \tag{4.3}$$

where X is total output, A is the coefficient matrix, Y is final demand, I is the identity matrix and $[I - A]^{-1}$ is the inverse Leontief matrix.

From the sectoral total output the primary inputs can be calculated:

$$L_i = l_i X \tag{4.4}$$

where L are the vector sectoral primary inputs and l are the coefficient primary inputs.

Dynamic Input–Output Analysis

Because of the mathematical handling, the period length of capital stock has to be given; MIS uses the following solution possibility:

$$X_t = [B^{-1}(I - A + B)]^t \times (X_0 - X_0^*) + X_t^* \tag{4.5}$$

where B is the capital coefficient matrix and X^* is the specific solution.

Investment

Investment is generated endogenously.

Net investment

$$I_{n,t} = \kappa_{t,i,j}(1 - 1/(1 + gv_j)) \times Va_{t,j} \tag{4.6}$$

where κ is the capital coefficient, gv is the growth rate of value added and Va is value added.

Replacement investment

The depreciation rate is determined by:

$$d = (d_1 + d_2)/2$$
$$d_1 = K/n$$
$$d_2 = d_1/(1 + gv)^n \qquad (4.7)$$

where d is depreciation rate.

$$I_r = d_t K_{t,i,j} \qquad (4.8)$$

where I_r is replacement investment:

$$I = I_r + I_n \qquad (4.9)$$

$$I_{i,t} = V_i I_{t,i} \qquad (4.10)$$

where V_i is the interconnected investment matrix. $I_{i,t}$ will be added to the IO table and then iteratively solved against the background of sectoral exogenous demand structure until an exactive benchmark is achieved.

Employment

Employment is determined as follows:

$$E_t = E_{t-1} \times gv_t/\pi_t \qquad (4.11)$$

where E is employment, π is the productivity growth rate and gv is the growth rate of value added.

The labour productivity (α and β) can be given externally, but must not exceed a maximal productivity growth:

$$\pi_t = \alpha + \beta gv_t \qquad (4.12)$$
$$\pi \leqslant \mu$$

where μ is maximal productivity growth.

Price model

Exogenous price changes induce a variation of all sectoral prices; this effect can be calculated with a price model. A tax, for example, increases the sum of primary inputs. A method to apply this effect shows the following formula:

$$p = rp' \times (I - A')^{-1} \qquad (4.13)$$

where A' is the transposed coefficient matrix and rp' is the sum of primary inputs times input prices.

NOTES

(1) The development of demand differs as follows: first, the user has to supply the global growth rate, then the sectoral development of demand (except investment) for the target periods.

(2) The user has to check the consistency of sectoral import and export assumptions.

(3) In accordance with empirical estimations, substitution elasticities between capital and energy lie between 0.4 and 0.6. The latest estimation (by AGEP, 1995) for the substitution elasticity between capital and energy is 0.6 for Germany in the period 1960–1993. This value means that more capital has to be inserted to substitute for energy; in this context, we use lower substitution elasticities.

(4) To describe price dependent energy policy changes and their consequences for macroeconomic sizes, it is necessary to simulate substitution effects.

(5) Calculations for the New Countries are being prepared within the IKARUS project.

(6) At the actual point of model development, the presentation of the "old" sectoral structure is used, which means that the aggregated transport sector comprises sectors 10 to 16.

(7) MIS shows other effects, e.g. for growth and capital stock, which are not described in this chapter.

REFERENCES

Kemfert, C. (1994) Energiereduktionen und CO_2 Emissionsminderungspotentiale im industriellen Bereich Westdeutschlands bis zum Jahre 2020.

Leontief, W. (1986) *Input–Output Economics*, Oxford University Press, Oxford.

OECD/IEA (1994) *Climate Change Policy Initiatives 1994*, update no. 1: *OECD Countries*, OECD, Paris.

Pfaffenberger, W. and Ströbele, W. (1994) *Projeck IKARUS: Makroökonomische Einbettung; MIS Modellhandbuch.*

FURTHER READING

Conrad, K. and Wang, J. (1994) Tradable CO_2 emission permits versus CO_2 taxes: economic impacts and costs by industry—an applied general equilibrium analysis for West Germany, in J.-Fr. Hake et al. (eds) *Advances in Systems Analysis: Modelling Energy-Related Emissions on a National and Global Level*, Konferenzen des Forschungszentrums Jülich, Vol. 15, pp. 241–261.

Davar, E. *The Renewal of Classical General Equilibrium Theory and Complete Input–Output System Models*, Avebury, Aldershot.

Deutscher Bundestag (1992) *Vorschlag für eine Richtlinie des Rates zur Einführung einer Steuer auf Kohlendioxidemissionen und Energie*, KOM (92) endg. Drucksache 12/2298.

Deutsches Institut für Wirtschaftsforschung (1995) *"Selbstverplichtung" für die Wirtschaft zur CO_2-Reduktion: Kein Ersatz für aktive Klimapolitik*, DIW-Wochenbericht 14/95, pp. 277–284.

Dornbusch, R. and Poterba, J.M. (1991) *Global Warming: Economic Policy Responses*, M1T Press, Cambridge, MA.

Duengen, H. and Schmitt, D. (1990) Konkurrierende Abgabenlösungen zur Reduzierung der CO_2-Emissionen, *Zeitschrift für angewandte Umweltforschung* **3**, 253–263.

Hake, J.-Fr., Jagodzinski, P., Kuckshinrichs, W., Markewitz, P., Martinsen, D. and Walbeck, M. (1994) IKARUS: a model system to reduce energy-related greenhouse gas emissions in Germany, in J.-Fr. Hake et al. (eds) *Advances in Systems Analysis: Modelling Energy-Related Emissions on a National and Global Level*, Konferenzen des Forschungszentrums Jülich, Vol. 15, pp. 213–238.

Proops, J.L.R., Faber, M. and Wagenhals, G. (1993) *Reducing CO_2 Emissions: A Comparative Input–Output Study for Germany and the UK*, Springer-Verlag, Berlin.

Schiffer, H.-W. (1992) *Der Richtlinienvorschlag der EG-Kommission zur Einführung einer CO_2-Energiesteuer, Braunkohle*, 11/92, 4–12.

Welsch, H. (1993) Carbon taxation in Germany: approaches and perspective effects, in C. Carraro and D. Siniscalco (eds) *The European Carbon Tax: An Economic Assessment*, Kluwer, Dordrecht.

Chapter 5

Benefits from Electricity Trade in Northern Europe under CO_2 Constraints

Tomas Larsson

INTRODUCTION

The electricity systems in northern Europe differ in wide respects between the countries. The high *per capita* consumption, particularly in Norway but also to some extent in Sweden and Finland, has been made possible through the high availability of low-cost hydro power. This chapter presents the first results of a study on the northern European electricity trade. So far, Denmark, Norway and Sweden have been analysed, with Finland and Germany still to be included. In Table 5.1 sources of electricity generation and *per capita* consumption of electricity are presented for the five countries.

In 1993 the exchange of electricity between Denmark, Finland, Norway and Sweden was 18 TWh, i.e. 5% of total generation (Nordel, 1994). Only a limited part of the trade is based on long-term contracts. The short-term trade takes place through bilateral agreements on the basis of declared marginal costs.

Restrictions on CO_2 emissions increase the cost effectiveness of electricity trading as a means for joint implementation. The marginal costs for reducing

Systems Modelling for Energy Policy. Edited by D.W. Bunn and E.R. Larsen

Table 5.1 Share of power generation (%) for fossil thermal power, hydro power and nuclear power, together with electricity consumption *per capita* for Denmark, Finland, Germany, Norway and Sweden. Values are for 1991

	Fossil thermal power	Hydro power and wind	Nuclear power	Consumption per capita (kWh/y)
Denmark	97	3	–	6 700
Finland	44	23	33	13 100
Germany	68	4	28	7 200
Norway	0	100	–	25 300
Sweden	5	43	52	17 000

CO_2 emissions in Norway and Sweden are very high (Kram, 1993), while an emission reduction of 20% by 2005 is included in the Danish energy plan.

Five scenarios concerning trade and nuclear power are analysed in this chapter:

- No trade allowed. Emission reduction through joint implementation.
- Trade limited to the existing transmission capacity. Emission reduction through joint implementation.
- Trade limited to the existing transmission capacity. Emission reduction country by country.
- New transmission capacity allowed for trade. Emission reduction through joint implementation.
- Trade limited to the existing transmission capacity. Emission reduction country by country. Swedish nuclear power phased out from 2000 to 2010.

These scenarios have been combined with four scenarios concerning CO_2 emissions:

- CO_2 emissions unconstrained.
- CO_2 emissions frozen at the 1990 level.
- CO_2 emissions for 1990 reduced by 20% over the period 2000–2010 and then kept constant.
- CO_2 emissions for 1990 reduced by 40% over the period 2000–2010 and then kept constant.

In the following section the methodology used here is discussed in relation to other possible approaches.

METHOD

In the method used here the energy system is represented as a *reference energy system* (RES; Marcuse et al., 1976). In Figure 5.1 a generic picture of an RES

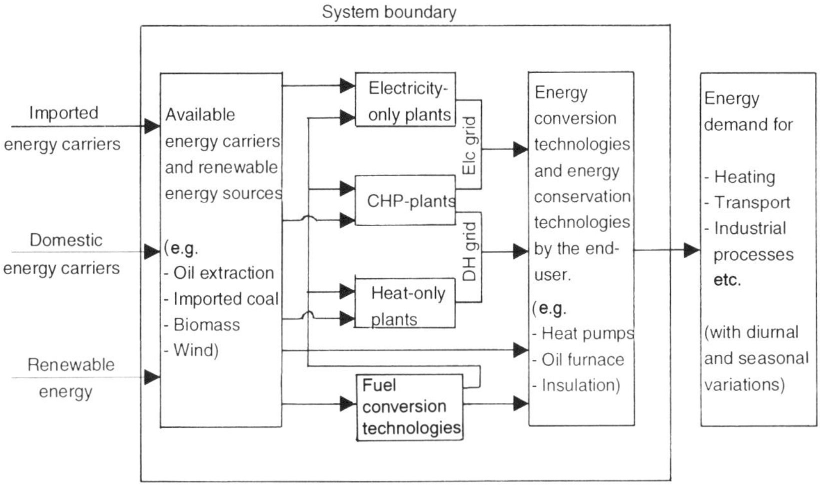

Figure 5.1 Generic description of a reference energy system

is presented. An RES describes the energy flows from extraction of energy via centralized conversion, transmission and distribution, to final conversion to useful energy. The existing energy system is described together with alternative technologies and energy flow paths. By describing the energy system as an RES, competition and synergies between different parts of the energy system can be studied, e.g. competition between energy supply and conservation, and combined heat and power (CHP) to supply both electricity and heat demand.

For this study the IEA–MARKAL model (Abilock et al., 1980; Altdorfer et al., 1980; Fishbone & Abilock, 1981; Fishbone et al., 1983) was used to describe the RES. MARKAL is a dynamic linear programming model of the technical energy system. Therefore, it is also possible to study the development of the energy system over time, including the timing of investments. Emissions and emission control technologies can also be modelled.

An RES gives *one* representation of the energy system, but does not provide *the* picture. The other methodological approaches that have been used to analyse Nordic electricity trade are discussed below, to describe what aspects of the energy system they represent.

The RES describes the energy system primarily in space. As mentioned above, MARKAL also describes the long-term perspective well. This makes MARKAL suitable for strategic analyses where investment programmes are evaluated. The short-term perspective, the focus for the operation of the energy system, is described with much less detail. The load curve is described with six steps over the year, and the differences between dry and wet years cannot be modelled.

A model of the electricity system alone can analyse these aspects in much more detail by basing the representation of the energy system on the *load curve*. These models are used to optimize the operation of the electricity system. Alternative technologies can be included in the model, thus investments in new generation technology can also be evaluated. The utilization of hydro power in the perspective of dry and wet years can be analysed. The description of the links to the rest of the energy system is limited or non-existent. Conservation measures and combined heat and power generation can be modelled to some extent. Models of this type are generally tailor-made for each power system. One model has also been developed for the Nordic electricity system, Samkjøringsmodellen. Models based on the load curve have more detail than models based on the RES, but they also have less scope. The implications of detail and scope for energy models are discussed in more detail in Wene (1989).

In models like MARKAL, useful energy demand is rigid. Only technical measures to reduce the energy demand can be studied. Energy conservation measures, such as insulation and more efficient appliances, can be included, but not changes in consumer behaviour caused by changes in energy prices. To study such effects the energy system has to be described as a *partial equilibrium model*. In this case the system boundary of the energy system is moved to the right in the RES. Consumers' sensitivity to changes in energy prices is modelled by own price elasticities and cross elasticities. One such model, Nordisk energimarkedsmodell (Bye & Johnsen, 1995), has been used to study the Nordic energy market. As the scope has been widened, some detail has been sacrificed. The model is, for example, static, i.e. no information is passed on between the time periods studied. Consequently all capital costs are annualized. A new model is currently under development, MICRO-MARKAL, which combines the detail of MARKAL with elastic demand.

More extreme scenarios, e.g. stringent CO_2 emission control policies, are likely to affect economic growth and the structure of energy demand. To study such effects a *general equilibrium model* is required. Only national general equilibrium models exist for the Nordic countries. The Norwegian model MSG-EE (Johansen, 1960) has been used to study the effects on the Norwegian economy of a CO_2 tax in the Nordic countries (Jarlset, Johnssen & Larsen, 1993). For the study the Samkjøringsmodellen was used to simulate a "world market" price for electricity. In general equilibrium models the scope is widened beyond the energy system to model the whole national economy. Thus, the energy demand can be dependent not only on energy prices but also on the relation between energy prices and prices of other commodities. As the scope is widened more technological detail is lost.

MARKAL has also been merged with a macroeconomic model to form MARKAL-MACRO (Manne & Wene, 1991). MARKAL-MACRO was developed to study the income effect on energy demand. As energy is the only commodity in the production function, the model is less suited for the study of structural change.

So far the representation of the energy system has been discussed in terms of detail and scope. All of these approaches consider consumers and producers as homogenous groups only reacting to external information. However, the electricity market is so concentrated on the supply side that some companies have the possibility of acting on the market to strengthen their position. Such behaviour can be modelled by a *game theoretic model*. One such model has been developed for the Swedish electricity market with the utilities as players (Andersson & Bergman, 1995). The utility model is developed to analyse what happens to energy prices and supply in the short term when the electricity market is deregulated. This is done for different assumptions on number of actors and degree of monopolistic behaviour. A similar study has been undertaken by Bjorvatn & Tjøtta (1993) but with the individual countries as players. The utility model is still under development and a dynamic version of the model is expected soon. Such a model would be valuable to investigate whether the present institutional structure of the electricity market can facilitate the cost-effective solutions derived from the other approaches.

Five different methodological approaches, including this one, have been used to analyse Nordic electricity trade. None of the approaches provides a complete picture of the situation. The focus of this study is the technical/economic implications of electricity trade under different scenarios concerning CO_2 emissions. Combined heat and power production has a large share of the electricity market, particularly in Denmark. Biofuels are also used to a large extent for CHP production. These phenomena are best represented by an RES-based model, which has been regarded as sufficient for the initial stage of this project. Later, the results from the RES-based analysis must be confronted with expert views from the other methodological approaches.

THE MODEL

MARKAL is a dynamic linear programming model of the technical energy system. It was developed within a multinational framework within the International Energy Agency (IEA). The model has frequently been used to aid decision makers at national, regional and local applications in a wide range of countries. Examples of national studies can be found in Kleeman, Seele & Wardiman (1991); Kram (1993), and ETSU (1994), of regional studies in Berger et al. (1990), and of local studies in Johnsson, Björkqvist & Wene (1992) and Rydén, Johnsson & Wene (1993). MARKAL has over the years been extended to include new features. To evaluate hedging strategies, sequential decision making has been included (Larsson & Wene, 1983; Larsson, 1986). To study trade of electricity between regions, multiple electricity grids have been introduced (Berger et al., 1988, 1990). It is the latter version of MARKAL that is used here, referred to as Extended MARKAL.

In the model the existing energy system is described together with all possible alternative energy technologies and energy flow paths. The model selects the combination of energy technologies and energy flow paths that best satisfies the specified system objective over the studied period (usually 25–40 years). The system objective is usually expressed as minimum total system cost with a real rate of discount of *r*%, with *r* typically 4–6%. Inputs are useful energy demand by sector region, quality, and duration; prices and availability of energy carriers in the market outside the system boundary; the cost and technical properties of new and existing energy technology; and constraints on the system's impact on the physical environment in the form of emissions, land use, etc. Similar models are EFOM (Finon, 1979; van der Voort et al., 1984) and Message (Agnew, Schrattenholzer & Voß, 1979). The ability of MARKAL to serve as a powerful decision support system has increased significantly through the development of a user interface (Goldstein, 1991).

DATA AND ASSUMPTIONS

The analysis covers the stationary energy systems of Denmark, Norway, and Sweden. Demand projections and technology assumptions are based on the MARKAL databases for Norway (Alm, personal communication) and Sweden (Vattenfall, 1994), and on the EFOM database for Denmark (Grohnheit, personal communication). For Norway and Sweden both supply and demand options are included, while for Denmark only supply options are included, i.e. only final energy demand is considered for Denmark. The total useful energy demand[1] for the three countries in 1990 is 550 TWh and is expected to grow at an average of 0.8% per year over the studied period, 1990–2030. Assumed costs for new transmission capacity are listed in Table 5.2.

In all scenarios it is assumed that each country maintains a capability to supply its own electricity demand. The Swedish nuclear power plants are only phased out in one scenario. In the other scenarios the existing capacity is retired

Table 5.2 Costs for new transmission lines

Transmission line	Cost (SEK/kW)
Norway–Sweden	800
Denmark–Sweden	2400
Denmark–Norway	3300
Jutland–Zealand	1000

over the period 2010–2020. Re-investments are allowed up the present capacity level of 10 GW at a cost of SEK 4500/kW.

The system objective is total discounted system cost with a real rate of discount of 5%. The result from the model is the system that best satisfies this objective from a technical/economic viewpoint. Organizational factors, e.g. degree of concentration on the supply side of the market, are not included.

RESULTS

Figure 5.2 shows the trade-off between total discounted system cost and accumulated CO_2 emissions. The differences in accumulated emissions between the scenarios depend on what happens at the early stages of the studied period. Figure 5.3 shows CO_2 emissions in the unconstrained cases. In all cases CO_2 emissions drop during the first years due to the introduction of cost-effective conservation measures and the export of electricity to Denmark. The latter option is not available in the case of no trade. Therefore, the initial drop in CO_2 emissions is smaller in this case.

Figures 5.2 and 5.3 When each country is obliged to meet the emission target, the problem becomes infeasible for Norway. Even keeping emissions constant is impossible for the years around 2000 as non-fossil technologies cannot be introduced at the rate necessary. For the 20% reduction case the

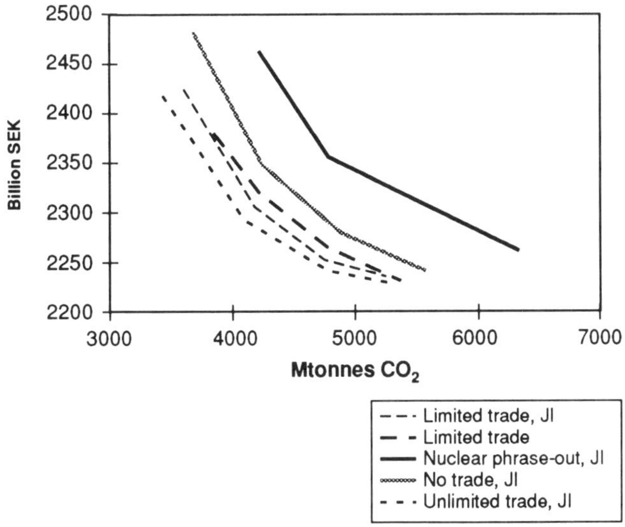

Figure 5.2 Trade-off between total discounted system cost and accumulated CO_2 emissions

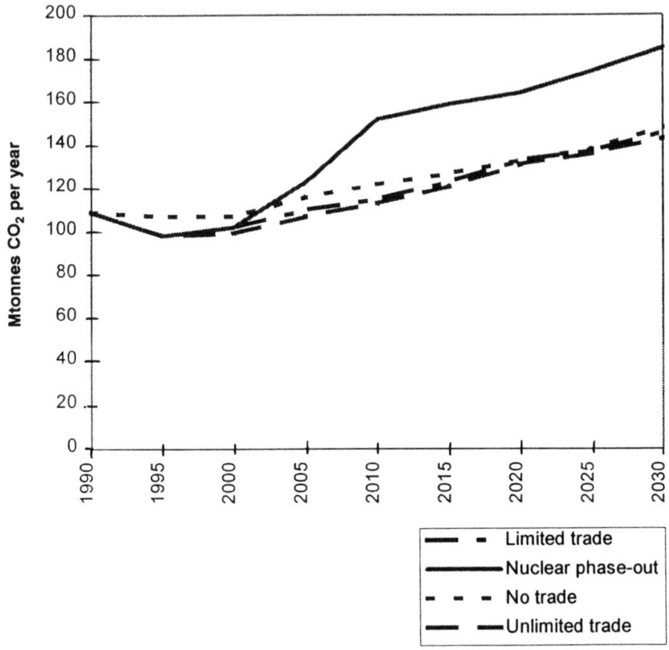

Figure 5.3 CO_2 emissions in the unconstrained cases

problem is infeasible for the years 2000–2020, and for the 40% reduction case from 2000 and throughout the period. To still be able to model the case of country by country emission reductions, a dummy CO_2 emission reduction technology has been introduced that reduces emissions at a cost of $2000/ton carbon. All options less expensive than this level are utilized. The cost of this dummy technology has then been deducted from the total system cost. This is also why the emissions are higher in this case compared to the case when joint implementation is allowed.

Figures 5.2 and 5.3 also show the importance of nuclear power to the Swedish electricity system. The cost of phasing out nuclear power is 30 billion SEK in the unconstrained case, 100 billion SEK in the CO_2 freeze case, and 150 billion SEK in the 20% reduction case. The 40% reduction case is infeasible when nuclear power is phased out and trade limited to the present transmission capacity. This is the reason why this trade-off curve has only three points.

The marginal cost of reducing CO_2 is illustrated in Table 5.3. The case where joint implementation has been excluded is omitted, as the marginal cost varies among the three countries.

Export of electricity can be made possible through three groups of technical measures: existing capacity can be utilized better, new capacity can be built,

Table 5.3 Marginal cost for reducing CO_2 ($/ton carbon; $1 = SEK 7.50)

Scenario	CO_2 constant	CO_2 reduced by 20%	CO_2 reduced by 40%
Limited trade	120	240	640
Unlimited trade	85	200	550
No trade	150	260	670
Nuclear phase-out	200	450	n.a.

Table 5.4 Hydro power production in Norway (TWh/year)

Scenario	Unconstrained CO_2	CO_2 constant	CO_2 reduced by 20%	CO_2 reduced by 40%
Limited trade	115	119	119	124
Unlimited trade	115	121	127	132
No trade	109	112	114	115
Nuclear phase-out	115	121	124	n.a.
No joint implementation	115	119	123	125

and domestic use can be reduced through conservation and fuel switching. For Norway a combination of these measures is undertaken. However, as the CO_2 constraint is tightened the construction of new hydro power plants is the dominating source. Table 5.4 shows electricity production from Norwegian hydro power plants.

The shadow prices of electricity become more equal among the countries as trade increases. In the case of free trade the difference is equal to the cost of transmitting the electricity from Norway to Sweden and Denmark. A Swedish nuclear phase-out would also influence the shadow prices of the other countries in the constrained cases, particularly for Denmark, as competition for fossil-free energy resources would increase. For Sweden, the nuclear phase-out leads to almost doubled electricity prices in the unconstrained case.

DISCUSSION

Here, the initial results of a study on joint implementation and electricity trade in Northern Europe are reported. Both trade and constraints on CO_2 emissions influence the marginal cost of electricity. Trade raises the shadow prices of electricity in Norway. This means that the energy-intensive Norwegian industries would lose in comparative advantages. Bye & Johnsen (1995) estimate the drop in domestic electricity demand for Norway to be 7.5 TWh for a rise of

NOK 100/MWh in the electricity price. Here trade alone causes a rise in shadow prices of about NOK 50/MWh. The structural changes caused by trade can therefore be expected to be relatively small.

When CO_2 is reduced, the effects are greater; in some cases the shadow price of electricity is almost tripled. Such price levels would for certain have an effect on useful energy demand. How great this effect would be, however, would depend on what CO_2 policies are implemented in the rest of the world.

ACKNOWLEDGEMENTS

I would like to thank Leif Alm and Fridtjof Unander of Institutt for energiteknikk in Norway and Poul Erik Grohnheit of Risø National Laboratory in Denmark. They made possible this study by supplying data on their countries' respective energy systems.

NOTE

(1) Although the analysis only considers final energy demand for Denmark, useful energy is used here for simplicity to describe the total demand for the three countries.

REFERENCES

Abilock, H., Bergström, C., Brady, J., Doernberg, A., Ek, A., Fishbone, L., Hill, D., Hirano, M., Kavanagh, R., Koyama, S., Larsson, K., Leman, G., Love, P., Sailor, V., Sato, O., Shore, F., Sira, T., Teichmann, T. and Wene. C.-O. (1980) MARKAL – a multiperiod linear programming model for energy system analysis, in R. Kavanagh (ed.) *Proceedings of an International Conference on Energy System Analysis*, 9–11 October 1979, Dublin, Reidel, Dordrecht, pp. 482–493.

Agnew, M., Schrattenholzer, L. and Voß, A. (1979) A model for energy system alternatives and their general environmental impact, WP-79-6, IIASA, Laxenburg, Austria.

Altdorfer, F., Blasco, M., Egberts, G., Finnis, M., Gundermann, J., Liemkuehler, K., Manthey, C., Pönitz, E., Rath-Nagel, S. and Tosato, G. (1980) Energy modelling as an instrument for an international strategy for energy research, development and demonstration, in R. Kavanagh (ed.) *Proceedings of an International Conference on Energy System Analysis*, 9–11 October 1979, Dublin, Reidel, Dordrecht, pp. 140–157.

Andersson, B. and Bergman, L. (1995) Market structure and the price of electricity: an ex ante analysis of the deregulated Swedish electricity market, *Energy Journal* **16** (2), 97–109.

Berger, C., Haurie, A., Lessard, E., Loulou, R. and Paskievici, A. (1988) MARKAL: development of materials, financial and multi-regional capability, École des Hautes Études Commerciales, Montreal, Canada.

Berger, C., Dubois, R., Lessard, E., Haurie, A. and Loulou, R. (1990) Assessing the

dividends of power exchange between Quebec and New York State: a systems analysis approach, *International Journal of Energy Research* **14**, 253–273.
Bjorvatn, K. and Tjøtta, S. (1993) Deregulation and common carriage in the Nordic power system, *Energy Journal* **14**, 57–74.
Bye, T. and Johnsen, T.A. (1995) Prospects for a common deregulated Nordic electricity market, Research Department, Statistics Norway, presented at Application of E3-Models, Göteborg, 8–9 May 1995.
ETSU (1994) *An Appraisal of UK Energy Research, Development, Demonstration and Dissemination*, HMSO, London.
Finon, D. (1979) Scope and limitations of a formalized optimization of a national energy system: the EFOM model, in A. Strub (ed.) *Energy Models for the European Community, An Energy Policy Special*, no. 1.
Fishbone, L. and Abilock, H. (1981) MARKAL, a linear programming model for energy system analysis: technical description of the BNL version, *International Journal of Energy Research* **5**, 353–375.
Fishbone, L., Giesen, G., Goldstein, G., Hymmen, H.A., Stocks, K.J., Vos, H., Wilde, D., Zölcher, R., Balzer, C. and Abilock, H. (1983) Users Guide for MARKAL (BNL/KFA version 2.0) report BNL-51701, Department of Applied Science, Brookhaven National Laboratory, Upton, NY.
Goldstein, G. (1991) PC-Markal and the Markal User Support System (MUSS), BNL-46319, Department of Applied Science, Brookhaven National Laboratory, Upton, NY.
Jarlset, T., Johnsen, T.A. and Larsen, B.M. (1993) Skatt på CO_2-utslipp i Norden – Virkninger for norsk krafteksport og bruk av olje til oppvarming i Norge, *Økonomiske analyser* no. 7-1993, Statistisk sentralbyrå, Oslo, Norway (ISSN 0800-4110).
Johansen, L. (1960) *A Multi-Sectoral Model of Economic Growth*, North-Holland, Amsterdam.
Johnsson, J., Björkqvist, O. and Wene, C.-O. (1992) Integrated energy-emissions control planning in the community of Uppsala, *International Journal of Energy Research* **16**, 173–188.
Kleeman, M., Seele, U. and Wardiman, D. (1991) National energy supply patterns for Indonesia: a computer aided energy planning and technology assessment study, in R.K. Pachauri, L. Srivastava and K. Thukral (eds) *Energy–Environment–Development*, Vikas, New Delhi, pp. 290–300.
Kram, T. (1993) National energy options for reducing CO_2 emissions. Vol. I: The international connection, a report of the Energy Technology Systems Analysis Programme/Annex IV (1990.93), Netherlands Energy Research Foundation, ECN-C93–101.
Larsson, T. (1986) GENFIX: providing for extra decision nodes in Markal, report A86-155, Department of Energy Conversion, Chalmers University of Technology, Göteborg, Sweden.
Larsson, T. and Wene, C.-O. (1993) Developing strategies for robust energy systems. I. Methodology, *International Journal of Energy Research* **17**, 503–513.
Manne, A.S. and Wene, C.-O. (1991) MARKAL-MACRO: a linked model for energy-economy analysis, report BNL-47161, Department of Applied Science, Brookhaven National Laboratory, Upton, NY (reprinted in J.-Fr. Hake et al. (eds) (1994) *Advances in Systems Analysis: Modelling Energy Related Emissions on a National and Global Level*, Konferenzen des Forschungszentrums Jülich, Vol. 15, pp. 647–667).
Marcuse, W., Bodin, L., Cherniavsky, E. and Sanborn, Y. (1976) A dynamic time

dependent model for the analysis of alternate energy policies, in K.B. Haley (ed.) *Operational Research '75*, North Holland, Amsterdam, pp. 647–667.

Nordel (1994) Nordel Annual Report 1993, Nordel's Secretariat, SK Power Company, Hellerup, Denmark.

Rydén, B., Johnsson, J. and Wene, C.-O. (1993) CHP production in integrated energy systems: examples from five Swedish communities, *Energy Policy*, **21**, 176–90.

van der Voort, E., Donni, E., Thonet, C., Bois D'Enghien, E., Dechamps, C. and Guilmot, J.F. (1984) Energy supply modelling package EFOM-12C, mark 1, CABAY, Louvain-la-Neuve, Belgium.

Vattenfall (1994) Beskrivning av markalmodellen och den svenska databasen, Vattenfall, Vällingby, Sweden.

Wene, C.-O. (1989) Exploring and mapping: a comparison of the IEA–Markal and CEC–EFOM technical energy system models and the ANL electric utility model, report BNL-52224, Department of Applied Science, Brookhaven National Laboratory, Upton, NY.

Chapter 6

Energy Planning and Multi-level Optimization: The Australian Energy Planning System Optimization Modelling Study

Sardar M.N. Islam

INTRODUCTION

Events in the international and domestic energy markets have generated several energy policy problems for the Australian economy. These problems include a lack of Australian self-sufficiency in energy, the appropriate pricing of energy products, the choice of fiscal instruments, the distributional equity of the benefits created in the energy sector, the conservation of energy (particularly of fossil fuels), the search for measures to manage supply disturbances, the optimum depletion rate of fossil fuels, and the determination of the exact boundary of energy policies (Weymans-Jones, 1986; Department of Primary Industries and Energy, 1988).

Several options are available to Australian policy makers to address these energy policy issues, and energy plans should consider these issues simultaneously. To formulate plans, energy models and planning studies have been conducted in Australia, but these studies have two important limitations

Systems Modelling for Energy Policy. Edited by D.W. Bunn and E.R. Larsen

(Centre for Applied Economic Research, 1985). Firstly, the models only have one level of optimization and therefore do not adequately represent the underlying multi-level policy system in the Australian economy: a clear and complete representation of the policy interactions between the government and economic agents is lacking in these models. Consequently, existing energy models cannot provide the results necessary to facilitate the decision making process, and cannot be satisfactorily applied to formulate multi-level energy planning policies.

In addition, operationalization of the policy planning problem via the formation of classified model variables as the model target and instrument variables has not been done.

Secondly, the existing energy model cannot produce results that will allow the formulation of a comprehensive set of energy policies consisting of energy technological and economic policy settings. The existing energy models produce results only for a partial set of the variables that are used to formulate policies. The mathematical programming models provide results mainly on energy technological policies and shadow prices, while the macroeconometric models produce results for taxes, subsidies and other economic policies. The deficiencies of existing energy models, when used in multi-level energy planning studies, necessitate the adoption of a multi-level optimization (MLO) approach to energy planning.

The objective of this study was to develop an energy policy planning model for the Australian economy that is appropriate in representing the underlying policy and institutional characteristics of the Australian economy and that can provide a set of useful energy policy strategies. It is argued in this study that an MLO model of the energy sector of the Australian economy is appropriate for energy policy formulation and modelling in Australia.

This chapter outlines the Australian Energy Planning System Optimization Model (AEPSOM) (Islam, 1995, and forthcoming) and a summary of the policy implications from the solutions of the model. AEPSOM is an MLO model that overcomes the two limitations of current energy models in Australia (as well as overseas).

MLO APPROACH TO ENERGY POLICY FORMULATION: JUSTIFICATION AND THEORETICAL FORMULATION

Government intervention in an economy as a means of achieving allocative efficiency is justified by the existence of market failures in a market or mixed economy. Generally, economic decisions in the mixed market economy are made by individual economic agents (the producers and consumers of energy), while policy makers only influence the behaviour of these economic agents.

Therefore, there are two types of decision makers: (i) the policy makers, who decide the optimum policy for the government; and (ii) the economic agents, who decide the optimum production and end use of the energy products, given the government's policy. The former decision is referred to as the *policy problem*, while the latter is a *behavioural problem*. It is envisaged that the behavioural problem is embedded in the policy problem, so that the economy is represented by a two-level policy system. Hence, an MLO modelling approach is necessary, particularly in a sectoral policy context such as energy planning.

The MLO modelling approach for energy planning developed in this study is consistent with Tinbergen's (1952) theory of economic policy planning, and provides a suitable operational framework. This study is also based on policy systems analysis, which is used to understand and identify the exact characteristics of the underlying policy planning problem.

THE AUSTRALIAN MODEL: AEPSOM

An empirical content to the theoretical approach developed in this study was given by AEPSOM. An abstract representation of the MLO model AEPSOM is reported below.

Policy objective function:

$$\underset{\{\pm T\}}{\text{Minimize}}\ WL = wG \tag{6.1}$$

subject to:

Definitional equations

$$G = I_1 Y_S + I_2 X_S + I_3 Z_S \tag{6.2}$$

Policy constraints

$$T_I \leqslant \pm T \leqslant T_p \tag{6.3}$$

Budget constraints

$$[(+T_1)Y + (+T_2)X + (+T_3)Z] - [(-T_1)Y + (-T_2)X + (-T_3)Z] \leqslant 0 \tag{6.4}$$

Behavioural objective function:

$$\underset{\{Y, X, Z \mid \pm T\}}{\text{Min}}\ C = (c_1 \pm T_1)Y + (c_2 \pm T_2)X + (c_3 \pm T_3)Z \tag{6.5}$$

subject to:

Demand constraints

$$Z \geqslant \overline{\mathbf{D}} \tag{6.6}$$

Intermediate energy balance constraints

$$Z = aX \tag{6.7}$$

Energy supply balance constraints

$$X = bY \tag{6.8}$$

Resource constraints

$$Y \leqslant \bar{\mathbf{Y}} \tag{6.9}$$

Capacity constraints

$$X \leqslant \bar{\mathbf{X}} \tag{6.10}$$

Non-negativity constraints

$$G, Y_S, X_S, Z_S, Y, X, Z \geqslant 0 \tag{6.11}$$

where WL is the value of the policy objective function; w is the vector of coefficient of the policy objective function $(1 \times e)$; G is the vector of energy target variables $(e \times 1)$; Y is a vector of primary energy $(p \times 1)$; X is a vector of energy products $(n \times 1)$; Z is a vector of end uses of the energy products $(m \times 1)$; $\bar{\mathbf{D}}$ is a vector of end uses in different sectors $(q \times 1)$; c_1, c_2, c_3 are the costs of supplying, converting and using energy $[(p \times 1), (n \times 1), (m \times 1)]$; $\pm T_1$, $\pm T_2$, $\pm T_3$ are vectors of different taxes and subsidies $[(p \times 1), (n \times 1), (m \times 1)]$ (lower case ts are the elements); a, b are matrices of technological coefficients $[(m \times n), (n \times p)]$; Is are the identity matrices (some of the elements of the diagonal matrix may be 0); Y_S, X_S, Z_S are subvectors of Y, X, Z; $\bar{\mathbf{Y}}$ is a vector of the fixed amount of available primary energy $(p \times 1)$; and $\bar{\mathbf{X}}$ is a vector of the fixed level of capacities $(n \times 1)$.

AEPSOM is an MLO partial equilibrium energy sectoral microeconomic planning model. The energy planning problem is modelled in such a way that the policy objective function is optimized subject to (i) the constraints on and consequences of the policy options, and (ii) the constraints imposed by the behavioural submodel on the degrees of freedom of the policy makers. The model has four main components: (i) a weighted policy objective function containing the objectives of the policy makers; (ii) the constraints on the choice of policy instruments; (iii) the objective functions of economic agents; and (iv) the constraints on the behaviour of economic agents.

The government determines a set of policy instruments and strategies that are constrained by the limits on the variation of taxes and subsidies and by the condition that net government revenue should be positive. The energy behavioural model represents the choice problem of economic agents subject to the energy resource supply, capacity and demand constraints.

AEPSOM was developed on the basis of the following specification of the

Australian energy policy planning problem:

(1) Energy policy objectives. Despite the relatively late formalization of Australian energy policy activities, the objectives of contemporary policies have become somewhat more apparent. Some energy policy objectives are commonly found in official government documents and academic work, and these suggest that the major objectives of the Australian energy policy are: the security of energy supply and import independence, the conservation of energy (particularly of oil), the efficient allocation of energy resources, and the equity in income and uses of resources in the energy sector.

(2) Energy policy instruments. The possible energy policy instrument alternatives in Australia are listed as follows:
   (a) Indirect control: (i) taxes and subsidies and (ii) government expenditures for energy conservation, research and development, and education and information.
   (b) Direct control: pricing of domestic crude oil.
   (c) Energy policy strategies/policy guidelines: (i) technological strategy, (ii) depletion policy and (iii) exploration and development strategy.

AEPSOM was specified for the years 1979/80 and 1989/90. Some of the data was estimated by the author and others were obtained from published sources.

SOLUTION ALGORITHM: MLP AND THE PPS APPROACH

The real test of an empirical model development is its numerical implementation. AEPSOM was numerically implemented as a multi-level programming (MLP) model (Candler and Norton, 1977). An MLP is considered as a collection of nested optimization problems at different levels. The main difficulty with an MLP model is its implementation by an algorithm. MLP is a recently developed mathematical programming technique, and research and development is continuing on the optimization of the MLP solution algorithm. In most of the existing MLP algorithms, the MLP solution is relatively difficult to obtain because in these algorithms some sort of transformation of the original problem is necessary. This makes the size of the transformed MLP large in comparison with the original problem.

To overcome these problems, the Parametric Programming Search (PPS) algorithm was developed in this study. In the PPS algorithm, the lower-level problem of a complete MLP problem is solved as a parametric programming problem. Alternative optimum (basic) solutions to the lower-level problem are searched to find the one that optimizes the policy objective function and satisfies the policy constraints. AEPSOM results produced by the PPS algorithm are close to the expected optimum results. Criteria such as efficiency in CPU time, and cost and efficiency in extension and transfer of the algorithm were applied to test the algorithm, and it was found to be satisfactory.

RESULTS: VALIDATION TESTS AND POLICY IMPLICATIONS

The expected outcome of the modelling work was a set of results for providing guidelines for the formulation of energy policies.

The AEPSOM model provides a set of results that allows for an analysis and understanding of the Australian energy policy problems. It also provided results that are capable of addressing energy policy issues in areas such as energy taxes and subsidies, pricing, energy technology, conservation, education and information, research and development, optimum depletion of exhaustible resources, and exploration and development activities of the government.

To test the reliability of the AEPSOM results, the following conventional validation tests were performed: *a priori* justifications about the relevance of the model, the underlying problems or systems, the usefulness of the output for achieving the objectives of the modelling study, accuracy of results, comparison of the model results with the results of other studies and intuitive judgements. AEPSOM results generated a numerical policy system in the Australian energy sector which has provided some insights into the characteristics of multi-level, multi-goal hierarchical policy formulation in the energy sector. The results have established some analytical aspects of the multi-level decision making process: the need for the determination of the optimum level of policy intervention ($\pm T$), the possibility of the formulation of an improved plan by the government, and the conflicting interests of the government and private sector economic agents.

AEPSOM results have been used to formulate a multi-level optimum energy plan for Australia, details of which are reported in the following section.

INDUSTRIAL DEVELOPMENT STRATEGIES FOR THE ENERGY SECTOR: A SET OF OPTIMUM INSTRUMENTS AND STRATEGIES

AEPSOM results have clearly demonstrated the need for changes in the relative price structure in the energy sector and have suggested deregulation of the energy market. AEPSOM results also indicate the need for rearrangement of fiscal instruments. For example, a tax of 4.63% on the cost of domestic crude oil can be implemented by using a package of fiscal instruments consisting of a resource rent tax, royalties and competitive bidding. Import duty on imported crude has been found necessary. Some other new excise taxes and subsidies will complete the energy sector fiscal instruments.

AEPSOM results imply a higher rate of extraction of coal and natural gas and lower rate for crude oil compared to their present rates. An active government exploration policy implemented through fiscal instruments can increase reserves of energy resources.

In order of priority, conservation activity in the transport sector followed by the agriculture and manufacturing industry sectors is implied by the results and suggests a rational prioritization of the conservation policy in Australia.

AEPSOM results also suggest the following energy policy implications: information about energy technologies and prospects, especially about conservation measures, should be one of the strategies within the government energy policies. As a market may fail to allocate socially desirable resources for research and development, government research and development activities will play an important role.

In energy technology areas, the desired options are: accelerated recovery from existing non-oil fossil fuel reserves, progressive reduction of the use of petroleum products, and the market penetration of renewable energy should be stressed in the government's energy policies. The exploration of the possibilities of cost savings in new technologies (specifically in producing liquid fuel from coal and natural gas) from the potential inventory of Australian energy resources and technologies should be explored.

AREAS FOR FURTHER RESEARCH

This study suggests three streams of further research. The theory of MLO itself should be investigated to provide insights into the interactions in a two-level decision process. Consequently, the numerical implementation and the implications of MLP energy models (either static or dynamic) should be evaluated. Lastly, further experiments on the PPS algorithm should be conducted to make the algorithm more widely used.

MAJOR FINDINGS

Various findings from this study are summarized in the following subsections.

Usefulness of the MLO Approach

The MLO model proposed and developed in this study accurately represents the energy policy system of a market economy. AEPSOM has predicted an improved energy plan and has generated an optimum energy system plan and an optimum energy policy plan simultaneously. Therefore, the methodological conclusion of the study is that an MLO model can provide an alternative methodology and framework for optimum multi-level energy planning.

Changes in Australian Energy Policies

AEPSOM has provided the following insights, guidelines and directions in Australian energy policy planning and issues: the necessity for deregulation of energy prices, reorganization of the existing taxes and subsidies in the energy sector, priority for conservation, education and information, and research and development policies, emphasis on the exploration and depletion policies, and the need for cost savings necessary for market penetration of new and renewable technologies. These changes will ensure reduction in energy use, and moves towards a safe, environmentally desirable and renewable energy technology dominated energy system for Australia.

Possibility for Sustainable Industrial Developments in the Energy Sector

The Australian energy system will undergo some significant changes in the long run if the policies suggested by AEPSOM are implemented and the assumptions in the model remain valid. Although pollution was not included in the model, the model results suggest two important policy strategies of sustainable development of the energy sector: reduction in the use of energy and the development of environmentally sound and renewable energy technologies.

These changes are in the directions required by the efficiency criteria of industrial economics and necessary for sustainable energy sectoral development, and are consistent with those discussed by Lowe (1994, pp. 196–212). For example, it was mentioned that the efficient use of end use energy products and the conservation of existing energy supplies could be attained by a variety of policy instruments, in addition to a tax/subsidy policy, used in AEPSOM (Lowe, 1994, pp. 196, 199). The role of new, and possibly renewable, energy sources can also provide alternative supplies (Lowe, 1994, pp. 205, 207). The institutional and financial aspects of the energy issues were also discussed, in particular the "decentralized management" of public utilities as reflected in a multi-level policy making process to achieve efficiency gains, and the impediments to new investment in new energy technologies.

CONCLUSION

The continuing search in Australia for a comprehensive energy policy justifies an interest in the formulation of such a policy (Saddler, 1981; Marks, 1986).

The events in the period 1979–1990 suggest that some energy policy objectives have not yet been achieved satisfactorily (Jones et al., 1991). This study has predicted an optimistic prospect for sustainable energy sector development in Australia. The optimum energy policies suggested by AEPSOM indicate the directions for the policy reformulation in Australia to move along a sustainable industrial development path.

ACKNOWLEDGEMENTS

The author thanks Fred Puno and Jim Gigas for their excellent research and editorial assistance in preparing this chapter.

REFERENCES

Candler, W. and Norton, R. (1977) Multi-level programming and development policy, bank working paper no. 258, World Bank, Washington, DC.
Centre for Applied Economic Research (1985) Energy modelling in Australia, University of New South Wales, Sydney.
Department of Primary Industries and Energy (1988) *Energy 2000: A National Energy Policy Paper*, Australian Government Publishing Service, Canberra.
Islam, S.M.N. (1995) The role of renewable energy in the energy system: an issue in the philosophy of renewal natural resources, *Energy Economics* **17**(2), 117–124.
Islam, S.M.N. (forthcoming) *Mathematical Economics of Multi-Level Optimization: Theory and Application*, Springer Verlag (Physica Verlag's contribution to Economics Series), Heidelberg.
Jones, B., Bush, S., Kanakaratnam, A., Leonard, M. and Gillan, P. (1991) Projections of energy demand and supply: Australia 1990–91 to 2004–2005, Bureau of Agriculture and Resources Economics, Canberra.
Lowe, I. (1994) Towards sustainable energy systems, in S. Dovers (ed.) *Sustainable Energy Systems: Pathways for Australian Energy Reform*, Cambridge University Press, Cambridge.
Marks, R. (1986) Energy issues and policies in Australia, *Annual Review of Energy* **11**, 47–75.
Saddler, H.D.W. (1981) *Energy in Australia: Politics and Economics*, Allen & Unwin, Sydney.
Tinbergen, J. (1952) *On the Theory of Economic Policy*, North-Holland, Amsterdam.
Weymans-Jones, T. (1986) *The Economics of Energy Policy*, Gower, London.

FURTHER READING

Asian and Pacific Development Centre (1985) *Integrated Energy Planning: A Manual*, 3 vols, Kuala Lumpur.
Ballenger, N.S. (1984) Agricultural policy analysis for Mexico: sectoral and macro impacts, PhD dissertation, University of California.

Blaug, M. (1985) *Economic Theory in Retrospect*, 4th edn, Cambridge University Press, Cambridge.
Bureau of Statistics (various volumes) *Australian Year Book*, Australian Government Publishing Service, Canberra.
Department of National Development and Energy (1981) *Australian Energy Statistics*, Australian Government Publishing Service, Canberra.
Department of National Development and Energy (1982) *Energy Demand and Supply Australia 1960–61 to 1980–81*, Australian Government Publishing Service, Canberra.
Department of Resources and Energy (1983) *Forecast of Energy Demand and Supply in Australia: 1982–83 to 1991–92*, Australian Government Publishing Service, Canberra.
Department of Resources and Energy (various issues) *Australian Energy Statistics*, Australian Government Publishing Service, Canberra.
Edwards, G.W. (1983) Economic aspects of energy policy, in *Energy '83: Towards an Energy Policy for Australia*, position papers, Australian Institute of Engineers, Sydney.
Gruen, I.H. and Hillman, A.L. (1981) A review of the issues pertinent to liquid fuel policy, *Economic Record* June, 111–127.
Islam, S.M.N. (1995) A sustainable economic growth model with endogenous technical progress: theory, numerical implementation methods and policy applications, CSES, Victoria University, Melbourne.
Intrilligator, M.D. (1971) *Mathematical Optimization and Economic Theory*, Prentice Hall, Englewood Cliffs, NJ.
Julius, D. (1981) Energy pricing in the developing countries: a review of literature, World Bank Energy Department paper no. 1, World Bank, Washington, DC.
Meier, P. (1984) *Energy System Analysis in the Developing Countries*, Springer-Verlag, New York.
Meier, P. (1986) *Energy Planning in Developing Countries: An Introduction to Analytical Methods*, Westview Press, Boulder, CO.
Munasinghe, M. and Schramm, G. (1983) *Energy Economics, Demand Management and Conservation Policy*, Van Nostrand Reinhold, New York.
Musgrove, A.R., Stocks, J.K., Essam, P., Le, D. and Hoetzl, V.J. (1983) Exploring some Australian energy alternatives using MARKAL, Division of Energy Technology, technical report TR2, CSIRO, Lucas Heights, New South Wales.
Treasury (1984) Energy markets: some principles of pricing, in P.J. Lloyd (ed.) *Mineral Economics in Australia*, Allen & Unwin, Sydney.

Chapter 7

Application and Limitations of Annual Models for Electricity Capacity Development

Poul Erik Grohnheit

INTRODUCTION

The type of modelling that is the topic of this chapter describes the energy system as a network of annual energy flows with capacity limits for energy conversion and energy transport.

The demands for energy services, final energy and materials produced by large energy consuming industries are exogenous. Using a linear programming (LP) algorithm, this demand is satisfied at minimum discounted costs.

The study period normally covers the lifetime of the equipment for energy extraction, energy conversion or consumer appliances for energy services. For large-scale equipment this period is some decades. This period is represented by a number of milestone years, typically every fifth year.

The cost of the optimal energy flow in each milestone year and the investment cost in the period since the last milestone year contribute to the discounted costs that are minimized. The initial equipment and the equipment that is installed as a result of the optimization remain available to the end of their technical lifetimes. This dynamics is referred to as *quasi dynamics*.

Systems Modelling for Energy Policy. Edited by D.W. Bunn and E.R. Larsen

All prices for fuels, investment in new equipment and operation and maintenance of existing equipment are exogenous, and subject to forecasts from sources outside the model.

Further constraints are added to reflect the infrastructure of the energy system or legal and political requirements, e.g. emissions constraints.

In this chapter the traditional applications of this model approach are described, with examples from international collaborative studies focusing on national strategies for energy and environment policies. These studies, developed over many years, offer modelling methods, energy system descriptions and data, and methods of analysis and presentation that are useful for other types of analysis, in particular in those areas where the weaknesses of models in the econometric and macroeconomic tradition are most obvious.

The practical use of the model approach is described with reference to modelling software, user interface, and computer and organizational resources.

Finally, applications of the model approach are described for different types of economic agents within the electricity sector.

THE BOUNDARIES FOR THE "ENERGY SYSTEM"

This modelling approach, with system boundaries describing a "national energy system", was used widely during the 1980s, both for national energy planning purposes by the national energy ministries and energy agencies, and for international comparative studies by international organizations. The driving forces in the development of such models have been the European Union (the EFOM model) and the International Energy Agency (the MARKAL model).

The main study objectives for using these models have been the penetration of new technologies, the impact of fuel price changes, and optimal emissions reduction strategies.

The same model approach has been used for regional or local energy planning studies covering the "energy system" within a smaller geographical area.

The model approach is also suited to sectoral studies. It has been useful for studies of the technology choice and assessment of investment options in the energy sectors or in large energy consuming industries, e.g. cement, steel, pulp and paper, etc. The market penetration of new consumer equipment may also be studied by this approach, e.g. more efficient cars or consumer appliances.

"BOTTOM-UP" VERSUS "TOP DOWN" MODELS

The main feature of the model approach is that the energy system is described by a limited number of technologies producing a single or a few well-defined physical outputs using well-defined energy carriers as inputs.

Each type of these technologies must be described by a few parameters:

- efficiency
- availability
- technical lifetime
- emission factors
- investment cost
- operation and maintenance cost
- initial and residual capacities

It follows that this "bottom-up" model approach is much less suited to describe the energy use in light industries that are producing goods that can be aggregated only in monetary terms, and using equipment that is not identifiable in a few physical terms. Nor is this approach suited to analysing the energy use that is most dependent on human behaviour.

The complementary "top down" approach from the econometric and macroeconomic tradition is based mostly on monetary variables and parameters estimated from long time-series of consistent national accounts statistics. Often such statistics are not available. Even in countries that have a long tradition of national accounts statistics, it is impossible to estimate, for example, long-term price elasticities for most large energy consuming industries.

The Energy Flow Optimization Model (EFOM), which was developed as the supply part of the energy model complex of the Commission of the European Communities, has been used for a number of studies since the 1970s.

The overall structure of the EFOM model is shown in Figure 7.1. The energy system is described as a network combining the extraction of primary fuels through a number of conversion and transport technologies to the demand for

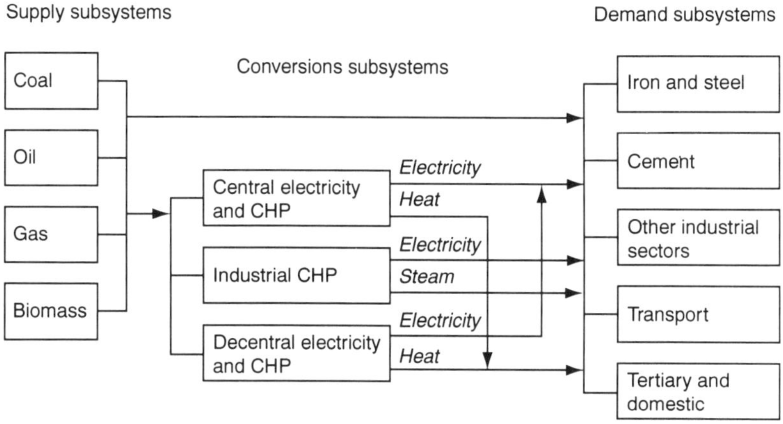

Figure 7.1 Modular structure of the EFOM model

energy services or large energy consuming materials. Some of the subsystems also contain emission abatement technologies for SO_2 or NO_x. Each of the subsystems may consist of a large number of links, which refer to a database that contains the network information (upstream and downstream nodes) and the set of parameter values mentioned above. The annual flows of electricity, district heating and natural gas are divided into four flows describing daily and seasonal variations over the year. Only the links representing primary fuels contain fuel price forecasts.

The power generating system is the central part of the energy system that is described here, including combined heat and power (CHP) for either industrial steam raising or district heating (DH). The model also allows technology substitution in the demand sectors, which may lead to different demands for electricity and heat (Grohnheit, 1993).

COST CURVES FOR EMISSIONS REDUCTION

An example of an application of the EFOM model is the construction of cost curves for the reduction of emissions of major atmospheric pollutants. Cost curves show the increasing value of the objective function for a series of scenarios for the same description of the energy system, where the constraint on one or more pollutants becomes more tight.

The EC cost-effectiveness analysis of CO_2 reduction options was part of a study under the JOULE Energy Research Programme of the Commission of the European Communities (1991) concerning options for CO_2 emissions reduction. The method of cost curves was developed for an earlier study on abatement strategies for SO_2 and NO_x emissions (Commission of the European Communities, 1989). The cost curve approach has since been used for several international collaborative studies, such as the UNEP Greenhouse Gas Costing Study (UNEP, 1994), which has developed a set of guidelines for the construction of cost curves.

Discounted Energy System Costs

The central definition of cost effectiveness was the discounted cost of the energy system within the limits set by the model description of the system for any feasible reduction level of a particular emission item within the period. A key result that complies with this definition is shown in Figure 7.2. The value of the objective function of the linear programming problem (i.e. the discounted energy system costs for the period 1985–2010) is shown for selected values of CO_2 reductions by 2005 compared to the 1988 level.

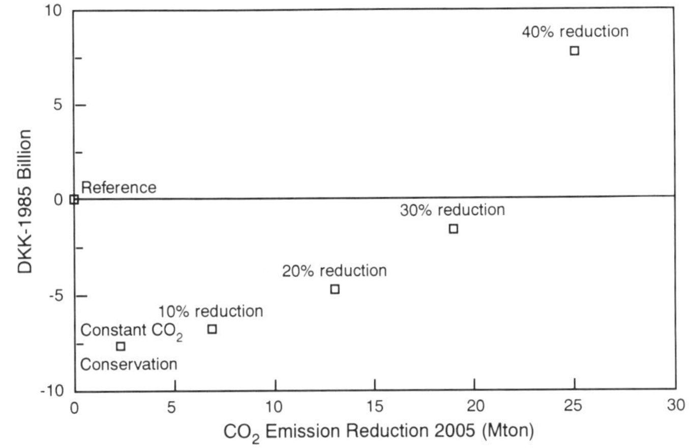

Figure 7.2 CO_2 emissions reduction and increase in discounted energy system costs for Denmark by 2005, 5% discount rate

Scenario Assumptions

A key issue for this type of study is the specification of a reference case. Here, the reference case contains the penetration of expected new technologies and the emissions abatement measures required by existing legislation, and some energy conservation measures that are assumed in the demand forecasts.

The conservation case is the reference case with penetration of further conservation measures that are options for the optimization, e.g. more efficient domestic electric appliances. It represents a "least cost planning" approach, assuming that conservation measures are subject to the same kind of rationality as utility planning.

The latter scenario is the starting point for introducing the emission constraints, "constant CO_2" through "40% reduction", in the optimization. Further reduction constraints would lead to a mathematically infeasible solution of the LP problem.

Assumptions for the Danish Study

The assumptions concerning the development of demand for energy services and fuel import prices are those of the harmonized multinational study.

Techno-economic data for energy technologies, including the technical options for emissions reduction, were based on earlier studies or detailed studies within the multinational study that is described here.

Pure national data describing the structure of the energy system and detailed assumptions for the demand for energy services were specified to comply with national statistics and the assumptions used for the national energy plan that were prepared at the same time. The most important policy assumptions are:

- The expansion of district heating grids connected to CHP stations, urban waste incineration plants and industrial surplus heat.
- The location of all new large power plants in connection to the major district heating markets.
- The build-up of the natural gas grid.
- No nuclear power plants within the foreseeable future.
- The energy conservation policy, in particular insulation of buildings.

Apart from excluding nuclear power, further political constraints were avoided in order get the truest optimization results. In consequence, there was no assumption concerning such issues as the promotion of indigenous energy sources and decentralized CHP, and no limit on natural gas import or penetration of the more expensive renewables.

In addition to the constraints that express important energy policy issues, a number of constraints are necessary to prevent unrealistic model results. Most of these quantify infrastructural constraints or long-term contracts. The most important of these are:

- Maximum total district heating supply and maximum heat supply from central CHP stations.
- Maximum supply from urban waste incineration to district heating and maximum supply of other biomass fuels.
- Levels of offshore gas extraction and gas exports.
- Import and export of electricity.
- Maximum capacity of wind turbines.

Although the penetration of the new renewable and fossil fuel technologies may be limited because of technological immaturity and social constraints, no arbitrary constraints were introduced to prevent the model from producing unrealistic results.

Overall Conclusion and Policy Recommendations

The main conclusion of the study for Denmark was that it would be possible to optimize the Danish energy system over a period to achieve a target for a substantial reduction of national CO_2 emissions, at extra costs no higher than could be outweighed by some cost-effective energy savings measures—often referred to as "no regret" measures. Such a conclusion is controversial, in particular the jump in cost from the reference to the conservation scenario.

The main elements in the CO_2 reduction policy should be energy conservation, intensive use of natural gas for electricity generation, and penetration of renewables. Further CO_2 reductions would also require that the use of natural gas be reduced and, thus, replaced by non-fossil fuel technologies (Morthorst & Grohnheit, 1992). The results for Denmark were similar to those of most of the other EU countries.

However, using these results as a general policy recommendation may be counterproductive for a target of cost-effective CO_2 emissions reductions. In this study the optimum would lead to a massive substitution of gas for coal in the power sector. This is actually happening, in particular in countries where the electricity supply industries have been liberalized. Sooner or later this substitution may lead to natural gas prices higher than the forecasts that were assumed for the optimization.

It should be emphasized that the outcome of any model study is understanding the behaviour of the system under specified conditions rather than recommendations to be implemented according to an optimal solution.

THE MODEL ENVIRONMENT AND SOFTWARE

The model results described in the previous section were some results from a multinational study which included both software model development and coordination of data acquisition. The development until that stage had required many years of software and model development, involving many national teams and international coordination.

Unfortunately, much of this work is now forgotten in reports or tape archives containing results of studies of issues that are obsolete for the time being. Much of this material may still be available and may be useful for new analyses. But it has been a discouraging experience for modellers working in different countries, that the benefits of modelling studies can be very limited because the personnel who were involved in a particular study have moved to other positions, and no institutional knowledge has been built up.

One of the lessons to be learnt from many years of experience in computer modelling is that it is most often the trivial tasks that are the most time consuming. So, before proceeding to the discussion of new applications of the modelling approach and new questions to be answered, it is valuable to focus on the trivial tasks of how to administer model data, software, documentation and presentations.

The Modular Structure of Optimization Models

Application of LP and other mathematical optimization techniques for large problems will require a model environment of the type shown in Table 7.1. The database may be common for several modelling or statistical purposes. The key module of the particular model is the interface, in which these data are

Table 7.1 The modular structure of an optimization model

General structure	EFOM model
Database	Energy network Techno-economic data Capacity data
Interface	Matrix generator
Optimizer	Linear programming software
Report writer	Energy and emission balances Capacity of new equipment Energy system cost

prepared for a general-purpose mathematical programming solver, or merely the brain and intuition of the user. Finally, the output and the model assumptions are presented by a report writer.

A properly designed structure of the various elements of the models will allow modular usage of the elements. Models are often presented by a brand name and promoted to increase their usage. However, no model can give the final answer to any issue. For many purposes it is far more interesting to analyse the same issue using different models, with the same assumptions and presentations that are comparable.

The Development of the EFOM Model

The EFOM model was developed as the supply part of the energy model complex of the Commission of the European Communities. Its origin was the reaction to the oil crisis in 1973: find an optimal mix of technologies that can be implemented to reduce dependence on imported oil. The model was used during the 1980s for reference projections of the energy systems in the member countries, and scenarios with assumptions concerning economic growth levels, oil import prices, or the role of solid fuels and nuclear power were studied (Guilmot et al., 1986). An extension of the model to include emissions of pollutants and abatement techniques was implemented for all member countries and for some countries outside the Community (Lefèvre et al., 1995).

This model, EFOM-ENV, was used for the assessment of national and Community emission reduction strategies under the JOULE Energy Research Programme. Although the policy conclusions of these studies were derived from the results of different model approaches, including macroeconomic models, in the practical model implementation work the EFOM-ENV model had often been used as a stand-alone tool for a collaborative study that involved national teams from most of the EU countries.

In the JOULE Framework Programme, which is now finished, further development of this tool is included in a Tool Package of the pilot phase of a new project called EURIO, which is part of an initiative for a European Energy–Environment–Economic Forum that has the objective of being a support for research and technology development (RTD) in the fields of energy planning and environmental protection. The Tool Package of the EURIO project contains models for energy supply, energy demand, macroeconomic impacts and technology data (Rentz et al., 1996).

The Software for the EFOM Model

The EFOM software was developed in the 1970s for large mainframes, when the media for data communication were punchcard input and lineprinter output. The Fortran code is now available in a portable version, which has been compiled by various compilers for mainframes, workstations and PCs. The compilation of the same source code (except a few system-specific features) using different compilers not only makes the software available to different systems, but is also an effective debugging method.

In addition to these enhancements of the classical EFOM software, EFOM was also developed in general model languages, most recently in GAMS (General Algebraic Modelling System) (de Kruijk, 1994).

The primary model interfaces for all of these systems are ASCII files in specified formats. ASCII files in an appropriate format remain the best choice for the model interface, because they can be treated by any software.

However, the appropriate *model* interface is not necessarily a good *user* interface. To handle a large amount of data or many scenarios of the same application model written in the EFOM software, it is necessary to automate this data communication. Many users of the EFOM model have developed their own user interface in their most familiar spreadsheet or database software. The version of the EFOM model that is described in the following sections uses a spreadsheet user interface to handle assumptions and results of many scenarios, which was developed over some years as a by-product of the multinational studies and additional national applications of EFOM. Because different users use different spreadsheets, it is necessary that user interfaces are portable among the different releases of existing and future spreadsheet software. For this reason, an early version of Lotus 1-2-3 is still being used, i.e. the single-sheet version for DOS (release 2.x). This will enable users to use the equipment and software they already have for other purposes.

Computer Resources and Organizational Requirements

The traditional application of the EFOM model covered most of the subsystems in Figure 7.1 for a single country and four to six 5-year periods. The linear programming problem for this system contains between 5000 and 10 000 rows and a similar number of columns. Solving this problem in the late 1980s would have required 1–2 CPU hours on a larger VAX system or 10–15 minutes on a large university mainframe. Today, an ordinary Intel 486 PC or a fast UNIX workstation is likely to require less time.

However, to maintain and manage a model of this volume of information will continue to require a large organization and many person-years.

For many types of studies much smaller models that cover only a single sector of the energy system will be far more useful. In the model for the electricity system, which is described below, the size of the LP matrix is less than 2000 rows, which is solved on a good standard PC in less than 2 minutes, and much faster on a workstation. This will allow running multiple scenarios for testing system boundaries and assumptions, parameter studies, and the analysis of the robustness of the results. With access to an existing database of techno-economic and structural parameters, it is possible for individuals to use this type of model for a wide range of purposes.

Dissemination of Model Software

The linear programming solver, LINPROG is available from Risø National Laboratory as a software package. It is distributed either as a stand-alone tool or included in a software package containing the EFOM model as source code that is portable to several computer systems (Kirkegaard & Grohnheit, 1995).

Since 1992 the price of a permanent licence for LINPROG has been 500 ECU (about $US650) for a single-user licence. We are practising the following licensing policy for multi-user and site licences: there is a basis price covering a single-user licence; two licences for the same system version allow installation at a multi-user environment (a PC network or a UNIX workstation); and three or more licences constitute a site licence for LINPROG in one or more system versions.

The LINPROG solver was developed and is completely owned by Risø. The software for the EFOM model was developed as a part of the Energy Research Programme of the European Commission. In collaboration with the Commission (DG XII/F/1), Risø has modified the code into a version that is portable to most Fortran systems. This version is available on a single diskette, labelled "EFOM Portable Source Code", which contains the source code, a compiled version by, for example, Lahey F77L-EM/32, and command files for running the model and for compilation by different systems, for example VAX/VMS or UNIX systems. The copyright of Risø for the "EFOM Portable Source Code" is thus limited to the structure of the diskette, some auxiliary code and the compilation, but not the contents of the source code itself.

Demonstration copies of both packages may be available on agreement for potential users as "invited shareware". "Shareware" means that users are invited to test the software, but those who want to use it regularly will need to pay a licence fee. "Invited shareware" is our suggestion for the distribution form within a limited group of users in order to support the dissemination of modelling tools and encourage further development.

A SECTORAL MODEL FOR ELECTRICITY

This model is written in the EFOM software and organized in modules as described in Table 7.1. There is a general database containing the energy flow network and the techno-economic values, which may be used for different countries or system boundaries. The model is run using national or regional specific data for initial capacities and infrastructural constraints, and a set of scenario assumptions, i.e. demand forecasts, energy prices, emissions constraints, etc.

The user interface for some of the sets of data is one or more spreadsheet models that also contain various tools for the documentation of scenario assumptions and graphic presentation of assumptions and results.

The model is developed from the study of CO_2 reduction strategies described in the previous sections, and most of the assumptions for the reference case are the same as those described for that study. This model is a sectoral model for the electricity sector, including CHP and district heating.

Figure 7.3 shows the principle of a sectoral model that produces two or more products in combined production, both subject to competition from outsiders. The market situations for the two products are very different. Electricity may be traded over long distances using the national or international transmission

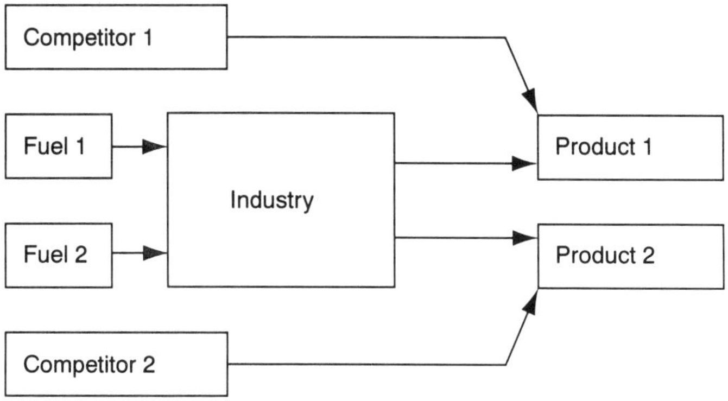

Figure 7.3 Bottom-up model for an industry with combined production and competition (source: P.E. Grohnheit, 1994)

grid, while heat must be used locally. The competitive situation for combined heat and power is described in detail by Grohnheit & Olsen (1994).

Danish Utilities in a Competitive Market

Traditional energy planning in most countries has been based on the idea of national autarky. Lack of indigenous resources of fossil fuels makes fuel import necessary in many countries, but the electricity demand is most often generated nationally, in many countries by nationalized or publicly owned utilities. However, electricity import to Denmark has been considerable in some years, whenever hydro power has been abundant in Norway and Sweden.

The recent development towards competition on a national and European scale for the electricity supply industry was studied in a project for the Danish Energy Research Programme with participation from the Local Governments' Research Institute (AKF), Roskilde University and Risø. This study included the task of adapting existing modelling experience to a quantitative analysis of a new organizational framework.

There is no single tool available for this analysis. The existing well-established methods of analysis will give only partial answers, and the methods of analysis of electricity markets and the behaviour of producers and consumers are only in their infancy.

To analyse the long-term consequences, the following analytical or modelling approaches, with increasing analytical sophistication, are being used or discussed:

- Short-term and long-term costs of competing technologies.
- An engineering model with an optimization of a network of energy flows.
- An econometric, partial equilibrium model for the electricity market.
- Models based on game theory that describe a market with few agents.

The first two approaches follow the logic of the traditional organization of the industry, where vertically integrated monopolies with reserve capacity traded according to some agreed principle. The econometric models are designed for the logic of the market-place to analyse the behaviour of many economic agents in order to find a clearing price through the interactions of supply and demand. There is a long tradition for these models, which are also supported by international cooperation on model development and elaborate international standards for national accounts statistics. The available econometric or macroeconomic models, however, do not offer a sufficient representation of the options for technology choices among the most important technologies within the energy sector. The game theory approach may be used to analyse strategic interaction between the agents in a market with oligopoly or monopoly power.

The Short-Term Marginal Costs for Electricity Generating Technologies

It is important to distinguish between the *short* and the *long* run. With the present prices there is too much capacity in Northern Europe. Therefore, it is likely that prices in a period after an opening of the market will decrease, possibly to the variable costs of the marginal technology. Demand growth and an increasing need for replacements will cause prices to go up in the longer run, possibly to a level where the construction of new capacity will become profitable.

Some price differences will, however, persist due to transport costs and limitations on transmission capacity. A continuation of market imperfections can, of course, also be the cause of price differences.

The competing technologies are presented in Table 7.2. Nuclear power has low marginal costs (Dkr0.06–0.07 per kWh; 1 Dkr = 0.16$US). As it is costly to change the load, nuclear power will be used as base load and will not be the price-setting technology under normal conditions. For political reasons it is not very likely that new nuclear capacity will be constructed in some countries.

Hydro power (with a water reservoir) is the most flexible generating technology. It can be stopped and started quickly and with little direct cost. Generators will therefore prefer to produce when demand and prices are high. In such periods, however, technologies with higher variable costs, such as condensing power, will determine the price (hydro power will be intra-

Table 7.2 Short- and long-term marginal cost of different technologies for power production (Dkr/kWh, 1990 prices)*

	Low fuel cost†		High fuel cost‡	
	Short term	Long term	Short term	Long term
Decentral gas-fired CHP	0.075	0.193	0.117	0.244
Central coal-fired CHP	0.062	0.194	0.070	0.212
Central gas-fired CHP	0.074	0.152	0.116	0.203
Coal condensing	0.122	0.243	0.137	0.279
Central gas condensing	0.122	0.194	0.191	0.279
Nuclear (France)§	0.057	0.224		
Nuclear (Germany)§	0.074	0.362		
Hydro (direct variable cost)	0.015	0.180		
Hydro (water value)	0.060	0.220		

* 5200 annual hours of production for all technologies, 5% discount rate.
† 12.43 Dkr/GJ for coal and 16.29 Dkr/GJ for gas.
‡ 16.29 Dkr/GJ for coal and 29.22 Dkr/GJ for gas.
§ Source: OECD/IEA (1993).

marginal). Direct variable costs are extremely low, Dkr0.01–0.02 per kWh. It can be argued, however, that opportunity costs are higher for a hydro plant with water reservoir, because of the possibility of storing the water and selling it at another time. Amundsen, Bjørndalen & Rasmussen (1994) quote an average value of Dkr0.06 per kWh.

Thermal (condensing) power is likely to become the marginal, price-setting technology most of the time on a Northern European market with competition. It is also the technology that is likely to be constructed to cover an increase in demand.

The production costs of CHP can be determined only by analysing the system of combined production of heat for local use and electricity for the national and international electricity grid. The power or heat grids are often fed from several generating units. The basis for scheduling these units for production in each time period will be their merit order, which is based on the short-term operating cost. In the case of combined heat and power production, however, the marginal cost of power also depends on the value of the co-generated heat. Thus, the merit order must be determined by considering the whole power and heat generating system. It then becomes reasonable to assume the marginal efficiency of the co-generated electricity, $\eta = 0.85$, while the efficiency of a modern condensing steam-generator producing electricity only is $\eta = 0.45$ at optimal conditions. These parameters are explained by Grohnheit (1993).

Table 7.2, however, represents traditional planning assumptions, used, for example, by international organizations for numerous studies. The main weakness is that it was necessary to assume the number of annual hours of production during the lifetime of each technology. The real long-term marginal cost for each technology is very sensitive to the utilization time, which may be very uncertain in a competitive market, in particular for the marginal technologies.

Sensitivity to Economic Parameters

To model the optimal investment strategies in new equipment the utilization times of the different types of generating units must be endogenous. For this purpose the engineering modelling approach is needed, in which the technological options are physically identified. A module of the EFOM model for the electricity sector with CHP is used as a tool for this analytical approach to study the consequences of changes in parameters that describe the different market frameworks.

Figure 7.4 shows some results of an application of this modelling approach for analysing the Danish electricity system until 2010, emphasizing the impact of changing market conditions on the choice of new generating capacity and fuels. Most of the assumptions for the reference case are the same as those of

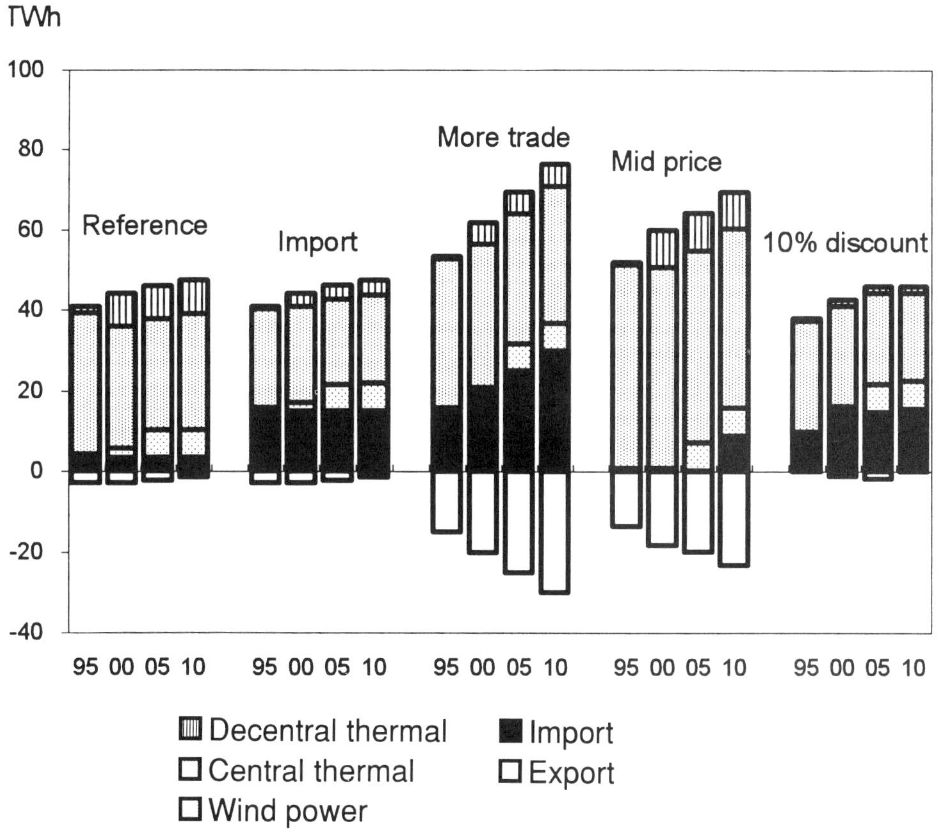

Figure 7.4 Electricity generation technologies in Denmark, 1995–2010

the multinational study of CO_2 reduction strategies described in the previous section.

The parameter variations concern:

- limits for electricity import and export
- import and export prices
- discount rate

In the reference scenario the Danish electricity system is optimized for the period 1995–2010 assuming only contracted import and export. In this model version import and export prices were set constant over the year. The import price was chosen at 0.15 Dkr/kWh for the whole period, and the export price should reflect long-term marginal costs for a reference technology (0.28–0.32 Dkr/kWh). The objective function is the total discounted costs for the period at

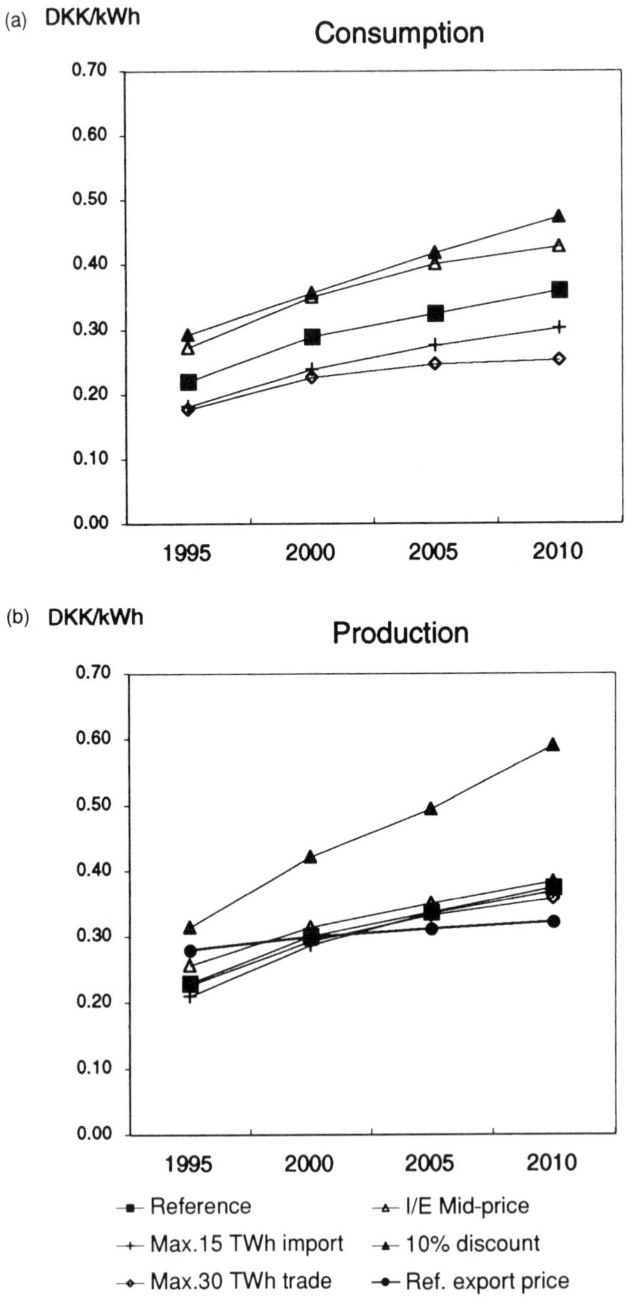

Figure 7.5 Average expenditure for (a) electricity consumption and (b) electricity production in Denmark, 1995–2010, at optimal development

5% discount rate. A set of constraints reflect the infrastructure of the Danish electricity and CHP system. International electricity trade is constrained at a minimum.

In the next scenario, "import", maximum import is set at the capacity of the transmission lines from 1994. The third scenario, "more trade", assumes further expansion of the transmission capacity. In the scenario "mid-price" the same price is set for both import and export. The last scenario is identical to "mid-price", but the discount rate is set at 10% in order to reflect the financial conditions for utility investment at a competitive market for electricity.

The optima for all of the scenarios with unconstrained trade volumes are very sensitive to changes of the import and export price assumptions. This shows that most electricity is generated at central thermal stations, either as combined heat and power or electricity-only generation. The volume of the latter will mirror the variations in import and export volumes.

Figure 7.5 shows the annual expenditure including investment in new capacity in each year per kWh consumed or produced in Denmark for the optimal development under the various scenario assumptions. Obviously, these expenditures will normally be increasing, when new capacity is required for an optimal long-term development of the industry, because no expenditures to meet existing financial obligations from existing equipment are included.

The cost concept shown in Figure 7.5 is undiscounted cost per unit for each period. These costs are calculated from optimal solutions, in which the discounted costs over all periods were minimized.

ELECTRICITY MARKET WITH SEVERAL AGENTS

In the scenario models there was usually only one economic agent: the Danish electricity supply industry. This agent enjoys a monopoly at the distribution side. However, when foreign trade is possible, there is competition at the production side, even without further liberalization. This leads to an industry with more agents with similar or different system boundaries, within which they will optimize, either by maximizing profits, minimizing costs or expanding their market share.

Table 7.3 shows some tentative results of modelling regional agents, which were obtained by using the model for different system boundaries. The Danish electricity and CHP system is divided into regional units supplying a metropolitan, city or town district heating grid and a share of the electricity market. At the production side, there is competition with outside electricity generators. For district heating, there is competition only within each interconnected district heating grid, i.e. between combined production, waste incineration, heat-only boilers etc.

Table 7.3 Results for 1995 of a model optimization of Danish CHP regions, 1990–2010

	Copenhagen 10%	Odense 10%	Midtown 10%	Residual 10%	Total 10%	Denmark 10%	Denmark 5%
Annual expenditure for electricity and district heating, Dkr (bn)							
Export revenue	−0.2	−0.4	0.0	−2.9	−4.5	−2.3	−2.4
Import	1.7	0.1	0.1	2.2	4.9	1.7	1.4
Fuel	1.4	0.8	0.1	7.3	13.3	11.7	11.6
New capacity	1.4	0.0	0.0	0.2	2.2	2.0	2.5
Total	4.3	0.5	0.2	6.8	15.9	13.1	13.2
Electricity generation by technology, GWh							
Export	−559	−1256	−50	−10242	−16137	−8414	−8503
Import	6175	290	225	7663	17585	5856	4998
Wind power	0	0	0	870	722	870	870
Nuclear	0	0	0	0	0	0	0
Central thermal	5369	3044	0	13596	31874	33054	31842
Decentral thermal	194	0	162	0	2789	5467	7587
Total	11180	2078	337	11887	36833	36833	36793
Primary fuels for electricity and DH, PJ							
Coal	42.3	22.7	0.0	96.7	235.8	242.7	242.3
Oil	2.2	1.2	0.0	6.0	13.1	14.9	14.3
Natural gas	10.5	2.4	2.5	16.0	75.8	110.6	108.9
Uranium	0.0	0.0	0.0	0.0	0.0	0.0	0.0
Biomass	5.0	0.0	0.3	0.0	9.8	18.7	18.7
Total	60.0	26.3	2.8	118.7	334.5	386.8	384.2
District heating, PJ							
Central CHP	19.3	9.2	0.0	0.0	65.1	65.2	65.2
Decentral CHP	1.0	0.5	0.5	0.0	11.3	29.6	38.9
Biomass CHP	0.1	0.0	0.2	0.0	2.6	4.2	5.6
Boilers, fossil	8.1	2.3	1.3	0.0	41.0	19.7	10.8
Boilers, biomass	7.4	0.0	0.0	0.0	8.1	8.6	6.9
Total	0.0	12.0	2.0	0.0	128.0	127.3	127.3
Emissions, 1000 t							
CO_2	4783	2381	141	10540	27705	30460	30290
SO_2	16.9	3.5	0.1	44.0	72.1	137.2	135.6
NO_x	14.7	3.5	0.4	21.0	55.9	66.1	63.9
Annual expenditure for electricity consumption at optimal investment							
	0.36	0.24	0.47	0.58	0.40	0.32	0.33
Annual expenditure for domestic electricity generation at optimal investment							
	0.44	0.25	0.63	0.52	0.41	0.32	0.32
Export price	0.28	0.28	0.28	0.28	0.28	0.28	0.28
New electricity generating capacity							
Renewables	0.0	0.0	0.0	0.0	0.0	0.0	0.0
Central thermal	0.0	0.0	0.0	36.3	30.2	62.7	62.7
Decentral CHP	0.0	0.0	27.8	0.0	445.4	958.1	1341.3
Biomass CHP	0.0	0.0	3.8	0.0	60.7	102.4	135.8

In contrast to most presentations of the results of an optimization model, which focus at the end of the planning period, Table 7.3 shows the results for the first period of an optimization over several periods. The results for the various agents may be conflicting, thus leading to a change in prices or investment plans.

Accepting the results shown in Table 7.3 as they are, the system will be far more expensive at the end of the first period, when optimizing within a number of units instead of the national system as a whole, even when using the same high discount rate at 10%. Import is much higher at the distributed optimization, and own production—in particular, from decentralized CHP—is much lower. There is the same kind of difference between the optimization results for the total system at the discount rates 10% and 5%.

ECONOMETRIC MODELS AND MARKETS WITH FEW AGENTS

Econometric models are designed to analyse the behaviour of economic agents in a market. There is a long tradition for econometric and macroeconomic models, and much effort has been devoted to international cooperation on model development, which is also supported by very elaborate international standards for national accounts statistics.

At a competitive market for electricity, prices and output are determined by the total interaction of demand and supply (including transmission costs). Norwegian researchers have developed models to determine equilibrium prices and output on a future competitive Northern European market. Amundsen, Bjørndalen & Rasmussen (1994) and Amundsen et al. (1994) have presented results from two equilibrium models for the study of prices, output and trade on an integrated Northern European market for power. One model analyses the short-term perspective whereas the other deals with the long-term perspective.

Future Market Prices in Northern Europe

The long-term model (Amundsen, Bjørndalen & Rasmussen, 1994) determines national consumption and volume trade in electricity in year 2000 based on existing and newly constructed transmission lines. The model takes account of existing production capacity in the various regions and calculates the optimal expansion of new production capacity. Calculations are based on the assumption of efficiency in production, free trade in electricity and the possibility of constructing new transmission lines. The resulting equilibrium prices for the Nordic Countries and West Germany (before 1990) are reported in Table 7.4.

The price in Denmark under the assumptions of free competition of the

Table 7.4 Calculated prices* in year 2000 on a Northern European competitive market for power (Dkr/kWh)

Country	Initial situation without competition (1988 data)	Free competition (year 2000)	CO_2 tax ($US10 per barrel oil) (year 2000)	Norwegian exports restricted to 10 TWh annually (year 2000)
Norway	0.197	0.199	0.209	0.167
Sweden	0.207	0.207	0.218	0.207
Finland	0.234	0.234	0.234	0.234
Denmark	0.309	0.248	0.222	0.303
West Germany	0.315	0.274	0.247	0.310

* Wholesale prices excluding transmission costs. Exchange rate 1 Nkr = 0.9 Dkr.
Source: Amundsen, Bjørndalen & Rasmussen (1994).

model (which includes average precipitation in Norway) will be 0.25 Dkr/kWh. This is similar to the unit cost of new coal condensing capacity and above the unit cost of new CHP capacity in Denmark.

The other model (Amundsen et al., 1994) calculates time-differentiated equilibrium prices under the assumption of existing generating and transmission capacity. The time periods are: summer day, summer night, winter day and winter night (see Table 7.5). The overcapacity mentioned above is reflected in lower equilibrium prices than in the long-run model.

Most of the resulting prices on the Danish market are above the short-term marginal costs of CHP reported in Table 7.2 (under the assumption of high fuel costs, the price on a summer night will be lower than the marginal costs of the two gas-fired CHP plants). The calculated price differences among the four

Table 7.5 Time differentiated prices on a Northern European competitive market for power (Dkr/kWh)

	Summer day	Summer night	Winter day	Winter night
Percentage of year*	31	13	39	17
Norway	0.114	0.114	0.114	0.114
Sweden	0.114	0.112	0.114	0.114
Finland	0.106	0.106	0.110	0.110
Denmark	0.131	0.106	0.151	0.114
West Germany	0.138	0.081	0.159	0.110

* Varies slightly among the five countries.
Source: Own calculations from Amundsen et al. (1994).

periods are not large. This reflects, among other things, the modifying influence of hydro power, which is not exposed to the same peak load problems as thermal production capacity.

The prices found by the equilibrium model are lower than the prices that were assumed when using the more traditional planning orientated methods (see Table 7.2 and Figure 7.5). This result is not surprising. In a situation with international overcapacity, the industry is warned against investment in further capacity, which would lead to financial difficulties.

FURTHER APPLICATION OF THE OPTIMIZATION MODEL

The modular structure of EFOM and the spreadsheet user interface that was used for the study of the electricity market in Table 7.3 allows two types of comparative analyses of the electricity sector:

- Different market conditions for the electricity system in a given region.
- The impact of given market conditions on different regions.

Examples of the first of these applications were described in "Sensitivity to economic parameters" (p. 102) and the second in "Electricity market with several agents" (p. 105). Both types of application require a modelling tool that is not necessarily interactive, but easy to use for describing, calculating, presenting and documenting many scenarios. The development of such a tool was described in "The model environment and software" (p. 95).

The data requirements for the descriptions of the two elements that are being combined are different. A proper description of a regional electricity system requires a large amount of data or data preparation to reach an appropriate level of aggregation. However, these data are stable over time, and most technology data may be applied for different regions or even different nations, assuming an international market for equipment. Only data for initial and residual capacities need to be different from region to region. At a reasonable level of aggregation these data are even available from international statistics. Such data need not be presented as part of the scenario description; they only need to be documented in the background.

Model Quantification of Market Conditions

Data for the market conditions, on the other hand, are highly uncertain and subject to changes and controversy, e.g. price and demand forecasts, discount rate, emission constraints, etc. The following description of market conditions for an optimization of a regional electricity system is developed on the basis of the experience gained from the analyses described in the previous sections:

- Monopolies: high prices for electricity trade. 5% discount rate.
- Overcapacity: low prices for electricity trade. 10% discount rate.
- Market: price increase for electricity trade. 10% discount rate.
- Regulation: price increase for electricity trade. 5% discount rate.
- Crisis: 1985 primary fuel prices. 5% discount rate.

For all of these market conditions, import and export prices are set at the same value. This is different from the previous Danish experience, where import prices for hydro and nuclear electricity from Norway and Sweden were much lower than export prices to Germany. The EU Transit Directive and new transmission lines in Northern Europe are likely to reduce this difference in the future. For the market conditions described above, "low prices" are import and export prices at short-term marginal cost of the marginal generating unit within the international transmission network. "High prices" are import and export prices at long-term marginal cost. It is assumed that market forces will lead to price increase from short-term to long-term marginal cost.

The high discount rate reflects the financial risk of investments at a deregulated market, while a regulated market or a monopoly will allow utilities to transfer the financial risk to the consumer.

The 1985 fuel prices were chosen to illustrate a possible new oil crisis. In 1985 the world market price of primary fuels was higher than practically all international forecasts for primary fuel prices for 2010.

Presentation Graphs for Several Scenarios

The presentation graphs shown in Figures 7.6 and 7.7 are a graphical improvement of a set of draft graphs from the spreadsheet user interface, into which the results from the EFOM model were copied. This set of graphs has been used as the basic presentation of new results for some years. A large number of scenarios was made using different combinations of the modules or the model. When updates are made to some of the modules, the fixed format of presentation is not only a first presentation of results, but also an important tool for debugging, because it is easy to compare the results from different versions of the model.

Explanation of Results and Model Limitations

The results for the Danish electricity sector subject to these different market conditions show that the volume of imports is considerable in the first two or three periods, because import prices are low or investment in new capacity is discourged by a high discount rate. In the last periods both imports and exports become very sensitive to the market conditions; higher prices on the international market will lead to more exports and even investment in order to meet those exports.

The latter result concerning investment, however, would involve a sub-

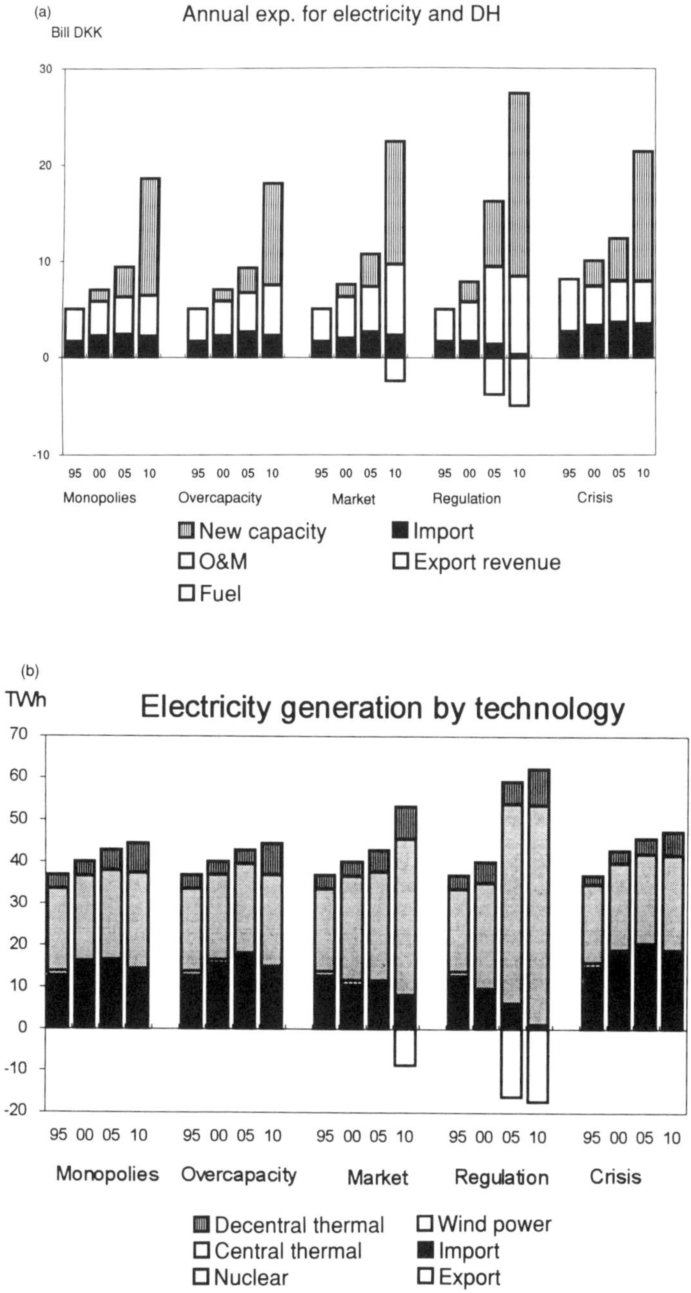

Figure 7.6 The Danish electricity and DH system in various market environments: energy system costs, production technologies and fuel requirements

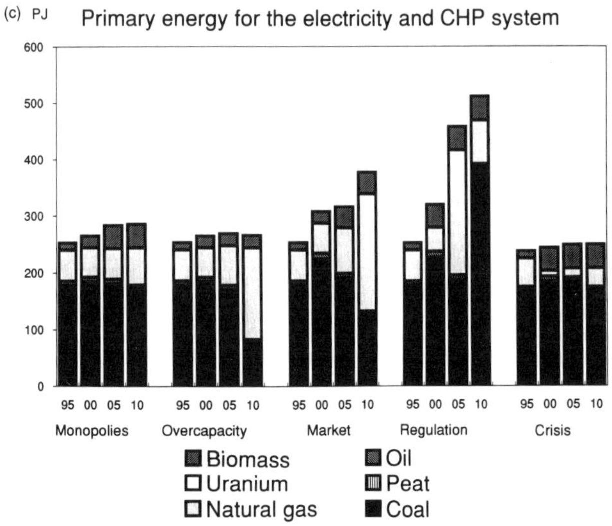

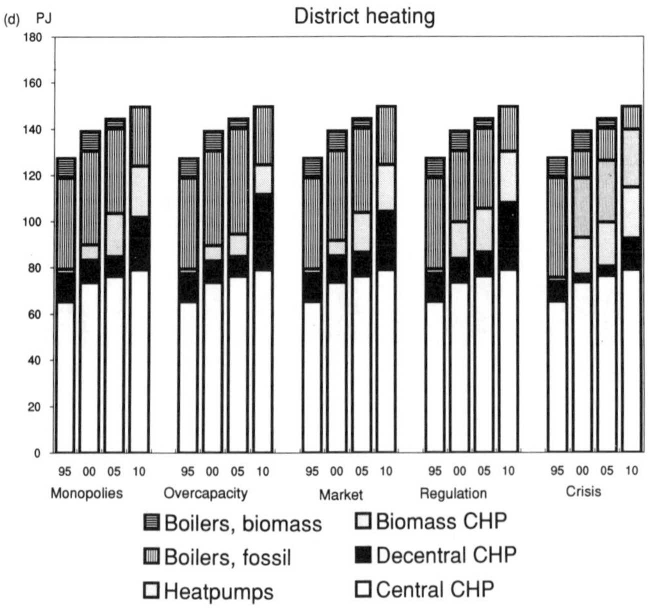

Figure 7.6 *Continued*

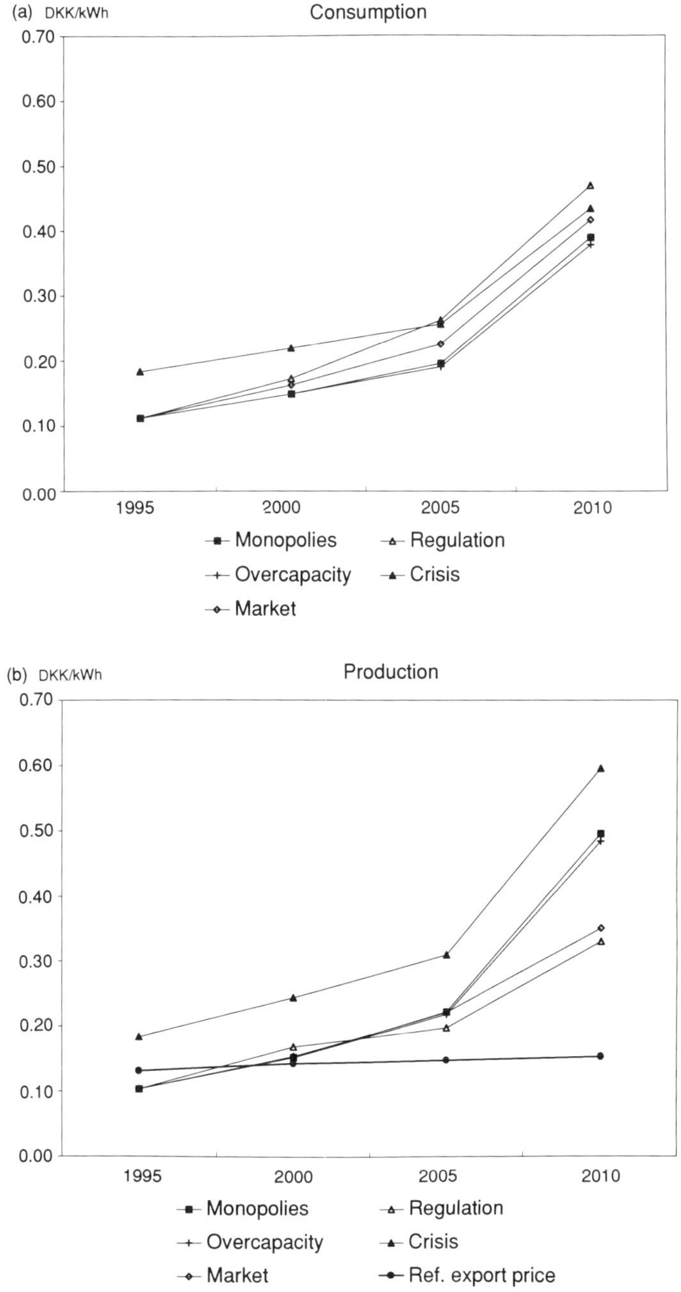

Figure 7.7 The Danish electricity and DH system in various market environments: unit costs and emissions

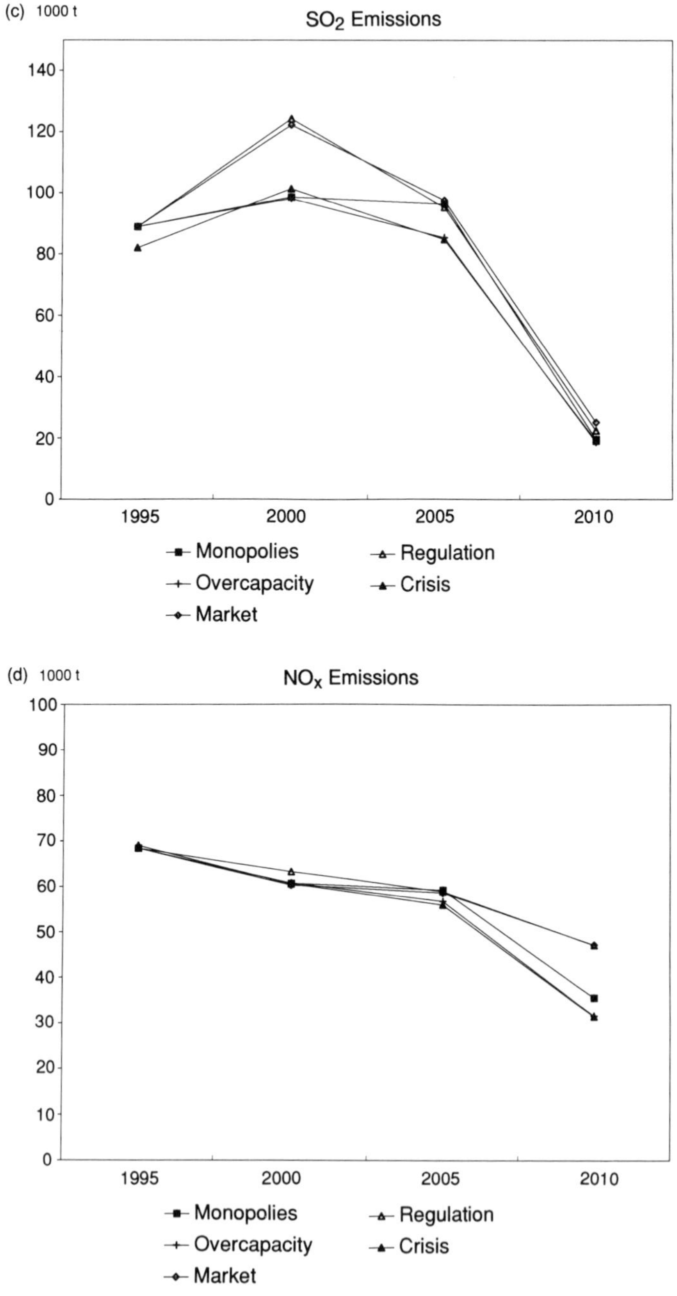

Figure 7.7 *Continued*

stantial financial risk. It may be justified, because the availability of a heat market, which may be supplied by natural gas fired combined cycle power plants, offers investment opportunities, which could be a basis for export, mainly in high and peak load periods.

In particular, this version of the model should not allow this conclusion to be drawn immediately. The market price is constant over the year. This may be a reasonable simplification, but it ought to be tested by a modification of this model that will allow for different prices during base and peak hours. This extension of the use of parameters is possible within the existing EFOM software. Such an extension of the application of the model would require a test of the validity of the decomposition of the annual flows. The method for such tests is to use the model assumptions in a model that is more detailed concerning the break-down of time.

International Cooperation on Modelling

As a contribution to project EURIO (Energy and Environment Modelling in Central and Eastern Europe) under the EU JOULE Energy Research Programme, the sectoral model for electricity was transferred to the GAMS version of EFOM that was developed by ECN, Netherlands (see de Kruijk, 1994). The same data and user interface can be used for both types of model software. This is a very important method for debugging both model versions.

The impact of given market conditions has been tested for five countries (Denmark, Sweden, Finland, Latvia and Lithuania) using the same technology options for all countries; only the initial structure of the electricity sectors of the various countries was different. This method of comparison isolates the impact of the initial structure on the optimal development of the sector. The market condition that was used for the test run was import and export prices set at long-term marginal costs for coal-fired power stations in condensing mode (electricity-only production). The results were conflicting: in all countries there was investment in new capacity in order to meet the maximum limit of export.

Using an internationally developed reference or sample model version with default techno-economic data is a first step in the development of a national model. This was done in most EU countries during the 1980s. A model to be used as a decision support tool for national ministries or agencies requires a large organization for data acquisition and scenario design.

The EFOM model was first designed for centrally planned national or international strategies to meet the challenges of, first, the oil crisis and later environmental concerns. The modelling approach and the tools that are developed to handle it are useful for wider purposes, including the analysis of the development of the electricity sector in a liberalized market. But it will require a lot of coordinated experience to design the simplifications that are valid for many different purposes.

ACKNOWLEDGEMENTS

Grants for this research were received from the JOULE Energy Research Programme of the European Commission, the Energy Research Programme of the Nordic Council of Ministers, and the Danish Energy Research Programme.

REFERENCES

Amundsen, E.S., Bjørndalen, J. and Rasmussen, H. (1994) Export of Norwegian hydropower under a common European regime of environmental taxes, *Energy Economics* **16**(4), 271–280.

Amundsen, E.S., Bjørndalen, J., Nyhus, H.F. and Tjøtta, S. (1994) A numerical model of an integrated European electricity market, SNF working paper no. 33/1994, Bergen.

Commission of the European Communities (1989) *Energy and Environment—Strategies for Acid Pollutant Reduction*, Brussels.

Commission of the European Communities (1991) COHERANCE: cost-effectiveness analysis of CO_2 reduction options, synthesis report/country reports, DGXII, JOULE Programme, Models for Energy and Environment, Brussels, May.

de Kruijk, H. (1994) The EU Energy and Environment model EFOM-ENV specified in GAMS: model description and user's guide, Netherlands Energy Research Foundation, ECN, Petten, ECN-C-94-021.

Grohnheit, P.E. (1993) Modelling CHP within a national power system, *Energy Policy* **21**(4), 418–429.

Grohnheit, P.E. and Olsen, O.J. (1994) The competitiveness of the Danish electricity supply industry in the European internal market for energy, in *IAEE 17th Annual Conference: Energy Markets in Transition*, Vol. III, Stavanger, Norway, 25–27 May.

Guilmot, J.-F., McGlue, D., Valette, P. and Waeterloos, C. (1986) *Energy 2000*, Cambridge University Press, Cambridge.

Kirkegaard, P. and Grohnheit, P.E. (1995) LINPROG—a linear programming solver: mathematical description and model applications, Risø-R-707(EN), Risø National Laboratory, Roskilde, Denmark.

Lefèvre, T., Nguyen, C.T., Bui, D.T. and Acaraya, S. (1995) EFOM-ENV, Energy Environment Planning in Developing Countries, Methodological Guide (European Commission, Asian Institute of Technology, Bangkok), United Nations ST/ESCAP/1531, New York.

Morthorst, P.E. and Grohnheit, P.E. (1992) UNEP greenhouse gas abatement costing studies—Denmark, Phase 1, Risø National Laboratory, Roskilde, Denmark.

Rentz, O., Lüth, O., Schöttle, H., Jattke, A. and Zundel, T. (eds) (1996) EURIO—support for Energy–Environment RTO Strategies (EC—DG XII research project JOU-CT93-0274), Final Report, Institute for Industrial Production (IIP), University of Karlsruhe.

OECD/IEA (1993) *Projected Costs of Generating Electricity. Update 1992*, OECD Nuclear Energy Agency/International Energy Agency, Paris.

UNEP (1994) Greenhouse gas abatement costing studies, Phase 2 guidelines, UNEP Collaborating Centre on Energy and Environment, Risø National Laboratory, Roskilde, Denmark.

Chapter 8

Energy–Economy–Environmental Interactions in a General Equilibrium Framework: The Case of Turkey

Y. Arikan, Ç. Güven and G. Kumbaroglu

INTRODUCTION

The effects of urban life and industrial activity on the environment create complicated policy decisions owing to the difficulty of not only measuring the environmental impact but also unravelling the intimate relationship and the trade-offs between economic welfare and environmental degradation. In industrializing countries like Turkey policy makers are far from sure even of the options that are available. Electricity demand is growing fast and it is not clear whether it is better to develop indigenous resources of doubtful quality, or to base growth on cleaner imported fuels, or whether a nuclear programme should be launched. Painful decisions are having to be made, when almost completed thermal plants are frozen for fear of endangering tourism potential. Huge multipurpose hydro and irrigation schemes are in progress but neither the extent of salination that can be caused nor its economic costs are well understood. Energy decisions are also intimately related to urban air pollution by way of alternatives available in transportation and residential heating.

Systems Modelling for Energy Policy. Edited by D.W. Bunn and E.R. Larsen

The analysis of economy-wide energy decisions requires in the first place detailed and fully disaggregated energy technology modelling, which allows representation of all complementarity and substitution relations among various energy activities. Of crucial importance at this stage is the quantification of environmental effects of all these activities, which poses several problems of formulation and measurement. Finally, the linkages between energy activities and the rest of the economy need to be quantified in a sufficiently disaggregated manner to allow for precise modelling in respect of economic sectors that depend on energy at varying degrees.

A first requirement of integrated modelling is to provide a combined interdependency framework which represents all of the interplay between energy, economy and the environment, but it is not at all clear according to what measure choices should be made. Energy–economy interactions have often been studied using a general equilibrium energy–economy model, seeking market clearing equilibrium solutions which explicitly or implicitly depend on maximizing some utility function. These models form a bridge between detailed multi-period energy-technology models that provide for comprehensive disaggregation of the energy sector, and economy models of the general equilibrium type, where an equilibrium solution is computed satisfying several linear and nonlinear equations. In the economy models the degree of detail provided is of necessity limited to disaggregating the economy into a number of uniform sectors. Each sector is assumed to produce a homogenous good, subject to a certain technology, often represented by a neoclassical production function which relates input factors such as capital and labour to homogenous output. In such models energy is represented by a few sectors at the most; since equalities are not allowed resource and capacity constraints cannot be written directly. Supply and demand equations should be provided to bring all energy activities into equilibrium. This cannot be done at the level of detail of energy-technology models and hence a comprehensive evaluation of all technological options cannot be carried out.

The discrepancy between energy and economy frameworks is moderated in energy–economy models to a certain extent by (i) defining as many energy related sectors as possible with the necessary minimum of non-energy sectors within a general equilibrium framework, or (ii) adopting the optimization framework with a detailed energy sector and representing the rest of the economy in a one-sector model. This second approach permits two-way linkages between energy and the economy and can provide an equilibrium solution.

In studying the energy economy and the environment all together, first priority must be given to a detailed representation of energy activities since these activities not only strongly interact with the economy but also produce the greatest environmental impact. However, the environment is also affected to varying degrees by non-energy economic activity. Hence, ideally, a suitable framework can be formed combining activity analysis (which handles the energy side) with multisectoral equilibrium (which takes care of the rest of the economy without too much aggregation). However, due to the large data requirements and calibration difficulties, a one-sector specification of the economy becomes more suitable for long-term analysis, and this is the approach adopted here. The following section provides a description of the model developed for Turkey.

THE MODEL

Turkey is an industrializing country with moderately high GDP growth rates that result in a considerable increase in energy demands. Therefore, the environmental situation is dominated largely by energy related factors. The Turkish energy supply policy has focused on the satisfaction of the fast growing energy needs. As a reaction to the oil crises the development of the large domestic hydro power and lignite resources has been promoted. In the near future, Turkey will be asked to fulfil the European emission control standards for integration into the European Union. This will imply the installation of emission control technologies with high investment and import costs. This is an issue that may have important implications for the economy of a country like Turkey, whose industry depends heavily on imports. Environmental considerations therefore cannot be separated from energy–economy interaction analyses. The model presented in this chapter takes these aspects into consideration while searching for optimal solutions under import restrictions.

The model developed is a general equilibrium model of interactions between the energy sector and the rest of the economy, where the economy is represented by an aggregate production function in which capital, labour and different energy forms are intermediate inputs to determine GDP. It is an ETA-MACRO (Manne, 1979) type model and is an extension of Güven's model (Güven, 1994), which also takes emission levels and costs into consideration (Arikan, Güven & Kumbaroglu, 1994).

The model comprises energy, economy and environmental submodels which are linked to produce equilibrium solutions through utility maximization over a planning horizon of T periods. The utility function is the discounted logarithm of consumption (C_t), compounded periodically, where each individual time period is identified by a representative year t. Hence the objective function is:

$$\max \sum_t \Delta_t \log C_t$$

The energy submodel divides energy consumption into the three groups of electricity (E_t), oil and gas (P_t) and solid fuels (S_t). Electric energy ($ELEC_{j,k,t}$) is obtained from hydroelectric and thermal plants, where the latter are further differentiated according to the type of fuel they use, i.e. lignite, hardcoal, natural gas, fuel oil and nuclear fuels. Primary energy carriers as well as hydroelectric energy are included in the set j, which differentiates between energy types, and the set k serves for further disaggregation into subgroups of j. P_t comprises petroleum products and natural gas while S_t includes hardcoal, lignite, wood, and plant and animal wastages, all being used as non-electric energy ($NE_{j,k,t}$) for several purposes other than electricity generation. Equations describing energy activities are then:

$$E_t = \sum_j \sum_k ELEC_{j,k,t}$$

$$P_t = \sum_j \sum_k NE_{j,k,t} \qquad (j = \text{gas, petroleum})$$

$$S_t = \sum_j \sum_k NE_{j,k,t} \qquad (j = \text{coal, lignite, wood, plant and animal wastages})$$

With regard to the required investment cost and the load factor, hydroelectric power plants are represented in six categories. Lignite reserves are grouped by their calorific costs and comprise four groups, one for non-electric use. Thus, the index for subenergy types, k, runs up to 6 or 4 and equals 1 for the energy types of hydro, lignite and all others, respectively. The sum of fuel consumption for electric and non-electric use gives the total fuel consumption ($CO_{j,k,t}$), hence:

$$CO_{j,k,t} = NE_{j,k,t} + e\ ELEC_{j,k,t}$$

where e is a parameter converting electricity units into the required calories of thermal fuel. Besides the availability of domestic fuels, imports of fuel oil, hardcoal and natural gas are allowed in the model so that total fuel consumption consists of the sum of domestic production ($DP_{j,k,t}$) and imported fuel ($FCO_{j,t}$)

$$CO_{j,k,t} = FCO_{j,t} + DP_{j,k,t}$$

Similarly, non-electric energy use is split into domestic ($DNE_{j,k,t}$) and imported ($FNE_{j,k,t}$) components:

$$NE_{j,k,t} = FNE_{j,t} + DNE_{j,k,t}$$

$$FNE_{j,t} \leqslant FCO_{j,t}$$

$$DNE_{j,k,t} \leqslant DP_{j,k,t}$$

Power plants are assumed to have a useful life of 30 years and therefore retirement is allowed:

$$ELEC_{j,k,t+1} = p\ ELECINC_{j,k,t+1} + ELEC_{j,k,t} - \frac{p}{30}\ ELEC0_{j,k} \qquad \text{for } t \leqslant \frac{30}{p}$$

$$ELEC_{j,k,t+1} = p\ ELECINC_{j,k,t+t} + ELEC_{j,k,t} - p\ ELECINC_{j,k,t-(30/p)} \qquad \text{for } t > \frac{30}{p}$$

where $ELECINC_{j,k,t}$ denotes annual incremental capacities and $ELEC0_{j,k}$ is the existing stock of capacity at the base year. p is the length (in years) of each subperiod of the planning period. Similarly, incremental variables are defined for non-electric use ($NEINC_{j,k,t}$) of oil and gas and solid fuels:

$$NE_{j,k,t+1} = p\ NEINC_{j,k,t+1} + NE_{j,k,t}$$

The economy submodel includes foreign trade and allows for the separation of domestic and foreign capital goods (KD_t, KF_t) resulting in two types of investment (ID_t, IF_t) according to the origin of capital, so that capital increments (KDN_t, KFN_t) during a p-year period are calculated as:

$$KDN_t = \frac{1}{2} p\ (ID_t + ID_{t-1})$$

$$KFN_t = \frac{1}{2} p\ (IF_t + IF_{t-1})$$

Gross investments during the terminal year must provide for expansion at the post-terminal growth rate (gd, gf) and replacement at the rate $(1 - \lambda)$ where λ is the annual survival factor:

$$ID_T = (gd + 1 - \lambda)\ KD_T$$

$$IF_T = (gf + 1 - \lambda)\ KF_T$$

The foreign exchange constraint limits total imports (M_t) so as not to exceed the sum of exports (X_t), factor incomes from abroad (F_t) and capital inflows from abroad (W_t):

$$X_t + F_t + W_t \geqslant M_t$$

Bounds are set as a fraction of the GDP, as follows:

$$F_t \leqslant \mu_t GDP_t$$

$$\phi_1 GDP_t \leqslant X_t \leqslant \phi_2 GDP_t$$

$$\eta_1 GDP_t \leqslant CG_t \leqslant \eta_2 GDP_t$$

GDP is computed as:

$$GDP_t = C_t + ID_t + IF_t + X_t - M_t \tag{8.1}$$

where imports of intermediates (INT_t), consumption goods (CG_t) and fuels (EF_t) are part of total imports:

$$M_t = IF_t + INT_t + EF_t + CG_t \tag{8.2}$$

The cost of fuel imports is found to be:

$$EF_t = \sum_j FCO_{j,t}\, PIF_{j,t}$$

where $PIF_{j,t}$ denotes the price of imported fuels. Since gross production (Y_t) is defined as:

$$Y_t = GDP_t + INT_t + EC_t$$

C_t can be obtained as:

$$C_t = Y_t - ID_t + EF_t - X_t + CG_t - EC_t$$

simply by substituting Equation (8.2) in Equation (8.1).

The survival factor is again incorporated to compute the surviving stocks in order to add to the newly added incremental quantities which have been defined by appending an *N* at the end of the variables. Total production is then:

$$Y_t = YN_t + \lambda^p Y_{t-1}$$

and similarly, the existing and the new stocks for each input (*KD*, *KF*, *L*, *INT*, *E*, *P*, *S*) are linked in the same way:

$$INPUT_t = INPUTN_t + \lambda^p INPUT_{t-1}$$

The incremental production aggregates the input factors of capital (KDN_t, KFN_t) and labour (LN_t) as well as energy (EN_t, PN_t, SN_t) according to a nested Cobb–Douglas function, and relates these primary and energy aggregates with a constant elasticity of substitution $[\sigma = 1/(1-\rho)]$ to the output:

$$YN_t = \gamma_t[\alpha_t[(KDN_t^{\text{skd}} KFN_t^{\text{skf}})^{\text{sk}} LN_t^{\text{sl}}]^\rho + \beta_t INTN_t^\rho$$
$$+ (1 - \alpha_t - \beta_t)[EN_t^{\text{se}} PN_t^{\text{sp}} SN_t^{\text{ss}}]^\rho]^{\frac{1}{\rho}}$$

The environmental submodel consists of a set of equations similar to those of the EFOM-ENV model (Rentz et al., 1992), which is used in all EC member countries harmonizing emission strategies. It differentiates fuel consumption for non-electric use according to the sector it is used in

($SNE_{j,k,t,l}$) by the subscript l, which distinguishes between the energy utilization sectors, transportation, households, and tertiary and domestic industry excluding cement production. Separate treatment is needed for cement production since the resulting emissions show great variations in comparison to other industries. The sectoral distribution of non-electric energy (sd_l) is assumed to remain constant over time, so $SNE_{j,k,t,l}$ is obtained as:

$$SNE_{j,k,t,l} = sd_l \, NE_{j,k,t}$$

But the focus is actually on SO_2 and NO_x emissions resulting from energy conversion and energy use, taking into account the fuel properties, combustion technologies and operating conditions. Lignites are split into two categories, one with low heating values and sulphur contents and the other with medium to high heating values and sulphur contents. Total emissions of type x, SO_2 or NO_x, are divided into two main groups of "electric" ($EPOL_{x,j,k,t}$) and "non-electric" ($NEPOL_{x,j,k,t,l}$), which are determined according to their origin of generation as:

$$EPOL_{x,j,k,t} = efe_{x,j,k} \, ELEC_{j,k,t}$$

$$NEPOL_{x,j,k,t,l} = efne_{x,j,k,l} \, SNE_{j,k,t,l}$$

where $efe_{x,j,k}$ and $efne_{x,j,k,l}$, the emission factors for electric and non-electric use, convert unit energy use into the resulting amount of emissions. Total emissions ($POL_{x,j,k,t}$) are then obtained as:

$$POL_{x,j,k,t} = EPOL_{x,j,k,t} + \sum_{l} NEPOL_{x,j,k,t,l}$$

Wet limestone and Wellmann Lord processes for SO_2 reduction, and primary measures and selective catalytic reduction for NO_x reduction, have been considered as feasible emission control technologies. For the transportation sector and for households, no emission control technologies are considered. The amount of pollutants filtered by abatement activities ($FILT_{x,j,k,t}$) is calculated from the installed abatement technologies in power plants ($REDE_{x,j,k,t,s}$) and industry ($REDNE_{x,j,k,t,l,s}$), in terms of power generation and primary energy as:

$$FILT_{x,j,k,t} = \sum_{s} REDE_{x,j,k,t,s} \, efe_{x,j,k} \, eye_{x,s} + \sum_{l} \sum_{s} REDNE_{x,j,k,t,l,s} \, efne_{x,j,k,l} \, eyne_{x,l,s}$$

where $eye_{x,s}$ and $eyne_{x,l,s}$ stand for the removal efficiency of control technology s in electric and non-electric use respectively. This amount is then deducted from the estimate of total pollution outputs. The resulting total net amount of air pollutants ($AIR_{x,j,k,t}$) is then:

$$AIR_{x,j,k,t} = POL_{x,j,k,t} - FILT_{x,j,k,t}$$

The total amount of SO_2 ($TOTSO2_t$) and NO_x ($TOTNOX_t$) diffused in the atmosphere in any one year is then calculated as:

$$TOTSO2_t = \sum_j \sum_k AIR_{x,j,k,t} \qquad \text{for } x = SO_2$$

$$TOTNOX_t = \sum_j \sum_k AIR_{x,j,k,t} \qquad \text{for } x = NO_x$$

Introduction of emission control technologies should be bounded from above so as not to exceed total consumption:

$$\sum_s REDE_{x,j,k,t,s} \leqslant ELEC_{j,k,t}$$

$$\sum_s REDNE_{x,j,k,t,l,s} \leqslant SNE_{j,k,t,l}$$

An economic life of 30 years, (the same as for power plants) has been used to write the depreciation equations for reduction technologies:

$$REDE_{x,j,k,t+1,s} = pREDEINC_{x,j,k,t+1,s} + REDE_{x,j,k,t,s} \qquad \text{for } t \leqslant \frac{30}{p}$$

$$REDE_{x,j,k,t+1,s} = pREDEINC_{x,j,k,t+1,s} + REDE_{x,j,k,t,s} - pREDEINC_{x,j,k,t-(30/p),s} \qquad \text{for } t > \frac{30}{p}$$

$$REDNE_{x,j,k,t+1,l,s} = pREDNEINC_{x,j,k,t+1,l,s} + REDNE_{x,j,k,t,l,s} \qquad \text{for } t \leqslant \frac{30}{p}$$

$$REDNE_{x,j,k,t+1,l,s} = pREDNEINC_{x,j,k,t+1,l,s} + REDNE_{x,j,k,t,l,s} - pREDNEINC_{x,j,k,t-(30/p),l,s} \qquad \text{for } t > \frac{30}{p}$$

The required investment and operation and maintenance costs for such technologies are aggregated on a per kWh ($REDECOST_{x,j,k,t,s}$) and per calorie ($REDNECOST_{x,j,k,t,l,s}$) basis for electric and non-electric energy to be added to the total energy cost (EC_t), which already includes the cost for electricity generation ($ECOST_{j,k,t}$) and for domestic and imported fuels ($PDF_{j,k,t}$, $PIF_{j,t}$)

for non-electric use:

$$EC_t = \sum_j \sum_k ELEC_{j,k,t}\, ECOST_{j,k,t} + \sum_j FNE_{j,t}\, PIF_{j,t} + \sum_j \sum_k DNE_{j,k,t}\, PDF_{j,k,t}$$
$$+ \sum_x \sum_j \sum_k \sum_s REDE_{x,j,k,t,s}\, REDECOST_{x,j,k,t,s}$$
$$+ \sum_x \sum_j \sum_k \sum_l \sum_s REDNE_{x,j,k,t,l,s}\, REDNECOST_{x,j,k,t,l,s}$$

With regard to the fact that technology imports are required for the installation of abatement technologies, the equation for fuel imports has to be updated in such a way that investment costs of abatement technologies are included:

$$EF_t = \sum_j FCO_{j,t}\, PIF_{j,t} + \sum_x \sum_j \sum_k \sum_s REDE_{x,j,k,t,s}\, REDECOST_{x,j,k,t,s}\, f_{x,j,k,s}$$
$$+ \sum_x \sum_j \sum_k \sum_l \sum_s REDNE_{x,j,k,t,l,s}\, REDNECOST_{x,j,k,t,l,s}\, f_{x,j,k,s}$$

where $f_{x,j,k,s}$ is the fraction of investment in total cost of abatement technologies.

Model estimation has been rather difficult owing to the functional form adopted for the production function. Some of the parameter values were obtained from existing literature while others were estimated by benchmarking to a base year, thereby calibrating the model to reflect the particular issues faced in Turkey.

The model has about 5560 constraints with 5749 variables. The solution of the model has been obtained using the nonlinear optimization program MINOS (Murtagh and Saunders, 1980).

RESULTS AND COMMENTS

Projecting the effects of reducing SO_2 and NO_x emissions is difficult because many assumptions must be made about how the state of the world will evolve over a very long period of time, with and without a control programme. Considerable uncertainty exists about both exogenous factors like population growth, energy prices etc., and behavioural factors like the response of energy demand and economic output to changes in energy prices and control programmes. Many projections are possible under different sets of assumptions. A base case scenario must be defined. For this study, the latest year for which all of the required initial data was available, 1991, was taken as the base year since no extraordinary political, social or economic events took place in that year. Model predictions cover a planning horizon of eleven successive periods, each lasting for five years. Consideration of such a long planning period is for eliminating the horizon effects, therefore only the results for the first six to seven periods are of real importance, and have been reported here. Some key assumptions used in the reference scenario, relying mainly on official predictions (Turkish Electricity Authority, 1991; World Energy Council Turkish National Committee, 1993), are shown in Table 8.1.

Regular upper bounds upon the extraction of domestic resources are set according to potential production capacities and special lower bounds are set for the consumption of natural gas, lignite and hardcoal in the first two time periods to satisfy the needs for agreed-upon projects signed by the government. Starting from 1995, import prices of oil and other fuels are assumed to increase by 2% and 0.5% per year, respectively, whereas the prices of domestic fuels remain constant at their 1991 levels.

In addition to the base case scenario, six other scenarios employing the same resource availability, macroeconomic and technology assumptions as in the base case, but considering different annual SO_2 and NO_x emissions targets from 2010 onwards, were studied:

- A: base case – no control.
- B: most restricted scenario—total NO_x emissions $\leq$ 1.2 million t/year; total SO_2 emissions $\leq$ 2.6 million t/year.
- C: 10% reduction in total NO_x and SO_2 emissions from the base case values.
- D: 20% reduction in total NO_x and SO_2 emissions from the base case values.
- E: EUM scenario—EU membership regulations: total NO_x emissions $\leq$ 1.2 million t/year; total SO_2 emissions $\leq$ 3.8 million t/year.
- F: European Union scenario—total NO_x emissions $\leq$ 1.8 million t/year; total SO_2 emissions $\leq$ 2.6 million t/year.

The last two scenarios have been adapted from a detailed study of the Turkish energy sector reported by Plinke (1992), which employed the EFOM-ENV model, with scope for the assessment of the impacts of applying the environmental standards of the European Union to the Turkish energy sector and the economy. In this study, as one of the possible strategies, it was argued that even if Turkey decides to develop its low quality lignite reserves, it is expected that it will be allowed some respite in fully applying the directive in the case of EU membership. In order to take this into account, Plinke defined a scenario assuming less stringent standards for SO_2 emissions. Total emissions limits in this scenario are a consequence of restrictions on individual plants or technologies. Scenario E of the present work follows from this scenario. Scenario F, on the other hand, is an adaptation of the cost-efficiency scenario of Plinke and is obtained by restricting the overall emissions directly. The following sections present a summary and a comparison of the solutions

Table 8.1 Assumptions used in the reference scenario

Parameters and macroeconomic assumptions		Energy prices ($\$/10^6$ BTU)	
Cost and utility discount rate:	8%/year	Lignite1:	29.67
Labour force growth rate:	1.5%/year	Lignite2:	5.46
Productivity growth rate:	1.5%/year	Lignite3:	2.29
Workers' remittances growth rate:	7.8%/year	Lignite4 (only for non-electric use):	5.46
Upper limit on exports as a percentage of GDP:	30%	Hardcoal:	3.34
		Natural gas:	3.76
Upper limit on imports as a percentage of GDP:	30%	Nuclear fuels:	2.08
		Oil:	8.49
Upper limit on foreign capital inflows:	4% of GDP	Fuel prices' growth rates:	
		Oil:	2.5%
Upper limit on consumer good imports:	4% of GDP	Natural gas:	2%
		Solid fuels:	1%

obtained with the model, with more focus on scenarios E and F, the EU scenarios.

Macroeconomic results for the base case scenario and the two EU scenarios are presented in Table 8.2.

Examination of Table 8.2 shows that with the imposition of control on SO_2 and NO_x emissions, the economy adjusts itself to slower growth with smaller GDP, investment, export, import and energy use values. In fact, the same effect on GDP is observable in all other scenarios in varying degrees. The reductions in GDP in each case measure the costs to the economy in terms of lost output resulting from the increase in the costs of energy inputs due to emission constraints: either less energy must be input to the economy or expensive emission control technologies must be coupled to energy consuming processes. Both these price and income effects take place simultaneously and lead to a relative slow-down of the economy. From Table 8.3 it is seen that GDP loss in the most restricted scenario, scenario B, increases from 0.2% in 2000 to 1.69% in 2025, compared to base case levels. Among the other scenarios, E and F cause increasing GDP losses over time but not exceeding 1.39 and 1.01%, respectively, in year 2025. The losses in scenario F are more than in E until 2015 and less than in E thereafter. This is explained by the emission limits of each scenario and the total emission trajectories followed in the solutions (Figures 8.1–8.3).

Table 8.3 shows that although in scenario F total emissions of NO_x were restricted to no more than 1.8 million t/year after 2010, this constraint never becomes binding, i.e. NO_x emissions remain below this limit. On the other hand, the SO_2 limit for the same scenario is more restricted than that for scenario E (2.6 versus 3.8) and it becomes binding as early as the year 2000.

Table 8.2 Macroeconomic results (10^9 \$US) for scenarios A (base case), E (EU, less restrictive) and F (EU, more restrictive)

	GDP			INV			EXP			IMP		
1991	70.80			25.07			13.59			21.05		
	A	E	F	A	E	F	A	E	F	A	E	F
1995	95.46	95.44	95.45	28.22	28.22	28.24	21.89	20.31	20.25	28.64	27.06	27.00
Growth (%)	6.2	6.02	6.15	2.39	2.39	2.40	10.00	8.36	8.30	6.35	5.15	5.10
2000	119.7	119.6	119.6	38.16	37.86	37.85	27.28	27.22	27.20	35.15	35.09	35.06
Growth (%)	4.64	4.62	4.61	6.22	6.05	6.03	4.50	6.04	6.08	4.18	5.34	5.37
2005	149.6	149.4	149.2	41.89	41.84	41.61	35.67	35.61	35.56	44.90	44.83	44.77
Growth (%)	4.56	4.55	4.53	1.88	2.02	1.91	5.51	5.52	5.51	5.01	5.02	5.01
2010	183.4	183.0	182.7	50.98	50.59	50.90	44.28	44.17	44.11	55.02	54.89	54.83
Growth (%)	4.15	4.13	4.13	4.00	3.87	4.11	4.42	4.40	4.40	4.15	4.13	4.13
2015	224.3	223.5	223.6	60.06	59.00	59.78	54.75	54.53	54.55	67.31	67.05	67.07
Growth (%)	4.11	4.08	4.11	3.33	3.12	3.27	4.34	4.30	4.34	4.11	4.08	4.11
2020	273.8	271.8	272.4	71.00	69.01	69.86	67.43	66.90	67.06	82.14	81.54	81.72
Growth (%)	4.06	3.99	4.03	3.40	3.18	3.17	4.25	4.18	4.22	4.06	3.99	4.03
2025	335.0	330.3	331.6	86.56	81.56	82.54	83.14	81.92	82.25	100.5	99.09	99.47
Growth (%)	4.12	3.98	4.01	4.04	3.40	3.39	4.28	4.13	4.17	4.12	3.98	4.01

Table 8.3 Summary solutions: alternative restrictions on total emissions

		GDP loss (% of base case value)	Energy cost (10^9 \$US)	Abatement cost (10^9 \$US)	SO_2 emissions (10^6 tons)	NO_x emissions (10^6 tons)	Total emissions (10^6 tons)
1995	A		7.53		2.46	0.67	3.13
	B	0.031	7.57	0.04	2.46	0.67	3.13
	C	0	7.53	0	2.46	0.67	3.13
	D	0.010	7.53	0	2.46	0.67	3.13
	E	0.021	7.56	0.03	2.46	0.67	3.13
	F	0.010	7.55	0.02	2.46	0.67	3.13
2000	A		10.50		3.34	0.78	4.12
	B	0.200	10.58	0.08	2.60	0.89	3.49
	C	0.060	10.53	0.03	2.94	0.78	3.72
	D	0.091	10.57	0.07	2.53	0.78	3.31
	E	0.100	10.53	0.03	3.34	0.78	4.12
	F	0.142	10.58	0.08	2.60	0.78	3.38
2005	A		13.43		4.11	0.93	5.04
	B	0.380	13.46	0.03	2.60	0.97	3.57
	C	0.067	13.46	0.03	3.62	0.92	4.54
	D	0.080	13.50	0.07	3.12	0.91	4.03
	E	0.147	13.44	0.01	3.78	0.90	4.68
	F	0.267	13.48	0.05	2.60	0.90	3.50
2010	A		17.40		4.64	1.05	5.69
	B	0.490	17.46	0.06	2.60	1.07	3.67
	C	0.049	17.45	0.05	4.08	1.04	5.12
	D	0.103	17.51	0.11	3.52	1.03	4.55
	E	0.223	17.46	0.06	3.80	0.97	4.77
	F	0.343	17.52	0.12	2.60	1.00	3.60
2015	A		22.79		6.16	1.23	7.39
	B	0.597	22.76	−0.03	2.60	1.20	3.80
	C	0.018	22.80	0.01	5.26	1.19	6.45
	D	0.142	22.82	0.03	4.52	1.17	5.69
	E	0.378	22.72	−0.07	3.80	1.08	4.88
	F	0.343	22.92	0.13	2.60	1.14	3.74
2020	A		29.88		8.12	1.43	9.55
	B	0.942	29.67	−0.21	2.60	1.20	3.80
	C	0.084	29.84	−0.04	7.20	1.39	8.59
	D	0.274	29.80	−0.08	6.27	1.36	7.63
	E	0.734	29.62	−0.26	3.80	1.20	5.00
	F	0.511	30.03	0.15	2.60	1.33	3.93
2025	A		39.35		11.11	1.73	12.84
	B	1.692	38.59	−0.76	2.60	1.20	3.80
	C	0.215	39.09	−0.26	9.88	1.66	11.54
	D	0.400	39.06	−0.29	8.63	1.63	10.26
	E	1.397	38.59	−0.76	3.80	1.20	5.00
	F	1.015	39.27	−0.08	2.60	1.57	4.17

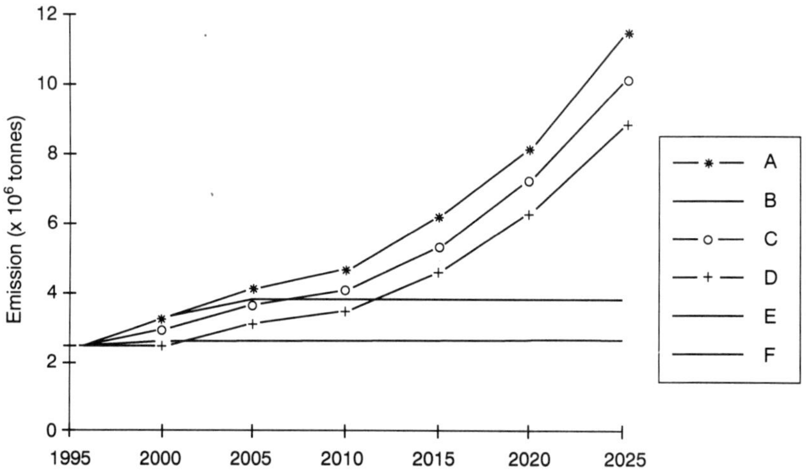

Figure 8.1 Total SO_2 emissions over time: alternative scenarios

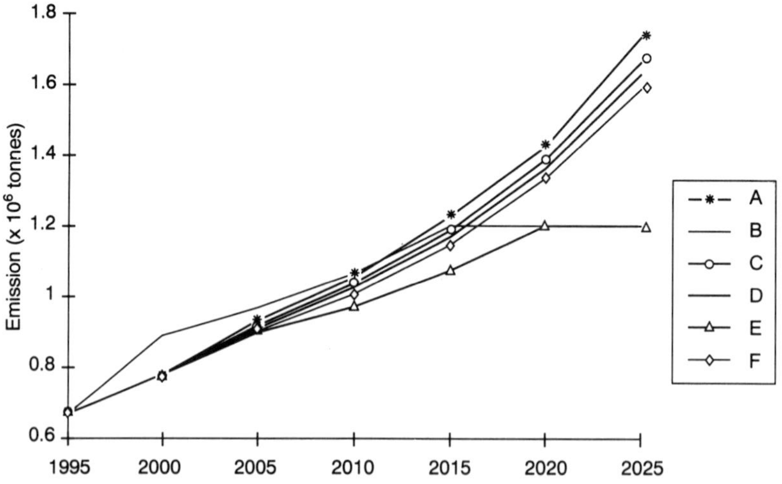

Figure 8.2 Total NO_x emissions over time: alternative scenarios

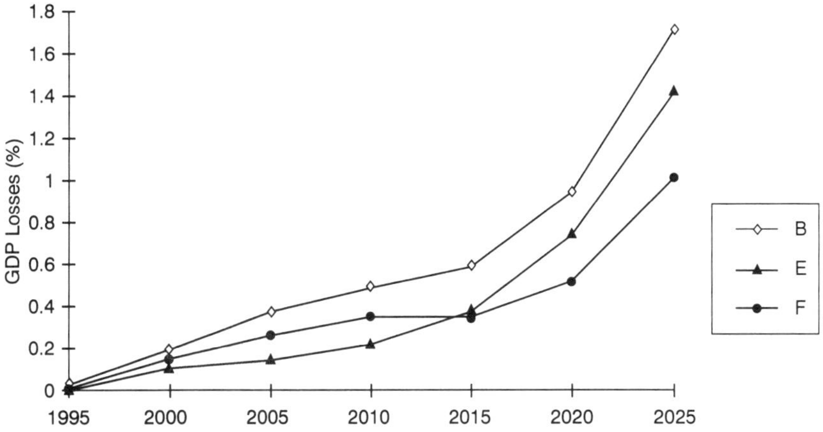

Figure 8.3 Percentage losses in GDP: alternative scenarios

After 2010, although the SO_2 limits of both scenarios become binding (corresponding to total SO_2 emissions increasing 54.4% and 5.69% relative to 1995 levels, respectively, for cases E and F), this time NO_x limits for scenario E are also reached, hence causing the observed results.

Among all scenarios the greatest GDP loss in year 2010 is 0.49% while it is around 0.6% in 2015, compared to base case levels. When expressed in constant US dollars they amount to $US0.9 billion and $US1.34 billion, respectively. On the other hand, the average growth rate of GDP between 2000 and 2010 is reduced from 4.45% per year to 4.28% per year only. Similarly, between 2000 and 2015 the same reduction is from 4.41% average yearly growth to 4.25%. Thus, it seems possible to reduce emissions significantly from their base case levels without drastically reducing the growth of the economy.

The energy costs reported in Table 8.3 are imputed costs. Since the model solutions are obtained through maximizing the discounted utility of consumption, these costs correspond to the imputed value of total expenditures on energy, which include, through energy prices, all of the costs of resource development and energy use. Therefore no direct comparison should be made with the costs obtainable with a cost minimizing model. Examination of the results shows that the differences with respect to the base case are not great and in some scenarios they are even negative. The third column of Table 8.3 shows cost differences in each period. It is understood that after 2010, when the emission constraints for SO_2 and NO_x become binding, there is a shrinking effect in energy use due to the emission limits: although it is possible to keep emissions within limits by control measures, still less energy must be consumed to compensate for the control costs and to keep macroeconomic balances. This is reflected in the negative cost differences.

Total discounted energy costs are displayed in Figure 8.4. It may be observed that in total the most expensive emission abatement scenario is scenario F. This scenario has the same 2.6 million tons of SO_2 emissions as scenario B but has a more relaxed limit on NO_x emissions. Therefore it must be the NO_x limit causing a different energy mix, coupled with different emission control technologies, resulting in the highest overall costs. However, when compared with the base case, percentage differences in total discounted costs are 0.124, 0.165, 0.29, 0.37 and 0.49, respectively for cases E, B, D, C and F, which shows that the highest is in the order of less than 0.5%.

Table 8.4 shows the results for energy aggregates for the base case and the two EU scenarios. Although there is in general a reduction in the use of all energy forms, the reduction in solids (mainly an aggregate of coal and lignite) is more pronounced than in oil and gas and in electricity. This is a result of the low quality of Turkish coal and lignite having high sulphur content. Electricity is reduced due to the escape from using low quality lignites as a supply option and because of the very high costs of emission reduction technologies. The reductions in the use of solids are 5.78% and 6.38%, respectively, for scenarios E and F in the year 2010. For the year 2015 these reductions become 9.86% and 10.23%, respectively. The corresponding reductions in electricity demand are 1.3% for both scenarios E and F in year 2010 and 1.26% and 1.63% respectively in year 2015. Oil and gas use decreases for scenario E and F by 1.31% and 0.07% in 2010 and increases by 1.08% and 2.22% in 2015. The column for oil and gas/solids ratio also reflects the shift away from solids.

Tables 8.5 and 8.6 display the tendencies to change in the composition of electric and non-electric energy from the base case of no control to control policies E and F, for SO_2 and NO_x emissions. In electric energy the shift away

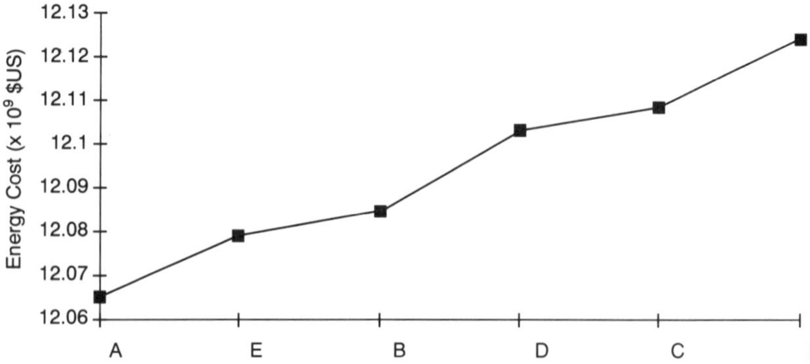

Figure 8.4 Total discounted energy costs: alternative scenarios

Table 8.4 Energy aggregates for scenarios A, E and F

	Oil and gas (mtoe)			Solids (mtoe)			Oil and gas/solid (mtoe)			Electricity (TWh)		
1991	27.12			24.96			1.09			60.16		
	A	E	F	A	E	F	A	E	F	A	E	F
1995	32.86	32.80	32.79	33.20	33.20	33.20	0.99	0.99	0.99	82.99	82.08	82.30
Growth (%)	3.91	3.88	3.87	5.87	5.87	5.87				6.64	6.41	6.46
2000	40.95	40.69	40.67	44.70	44.70	44.70	0.92	0.91	0.91	126.0	124.5	124.8
Growth (%)	4.50	4.40	4.40	6.13	6.13	6.13				8.71	8.69	8.69
2005	52.46	51.88	52.06	50.48	49.33	48.71	1.04	1.05	1.07	187.0	184.5	184.7
Growth (%)	5.08	4.98	5.06	2.46	1.99	1.73				8.21	8.19	8.16
2010	63.26	62.43	62.82	55.63	52.41	52.08	1.14	1.19	1.21	268.7	265.2	265.4
Growth (%)	3.81	3.77	3.83	1.96	1.22	1.35				7.53	7.52	7.52
2015	73.80	74.60	75.49	65.19	58.76	58.52	1.13	1.27	1.29	378.9	372.7	374.1
Growth (%)	3.13	3.63	3.74	3.22	2.31	2.36				7.11	7.04	7.11
2020	88.79	88.59	90.29	77.45	69.29	69.35	1.51	1.28	1.30	531.5	519.7	523.5
Growth (%)	3.77	3.50	3.64	3.51	3.53	3.45				7.00	6.88	6.95
2025	105.4	105.4	108.3	100.5	84.62	85.49	1.05	1.25	1.27	736.2	711.8	719.1
Growth (%)	3.49	3.54	3.70	5.36	4.08	4.28				6.73	6.50	6.56

Table 8.5 Composition of electric energy (TWh) for scenarios A, E and F

	1991	1995	2000	2005	2010	2015	2020	2025
Hydroelectric A	22.68	28.8	38.38	59.60	86.23	98.18	109.6	118.4
%	37.7	34.8	30.4	31.9	32.1	25.9	20.6	16.1
Hydroelectric E		28.86	38.38	65.01	86.23	98.32	109.7	118.8
%		34.8	30.5	33.0	32.2	26.0	20.7	16.2
Hydroelectric F		28.8	38.38	65.01	86.23	98.18	109.6	118.7
%		35.1	30.8	35.2	32.5	26.2	20.9	16.5
Lignite A	20.56	30.86	45.46	52.98	58.01	79.12	109.2	170.9
%	34.2	37.2	36.1	28.3	21.6	20.9	20.5	23.2
Lignite E		30.68	45.29	50.69	54.53	68.51	97.57	148.2
%		37.0	36.0	27.2	20.3	18.1	18.4	20.2
Lignite F		30.16	44.77	45.85	44.53	54.07	80.86	124.78
%		36.6	35.9	24.8	16.8	14.4	15.4	17.3
Fuel Oil A	3.33	2.85	2.30	1.75	1.21	0.66	0.11	0
%	5.5	3.4	1.8	0.9	0.4	0.2	0.02	0
Fuel Oil E		2.85	2.30	1.75	1.21	0.66	0.11	0
%		3.4	1.8	0.9	0.4	0.2	0.02	0
Fuel Oil F		2.85	2.30	1.75	1.21	0.66	0.11	0
%		3.4	1.8	0.9	0.4	0.2	0.02	0
Hardcoal A	1.00	1.86	1.70	1.53	1.37	1.20	1.03	0.87
%	1.7	2.2	1.3	0.8	0.5	0.3	0.2	0.1
Hardcoal E		1.86	1.70	1.53	1.37	1.20	1.03	0.87
%		2.2	1.3	0.8	0.5	0.3	0.2	0.1
Hardcoal F		1.86	1.70	1.53	1.37	1.20	1.03	0.87
%		2.2	1.3	0.8	0.5	0.3	0.2	0.1
Natural gas A	12.59	18.56	38.19	71.09	103.99	136.77	191.7	253.60
%	20.9	22.4	30.3	38.0	38.7	36.1	36.1	34.4
Natural gas E		18.56	37.57	70.47	103.38	146.28	201.2	271.58
%		22.4	30.2	38.0	38.7	38.0	37.4	36.7
Natural gas F		18.56	37.67	70.57	103.47	146.37	201.3	271.7
%		22.5	30.2	38.2	39.0	39.1	38.4	37.8
Nuclear A	0	0	0	0	17.94	62.94	119.9	192.44
%	0	0	0	0	6.7	16.6	22.6	26.1
Nuclear E		0	0	0	20.96	65.96	123.0	195.46
%		0	0	0	7.8	17.4	25.2	26.7
Nuclear F		0	0	0	28.59	73.59	130.6	203.1
%		0	0	0	10.8	19.7	24.9	28.2

Table 8.6 Composition of non-electric energy (mtoe) for scenarios A, E and F

	1991	1995	2000	2005	2010	2015	2020	2025
Lignite A	4.90	8.72	14.50	15.82	15.82	15.82	15.82	15.82
%	18.5	20.3	27.5	26.1	23.4	21.1	19.4	17.8
Lignite E		9.04	14.82	17.17	19.77	22.95	23.66	23.66
%		20.9	20.8	28.2	29.2	30.9	29.6	27.7
Lignite F		8.96	14.75	16.52	19.21	22.07	22.07	22.07
%		20.8	27.9	27.4	28.6	29.8	27.5	25.5
Oil A	16.17	23.20	23.70	28.21	32.00	35.57	38.72	41.78
%	61.3	54.1	44.9	46.5	47.3	47.5	47.4	47.0
Oil E		23.14	23.44	27.62	31.17	34.15	36.31	37.76
%		53.6	44.3	45.4	46.1	46.0	45.4	44.2
Oil F		23.14	23.42	27.80	31.57	35.04	38.00	40.60
%		53.7	44.3	46.1	47.0	47.4	47.3	47.0
Hardcoal A	2.18	6.49	6.55	8.57	11.81	15.40	19.14	23.35
%	8.3	15.1	12.4	14.1	17.5	20.6	23.4	26.2
Hardcoal E		6.49	6.55	7.93	8.58	8.97	11.94	15.85
%		15.0	12.4	13.0	12.7	12.1	14.9	18.5
Hardcoal F		6.49	6.55	7.93	8.25	8.74	12.12	15.67
%		15.1	12.4	13.1	12.3	11.8	15.1	18.1
Natural gas A	3.13	4.50	7.99	7.99	7.99	7.99	7.99	7.99
%	11.9	10.5	15.1	13.2	11.8	10.7	9.8	9.0
Natural gas E		4.50	8.13	8.13	8.13	8.13	8.13	8.13
%		10.4	15.3	13.3	12.0	10.9	10.1	9.5
Natural gas F		4.50	8.11	8.11	8.11	8.11	8.11	8.11
%		10.4	15.3	13.4	12.1	10.9	10.1	9.4

from lignites is very pronounced. This is partly compensated by the development of hydro potential at a greater level and increasing natural gas plant capacity, and after 2010 by the penetration of nuclear plants at a much faster rate than before. The results for non-electric energy seem surprising in the sense that a large increase is observed in the share of lignites and a large reduction in the share of hardcoal when emission limits become binding after 2010. There is also a relatively moderate reduction in the share of oil and a moderate increase in the share of natural gas. These developments are a consequence of the different efficiencies and the relative costs of emission control technologies. For SO_2 emissions the two technologies, wet limestone and Wellmann Lord processes, have 95% and 98% efficiencies whereas the efficiencies for NO_x reduction with primary measures and with selective catalytic reduction are 40% and 88%, respectively. It is understood that allowing more lignite use for non-electric purposes and employing high efficiency reduction technologies become less costly in the overall picture.

However, it should be noted that total use of lignite for all purposes in the years 2010 and 2015 are 15% and 22% less than in the base case, respectively.

CONCLUSIONS

Regardless of the shortfalls stemming from the necessity to aggregate quantities and technologies, the framework developed here appears useful in explaining the interactions between energy, economy and environment.

The model, which is the first of its kind for Turkey, is suitable for making a range of policy analyses. The representation of the economy through a macroeconomic submodel linked with submodels of the energy sector and the environment adequately allows for possible feedback effects, both from the energy sector and the environment to the economy as well as the effects of macroeconomic constraints, such as import restrictions, on the energy sector and environment. With this structure it becomes possible to measure the economy-wide implications as well as the costs or benefits of different energy and environment scenarios. This is a capability that integrated models of the energy sector and the environment do not always possess because they treat macroeconomic aspects through exogenously prespecified parameters. Although calibration of the present model is very difficult, once it is tuned, the model constitutes a consistent framework for the elaboration of interdependencies between energy policies, the economy and the environment.

Various solutions of the model point out that the impact of emission controls on economic output over the next 25 years may be quite limited. In the most tightly controlled scenarios, B and F, the imposed limits on SO_2 emissions meant keeping the emissions at a level that is only 5.65% more relative to the 1995 level for the rest of the planning horizon. However, it is not the SO_2 limit that determines the GDP losses relative to the base case. The combined effects of SO_2 and NO_x limits are readily observed from the results of scenarios E and F when the tighter limit on NO_x emissions is reached; even the looser limit on SO_2 emissions in scenario E cannot prevent the greater GDP losses compared to scenario F after 2015. GDP losses increase with time as control programmes continue to be effective.

The GDP losses also include the impact of emission control measures on investments. The tighter the control limits the slower the growth of investments becomes, as observed from comparing scenarios A, E and F.

In conclusion, Turkey can reduce emissions considerably without slowing economic growth too much. Achieving this will require following a carefully selected mix of energy alternatives coupled with appropriate emission abatement technologies. Results of the solutions indicate that such choices are feasible. It should be noted, however, that these conclusions are based on the input assumptions, and a different set of costs and supply technology assumptions may lead to different quantitative results. Sensitivity analysis is necessary to check the robustness of the conclusions derived from these early results of the model. If it turns out that the conclusions are still valid, in other words the GDP losses associated with reasonably strict emissions control strategies are not more than 1.5–2% of GDP, then the resulting policy implications for Turkey will be quite important.

ACKNOWLEDGEMENTS

This work was supported by the Turkish Scientific and Technical Research Council (TUBITAK) and the Middle East Technical University.

APPENDIX: NOTATION

Parameter values of the production function, taken from official predictions or determined through first order optimality conditions, are as follows:

sl = 0.28	se = 0.45	$\gamma = 0.43$
sk = 0.72	sp = 0.31	$\alpha = 0.841$
skd = 0.70	ss = 0.24	$\beta = 0.00019$
skf = 0.30	$\lambda = 0.97$	$\sigma = 0.30$

Variables

All variables, indices and parameters which appear in the text are defined as follows:

$AIR_{x,j,k,t}$	Total net amount of x type pollutant diffused in the atmosphere by j, k type fuel use
CG_t	Imports of consumption goods
$CO_{j,k,t}$	Total consumption of fuel type j group k
C_t	Consumption
$DNE_{j,k,t}$	Domestic amount of $NE_{j,k,t}$
$DP_{j,k,t}$	Domestic production of j, k type fuel
EC_t	Total energy cost
EF_t	Cost of fuel imports
$ELECINC_{j,k,t}$	Annual incremental electricity consumption which is produced by j, k type fuel fired power plant
$ELEC_{j,k,t}$	Electricity consumption which is produced by the kth group of j type fuel fired power plant
$EPOL_{x,j,k,t}$	Part of $POL_{x,j,k,t}$ resulting from electricity generation
E_t	Total electricity consumption
$FCO_{j,t}$	Consumption of imported j type fuel
$FILT_{x,j,t}$	Amount of x type pollutant from j, k type fuel use filtered by abatement activities
$FNE_{j,k,t}$	Imported amount of $NE_{j,k,t}$
F_t	Capital inflows from abroad
GDP_t	Gross domestic product

ID_t	Domestic money component of investments
IF_t	Investments made with foreign capital goods
INT_t	Imports of intermediate goods excluding fuels
KD_t	Domestic capital stock
KF_t	Foreign capital stock
L_t	Labour stock
M_t	Total imports
$NEINC_{j,k,t}$	Annual incremental j, k type fuel consumption for non-electric use
$NE_{j,k,t}$	Consumption of the kth group of j type fuel for non-electric use
$NEPOL_{x,j,k,t,l}$	Part of $POL_{x,j,k,t}$ resulting from l type non-electric use
$POL_{x,j,k,t}$	Amount of x type pollutant produced through use of j, k type fuel
P_t	Oil and gas consumption
$REDE_{x,j,k,t,s}$	Installed abatement activity s in power plants
$REDNE_{x,j,k,t,l,s}$	Installed abatement activity s in industry
S_t	Solid fuel consumption
$TOTNOX_t$	Total NO_x emissions
$TOTSO2_t$	Total SO_2 emissions
W_t	Factor incomes from abroad
X_t	Total exports
Y_t	Gross production
$EN_t, INTN_t, KDN_t, KFN_t, LN_t, PN_t, SN_t$	Incremental $E_t, INT_t, KD_t, KF_t, L_t, P_t, S_t$

Indices

j	Index for energy types
k	Index for sub-energy types
l	Index for sectors
s	Index for emission control technologies
t	Index for years which represent periods
T	Terminal period
x	Index for pollutant types

Parameters and Constants

λ	Annual survival factor; $(1 - \lambda)$ denotes stock depreciation.
$\mu_t, \phi_1, \phi_2, \eta_1, \eta_2, \gamma_t, \alpha_t, \beta_t$	Scale parameters
Δ_t	Discount factor
e	Required thermal fuel for unit electric production
$efe_{x,j,k}$	Emission factor of x type pollutant resulting from j, k type fuel use for electricity generation
$efne_{x,j,k,l}$	Emission factor of x type pollutant resulting from l type non-electric use of j, k type fuel
$eye_{x,s}$	Removal efficiency of control technology s for x type pollutant from power plants
$eyne_{x,l,s}$	Removal efficiency of control technology s for x type pollutant from l type non-electric use
$f_{x,j,k,s}$	Investment-fraction of total cost of x type pollutant and j, k type fuel related s type abatement technology

gd	Post-terminal domestic growth rate
gf	Post-terminal foreign growth rate
p	Period length in years
sd_l	Sectoral distribution of non-electric energy

REFERENCES

Arikan. Y., Güven, Ç. and Kumbaroglu, G. (1994) Energy–economy–environment interactions in Turkey, technical report, no. 94-14, Middle East Technical University, Industrial Engineering Department, Ankara, Turkey, August.

Güven, Ç. (1994) Energy planning under import restrictions, *European Journal of Operational Research* **72**, 519–528.

Manne, A.S. (1979) ETA-MACRO: A model of energy–economy interactions, *Advances in the Economics of Energy and Resources*, **2**, 205–233.

Murtagh, B.A. and Saunders, M.A. (1980) MINOS/AUGMENTED users guide, Department of Operations Research, Stanford University, Stanford, CA.

Plinke, E. (1992) Energy and emission control strategies for Turkey, PhD thesis, University of Karlsruhe.

Rentz, O., Haasis, H.D., Jattke, A., Ruß, P. and Wietschel, M. (1992) Impacts of energy supply structure on national emission reduction potentials and emission reduction costs, research report, UFOPLAN, no. 104 04 006, Karlsruhe, October.

Turkish Electricity Authority (1991) Wasp model usage in Turkey, Research and Planning Department report, TEA RPD-352 no. 10/1-195.

World Energy Council Turkish National Committee (1993) 1991 energy report.

Chapter 9

The Economics of the CO_2 Problem: What about the Supply Side?

Jürgen E. Blank and Wolfgang J. Ströbele

INTRODUCTION

By now, every energy economist has understood what we call the greenhouse gas (GHG) problem. And facing the severity of the problem, the scientific community has several computable general equilibrium models, which more or less uniformly calculate some optimal tax rate, especially for different sets of CO_2 reduction goals. Most of these models belong to the class of so-called top-down models. These models dominate the discussions at international conferences and in the journals. On the other side, very many energy economists point to existing, more or less obvious, inefficiencies in real-life energy systems: by reducing these inefficiencies, they offer a "free lunch" for CO_2 reduction. From this bottom-up analysis some types of "no-regret strategies" are developed. Since both kinds of model builders meet at the same conferences, it is astonishing to observe that communication problems exist between them.

For a cost–benefit analysis we assume for the moment that both the damage function and the cost function for reducing CO_2 emissions can be found somehow. Most energy experts would agree that in qualitative terms we may have a situation like that shown in Figure 9.1.

Systems Modelling for Energy Policy. Edited by D.W. Bunn and E.R. Larsen

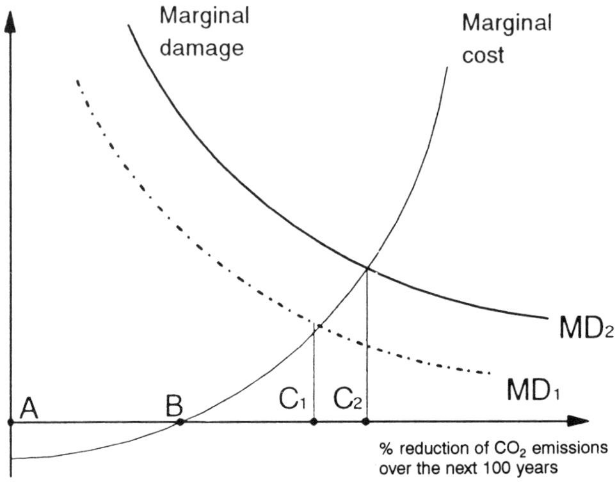

Figure 9.1 Qualitative structure of cost–benefit considerations

The *free lunch option* takes into account the considerable net-benefit improvements in an interval of some 10–20% reduction of CO_2 emissions (a move from A to B). These improvements stem from:

- Reductions of individual inefficiencies in firms and households.
- Reductions of governmental subsidies and other policies, which currently enforce the use of CO_2-intensive fuels.
- Additional positive external effects like reduced SO_2 or NO_x emissions if the burning of fossil fuels is reduced.

This type of improvement is the central concern of most bottom-up models, showing in detail how and where such inefficiencies may be removed. Therefore, typically, bottom-up models come up with rather "cheap" CO_2 reductions. This may be true for the first 10–20% and for an intermediate time horizon of CO_2 policy.

The reduction of CO_2 emissions from B to C is costly in terms of GDP loss, consumption loss or other economic categories. Global GHG models compute some *optimal shadow prices* for necessary reductions. The shape and the position of the marginal damage curve depend strongly on the assumed discount rate and the time horizon of the calculation, and on the fact that severe nonlinearities (such as a significant and rapid change of the Gulf Stream in the northern Atlantic) do not prevail due to the climate changes.[1] Still, there is a lot of uncertainty about the long-run damages: in economic terms we might therefore argue whether the optimal point of CO_2 reduction is C_1 or C_2 (for a discussion of cost–benefit arguments, see, for example, Nordhaus (1991) or Fankhauser (1994)).

SPECIFICITY OF CO_2 AS AN ENVIRONMENTAL PROBLEM

Nowadays, it is well-accepted knowledge that the CO_2 problem is a very specific environmental issue. The main characteristics are given by:

- There is *no (reasonable) end-of-pipe technology* preventing CO_2 emissions. Therefore, the only way to reduce CO_2 emissions is by substitution either of fossil fuel by capital (including know-how) or by fuels that are less CO_2 intensive (natural gas, nuclear power, renewables).
- The damage is not caused by the flow of emissions but merely by the *global accumulated stock* in the atmosphere. Therefore neither the region nor the time of emissions (within a range of some decades) is relevant for the damage. This gives the theoretical advantage that a CO_2 emission tax or some kind of certificates are reasonable instruments for environmental policy.[2] On the other hand, this peculiarity creates all possible problems of incentives of international cooperation: the individual free-rider position *vis-à-vis* both other regions and future generations is a severe impediment for any environmental policy.
- There is a one-to-one function between CO_2 emissions and the burning of certain types of *fossil fuels* (hardcoal, brown coal, oil, natural gas). This allows for a very simple form of taxation: one need not put a tax on some measured CO_2 emission but instead on the specific fossil fuel.[3]
- Since the damage is caused by the stock of CO_2 we have very *long time horizons* for the calculation of damage done by CO_2.

This raises three further problems. First, since *all* of these future damages can only be calculated by means of some long-term simulation models there is a high degree of *uncertainty* stemming from the unavoidable complexity of the models and the large number of very uncertain parameters. Today we have reliable information only on a certain range of substitution possibilities, technologies, demand reactions, etc., but the long-term adjustments will probably go far beyond that range of known parameters.

The second consequence of the very long-term structure of the problem is the economic evaluation of costs of the reduction of CO_2 emissions today (whose costs have to be carried by mankind in the next decades) and the benefits in the form of avoided catastrophes or other losses far in the future. In other words: the *rate of discount* and/or the degree of *risk aversion vis-à-vis* potential catastrophic events determine any good cost–benefit argumentation.

The third problem is the degree of *confidence in the quality of modelling*: whereas in many other areas, modelling can be understood as one helpful method among others to a better understanding of complex problems, there is in fact no alternative to modelling with respect to the very long run CO_2 cost–benefit considerations.

From our point of view, we realize that the well-trained "environmental economist" makes use of a certain chain of reasoning, where he or she does not take into account the following second aspect of the CO_2 problem (for an excellent example, see Barrett (1995)).

SPECIFICITY OF CO_2 AS A RESOURCE PROBLEM

Since the CO_2 emissions to a large extent can be attributed to the burning of fossil fuel, it is natural to ask: What determines fuel prices without a CO_2 restriction and what will be changed if some CO_2 restriction has to be taken into account? This question may be looked at from the point of view of the benevolent world-economy planner[4] or alternatively by looking at the mechanisms that work on the markets for oil, gas and coal in the long run. The main problem of any modelling effort will then be to put these two aspects—environmental policy recommendations on the one hand and resource pricing on the other—together.

Since all computable economic greenhouse gas models share one common feature—namely, the ultimate transition to a CO_2 free backstop technology—we will use this elementary approach to clarify some theoretical arguments. We will start with a very basic model where we assume, for analytical purposes, that there is a very large potential of the atmosphere to accumulate CO_2 without any damage to the climate. Then the resource pricing is solely determined by the scarcity of the resource from a finite stock. We assume that there is a net utility derived from consumption, which is either C from fossil fuel or Z from a backstop technology. The latter may be given by a large-scale nuclear or solar technology with constant unit cost k. Then the planner's optimization problem is given by:[5]

$$\max \int_0^\infty e^{-rt} \cdot [U(C_t + Z_t) - k \cdot Z_t]\, dt \tag{9.1}$$

s.t.

$$\begin{aligned} &\dot{S}_t = -C_t && \text{stock of resource} \\ &S_0 > 0 && \text{given, } S_t \geq 0 \\ &k > 0 && \text{given backstop cost} \end{aligned}$$

The Hamiltonian approach of dynamic optimization yields:

$$H = e^{-rt} \cdot [U(C_t + Z_t) - k \cdot Z_t] - \mu_t^R e^{-rt} \cdot C_t$$

$$\frac{\delta H}{\delta C_t} = 0 \quad : U'(C_t + Z_t) = \mu_t^R \tag{9.2}$$

$$e^{rt} \cdot \frac{\delta H}{\delta S_t} = -\dot{\mu}_t^R + r\mu_t^R : \mu_t^R = \mu_o^R e^{rt}$$

Boundary condition: $S_T = 0$, $\mu_T^R = k$.

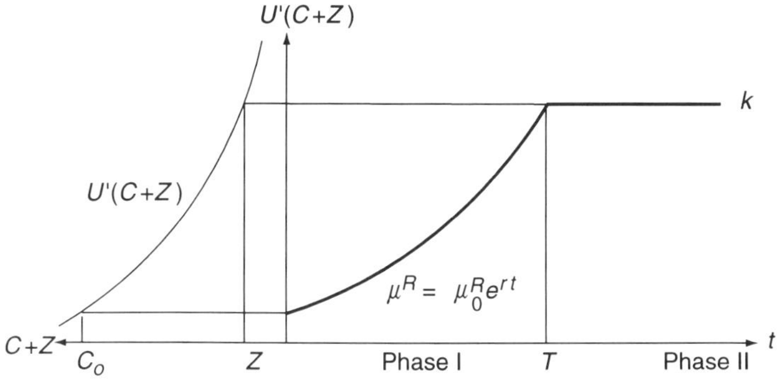

Figure 9.2 Price path with transition to a backstop technology without environmental constraint

The solution of this simplest type of model in the world of natural resource economics is the well-known transition problem to a backstop technology. The price path in a competitive world of general equilibrium corresponding to the optimal solution μ is shown in Figure 9.2. The solution has two phases. First, *all of the resource stock* is used up in finite time on a so-called Hotelling path ($C>0$, $Z=0$). Thereafter, we have the sudden transition at time T to the use of the *backstop technology in phase II* ($C=0$, $Z>0$). In this simple model we do not have research and development, putty-clay capital stock in energy production, exploration activities to enlarge the initial (known) resource endowment, etc. All of these aspects have been dealt with separately in resource economics during the past 20 years. But the main message is clear already in this basic analytical framework: There is a *resource* (*scarcity*) *rent* μ^R that is equal to the price of the resource in a perfect competitive market. In the more sophisticated resource models this rent also stimulates further exploration for widening the resource base. The rent then not only equilibrates supply and demand over time, but also has a dynamic incentive function. Of course, this ideal Hotelling logic is distorted by many factors in real-life energy markets.[6] But for a long-run perspective most economists accept the basic logic of the Hotelling principle. Even if we allow for exploration activities or some monopolistic or oligopolistic elements, the main arguments still hold in qualitative terms, even if we thereby get a modified price path over time.

RESOURCE PRICING WITH AN ENVIRONMENTAL CONSTRAINT

Let us again for simplicity assume that we have a clear cut given *upper limit* for the total acceptable accumulated stock of CO_2, which will be given the letter M^* in the following model. If the fictitious UN planner has to take this upper limit into account as an additional constraint, he or she has to solve the following optimization problem:

$$\max \int_0^{\infty} e^{-rt} \cdot [U(C_t + Z_t) - k \cdot Z_t] \, dt \tag{9.3}$$

s.t.

	$\dot{S}^t = -C_t$	stock of resource
	$\dot{M}_t = -\gamma \cdot M_t + C_t$	stock of accumulated $CO_2 = M$
	$S_0 > 0,\ M_0 > 0$	given, $S_t \geq 0$
	$k > 0$	given backstop cost
	$M_t \leq M^*$	CO_2 constraint

The Hamiltonian approach of this modified optimisation problem is given by:

$$H = e^{-rt} \cdot [(C_t + Z_t) - k \cdot Z] - \mu_t^K e^{-rt} \cdot C_t + \lambda_t e^{-rt} \cdot (-\gamma \cdot M_t + C_t)$$

$$\frac{\delta H}{\delta C_t} = 0 \quad : U'(C_t + Z) = \mu_t^K - \lambda_t$$

$$e^{rt} \cdot \frac{\delta H}{\delta M_t} = -\dot{\lambda}_t + r\lambda_t \quad : \lambda_t = \lambda_0 \cdot e^{(r+\gamma)t} \tag{9.4}$$

$$e^{rt} \cdot \frac{\delta H}{\delta S_t} = -\dot{\mu}_t^K + \delta \mu_t^K \quad : \mu_t^K = \mu_o^K e^{rt}$$

Only if the shadow price for using the atmosphere as a garbage storage (for CO_2) is zero ($\lambda = 0$ for all times), i.e. we have no scarcity from there, we have the coincidence of this modified model with the standard backstop model above. If the additional constraint ($M \leq M^*$) is binding, then we have the situation drawn in Figure 9.3. From the point of view of the UN planner, we

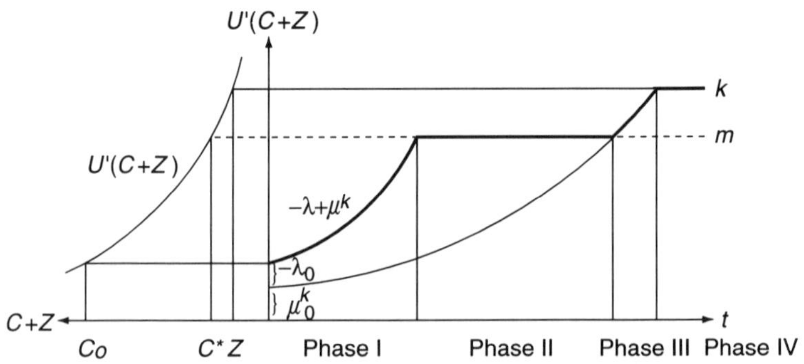

Figure 9.3 Changing the optimal price path via an environmental constraint

get a quite different price path (the bold line in Figure 9.3). The initial price $(-\lambda_0 + \mu_0^K)$ now consists of two components, which actually describe *two quite different "scarcities"* : while there is still a scarcity rent for the resource (given by μ^K), this is now smaller than μ^R in the case above. This stems from the fact that this scarcity rent μ^K is derived from the resource use in phase III, where there is no longer an atmospheric CO_2 restriction. In phases I and II the price path is dominated by the CO_2 restriction. This restriction is reflected in the dynamics of λ.

The institutional requirements for "pricing" of a global good like the use of the atmosphere pose a difficult problem of their own and are quite different from resource pricing. Contrary to stocks of resources, which are either owned by countries or by companies, there are no individual property rights of the atmosphere, or any other institutional framework to cope with the problem. This raises the question: If we do not have the benevolent UN planner but instead decision makers in the resource producing countries (oil and gas producers, coal producers) and the energy consuming countries, how can we implement a regime of resource management over time, that to a large extent reflects the message of model II? In fact, this message requires that the evaluation of fossil fuels is drastically changed by the CO_2 restriction: for all owners of fossil fuel stocks this additional restriction means a heavy loss in terms of rent $\mu^R - \mu^K$. To some extent the discovering of the CO_2 restriction means a partial expropriation of resource owners.[7] *The supply side of fossil fuels matters essentially for the CO_2 problem.*

The discussion has shown that the economics of greenhouse gases necessarily must simultaneously look at the problem from three perspectives:

- Economics of natural resources tells us what will happen with the price path of resource supply under a certain environmental restriction.
- Environmental economics tells us how to implement schemes to prevent free-riding or leakage effects.
- Energy economics tells us what will happen in the energy markets with respect to shifts between the different types of energy sources and what kind of sectoral effects will come up when a certain CO_2 scheme is implemented.

It will not be adequate to look on some induced substitution process on the demand side of energy, but similarly important is understanding possible reactions of resource suppliers.

As a next step we therefore look into the different modelling efforts that were developed in recent years to cope with the CO_2 problem, and how they solve this problem. Some of these models are optimization models, using some kind of UN planner: here we may ask whether the supply side is modelled adequately. Other models try to give a real-life picture of energy demand and supply mechanisms: here we may ask whether, and eventually in which way, the rent-depressing effect of a CO_2 policy is modelled.

Table 9.1 Comparison of three GHG models*

	GREEN	Global 2100, 5R	ERB
Authors	OECD (1994)	Manne–Richels (1992)	Edmonds & Reilly (1985)
Type	Computable general equilibrium	Computable general equilibrium	Computable partial equilibrium
Version	GREEN 1.0		Version 4.01
Time periods	1985 (base year), 1990, 1995, 2000, 2005, 2010, 2030, 2050	1990–2100 10-year periods	1990–2095, 15-year periods
Dynamic structure	Myopic	Forward looking	Myopic
Regions[14]	*USA* *Japan* (JPN) *EU* (EEC) *Other OECD* (OOE) *Central and Eastern Europe* (CEE) *Former USSR* (FSU) Energy-exporting LDCs (EEX) China (CHN) India (IND) Dynamic Asian Economies (DAE) Brazil (BRA) Rest of World (RoW)	*USA* *Other OECD* China Former USSR RoW	*USA* *Western Europe and Canada* Asian OECD *Central and Eastern Europe* Far East Middle East Africa Latin America South and East Asia
Fuel sources	Coal Crude oil Natural gas Refined oil Electricity (including gas and water distribution)	Non-electric energy supplies: Net oil import Coal direct use Oil (low cost) Oil (high cost) Natural gas (low cost)	Primary energy: Oil Conventional Nonconventional Gas Conventional

	Carbon-based backstop Carbon-free backstop Carbon-free electric backstop	Natural gas (high cost) Renewables Synthetic fuels Non-electric backstop Electricity technologies: Existing: Hydro (including geothermal, and other renewables) Gas fired Oil fired Coal fired Nuclear New: Gas fired Coal fired High-cost carbon free Low-cost carbon free	Nonconventional Solids Coal Biomass Nuclear Solar Hydro Secondary energy: Liquids Gases Solids Electricity
Tradable goods	Coal Gas Oil Refined oil Electricity Agriculture Energy intensive good Aggregative other good	Oil Carbon emission rights	Fossil fuels
Backstop	Leasing concept	Capital accumulation	Leasing concept

* Annex 1 countries are printed in italics. In Global 2100 Central and Eastern Europe are part of RoW.

GLOBAL GHG EMISSION MODELS AND THEIR BASIC ASSUMPTIONS

The Models

This section gives a general overview of the resource supply side of three global energy/greenhouse gases (GHG) models. The models considered are GREEN, Global 2100 and the ERB model.[8] Major features of the models are given in Table 9.1.

In all three models oil and natural gas are treated as exhaustible natural resources. The initial endowments of proven and unproven reserves are given. Coal, due to its very large endowment within the relatively short time periods considered, is treated as a non-exhaustible resource.[9] The theory of exhaustible resources sketched above shows that the price for a unit of a depletable resource (here, oil and gas) consists of marginal cost of extraction and the (positive) shadow price, which represents the scarcity of the resource and the (negative) shadow price, representing the externalities. The resource producers' optimal extraction programme must be orientated along the shadow prices and must take into consideration the existence of current and future backstop technologies.

Of the three models, only Global 2100 selects such an extraction path due to the forward looking behaviour of resource producers. In ERB and GREEN the resource producers, like all other agents, are myopic.

As in the sections on the specificity of CO_2, one may model the supply of a resource substitute in terms of a constant unit cost for the backstop. There is no separate capital stock for the backstop.[10] Only in Global 2100 is there a capital stock necessary for the production of a (electric and non-electric) substitute for fossil fuels.

Global 2100

The current supply of an exhaustible resource in a period is a simple linear function of current proven reserves. Therefore the process of reserve extension is the driver for current resource supply decisions and represents the link between different time periods.

Proven reserves are determined by the pre-period proven resource plus a time weighted annual average of gross reserve additions from undiscovered resource less current production. Undiscovered resources are determined by those of the pre-period minus the time weighted average of gross additions to proven reserves. The gross reserve additions are determined within the dynamic optimization process in such a way that the scarcity rents on the exhaustible resources cannot rise more rapidly than the marginal productivity of capital. Indeed, the scarcity rent may rise at a rate less than the marginal productivity of capital. This allows the simultaneous production from different resource cost categories.

Oil is the price leader in the exhaustible resource market. The international oil price is set by the RoW region in such a way that the demand by OECD countries is met. The imports from the USSR and China are determined exogenously. Within the OECD countries the oil price level is given. The supply price for natural gas is determined by a given oil–gas price differential (i.e. $US1.25 per GJ). In no case is there any premium on the gas supply price due to its lower content of carbon compared with that of oil.

GREEN

Contrary to Global 2100, where agents are optimizing over the total planning period, in GREEN (and in ERB) agents are optimizing only for the current period, ignoring totally the past and the future. Since the initial endowments of the exhaustible resources are given, the totals of proven and unproven reserves for oil and natural gas have to be predetermined in each period. A price-sensitive conversion rate converts unproven reserves into additional proven reserves. The level of the conversion rate depends on the relative real price of the exhaustible resource compared to a reference price scenario. A higher/lower supply price induces increasing/decreasing rates of conversion. The resulting level of the conversion rate is calibrated to meet output projections from energy experts, including IEA[11] (see Figure 9.4).

The potential supply in the current period is a simple linear function of current proven reserves.

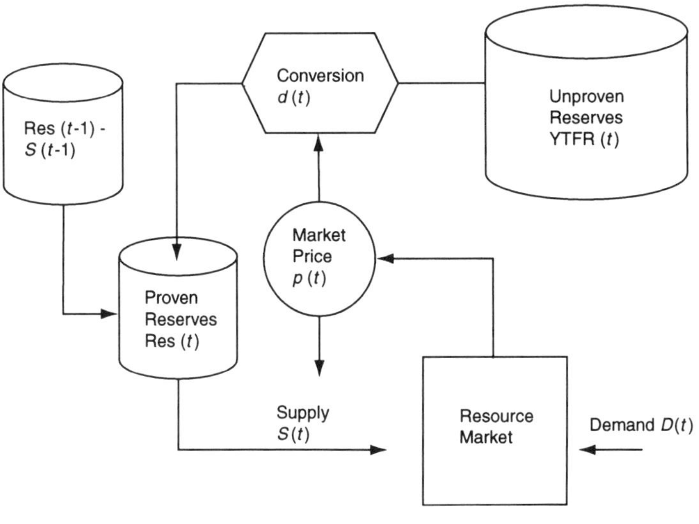

Figure 9.4 Resource depletion in GREEN

For all but the EEX region (mainly OPEC countries), potential supply is identical to actual supply. Only current production in the EEX region responds to price signals by a given supply elasticity. The potential supply is the corresponding upper bound.

Since all OPEC suppliers are part of EEX, GREEN generates an endogenously determined oil price by EEX's behaviour as the world's dominant oil supplier. This endogenous world oil price (PW_{Oil})[12] is the result of a *tâtonnement* procedure equating global supply with global demand in the short-term market equilibrium.

Natural gas resources in the EEX region are assumed to be infinite and production will be determined by market equilibrium conditions. Therefore, the price elasticity of natural gas supply for the EEX region is infinite. Unlike the case of oil, there is no uniform international natural gas price in GREEN, because of the Armington good character of natural gas. Imported gas is not a perfect substitute for domestic natural gas. From our point of view, this assumption seems very doubtful, because advances in exporting liquefied natural gas (LNG) and the construction of new pipeline systems, mainly from the Middle East (with 40% of natural gas reserves) and Russia to Europe, makes natural gas a homogenous good. We propose that, as for oil, natural gas should be modelled as a homogenous good.

ERB

Each region is endowed with an initial amount of the exhaustible resources oil (conventional and unconventional), natural gas and coal, differentiated by five resource grades characterized by extraction costs. In each period the profitable resource grade is determined. From each actual resource grade i one gets the accumulated production Q_t in period t:

$$Q_t = \sum_{j=1}^{i} S_j - S_i \cdot \frac{c_{i+1} - p_t}{c_{i+1} - c_i} \cdot \theta \tag{9.5}$$

S_j stands for the resource endowment of grade j, c_i is the minimum production cost (in the interpretation of ERB c_i should include taxes and environmental costs)[13] of the actual resource grade i, p is the domestic supply price and θ is a variable representing the rate of technical progress in extraction technologies. Cumulated production is price sensitive: a higher supply price results in higher cumulated production. High production cost differentials result in less actual production.

The rate of current production R_t is derived from cumulated production by:

$$R_t = 2 \cdot \frac{Q_t - Q_{t-1}}{n} - R_{t-1} \tag{9.6}$$

where n is the number of years between two periods.

An initial price for the fossil fuels is given exogenously. Since demand and supply for fossil fuels are determined independently it is not guaranteed that demand equals supply at this price. To close the potential gap between supply and demand, new prices are calculated by demand and supply elasticities until the gap becomes sufficiently small.

The production of fossil fuels is not based on strategic economic considerations over time. All fossil fuel producers act as price takers. Even large suppliers and exporters, like the Middle East region in the case of oil, behave as any small producer.

Comparison of the GHG Models

This section compares the outcome of the three models described above with respect to different types of GHG reduction scenarios. The following scenarios are considered.

Reference case

For simplicity we take the reference or baseline case from the original models. Therefore, each baseline scenario is different in its basic assumptions. All three models have in common that they assume some rate of autonomous energy improvement (AEEI). This serves as a catch-all variable for successive advances in using the free-lunch options mentioned in the introduction section, for structural change towards less energy intensive sectors and for better information and institutional settings to use less energy. For our comparison, AEEI is set at 1% annually until the year 2005 and thereafter at 0.25%.[14]

Toronto-type agreement

Based on each reference case, carbon emissions in Annex 1 countries (OECD, former USSR, Central and Eastern Europe) have to be reduced to 65% of their 1990 levels by the year 2050. All other countries are allowed to emit up to 125% of their 1990 levels.

The tax imposed in the models is a carbon tax. Each region imposes an individual tax on carbon emitted to fulfil its regional emission quota. There is no trade in carbon emission rights. In GREEN and ERB the carbon tax is an excise tax, applied at the level of consumers of primary fuels.[15] In Global 2100 the marginal or dual value of the annual carbon limit is the value of regional carbon emission rights. In the optimum the value of a carbon emission right is equivalent to a carbon tax imposed on the consumption of fossil fuels.

Market Shares of Energy Sources

Global 2100

The imposition of a carbon tax shows almost no effect on the amount of total energy consumption. That means that all relevant substitution processes run on the supply side of the economy. The large-scale use of coal and coal- and shale-based synthetic fuels in the reference case are substituted by non-electric backstops and low- and high-cost advanced electricity technologies in the CO_2 case. Within the RoW region the low-cost oil production pattern is unchanged with respect to the different carbon reduction scenarios. As a result of the carbon taxation, there is a decline in oil export by the year 2030, but thereafter the RoW is no longer engaged in international oil trade.

There is even no international trade in oil in the baseline scenario after the year 2030. This seems very dubious,[16] and it can only be explained by the massive regional production of coal and synthetic fuels in the baseline scenario and of non-electric and electric backstop in the carbon reduction scenario. Hence there will be *no* need for *oil imports in the OECD regions after 2030* in both cases. Oil price in the baseline case rises up to the unit cost of synthetic fuels, whereas in the carbon reduction scenario oil price rises up to the unit cost of the non-electric backstop (see Figure 9.5).

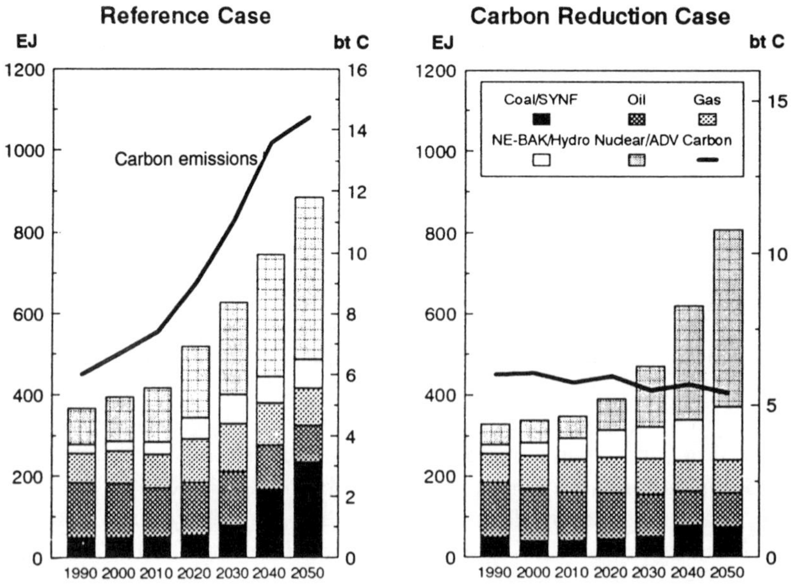

Figure 9.5 Composition of global energy consumption and carbon emissions in Global 2100

Our critique with respect to this point is that this approach in modelling neglects rather simple responses from the oil suppliers. The Middle East suppliers will still have a huge oil reserve base left after 2030. By setting an oil price just below the backstop price, the backstop technology will at least partly be priced out of the market.[17] Hence, oil exporters can gain market shares and raise their profits. In the model, there is an additional problem: "unknown resources", which feed the process of increasing known reserves in the RoW region, become almost zero, i.e. there will be no gross reserve additions. Due to the price-independent linear extraction function, in RoW the production of oil is in fact price-independent. Although in general agents are characterized by forward looking behaviour, in fact oil producers (note, again, that oil is the only tradable good) are heavily myopic. We doubt whether any oil manager in Saudi Arabia or Kuwait will agree with such an approach.

Figure 9.6 shows the production paths for oil and natural gas in the RoW region, which includes OPEC. As one might expect, oil production declines (after 2010),[18] and gas production will rise, due to its lower content of carbon. After 2030 oil production declines. Again, there is a puzzling point: in the case of a carbon reduction strategy, low-cost oil production is always identical to the reference figures.[19] What about supplying the rest of the (apparently cheap) reserves still in the ground in the Middle East countries after 2030/40 instead of producing high-cost oil and ceasing oil exports?

GREEN

In GREEN the imposition of a carbon tax has a significant effect on the amount of total energy consumption. Total energy consumption drops down from 899

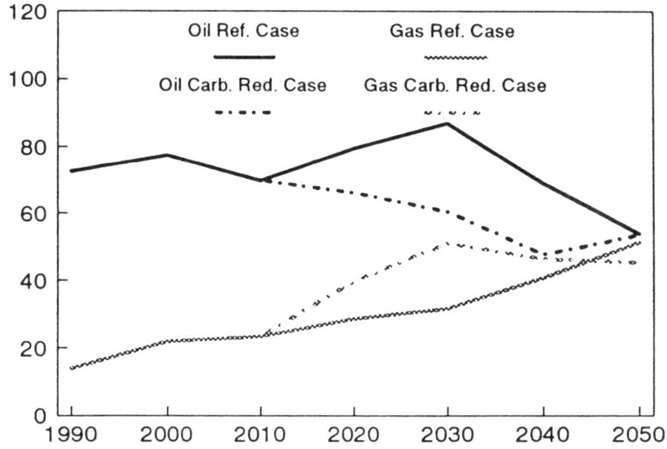

Figure 9.6 Oil and gas production (EJ) in the RoW region in Global 2010

EJ in the baseline case to 385 EJ in the reduction case in the year 2050. The main burden falls on the consumption of coal and the carbon-based backstop (Figure 9.7). The contribution of coal declines from 423 EJ in 1990 to 159 EJ in 2050, i.e. there is a decrease in the market share for coal from 47% to 12.7%. Carbon-free backstops (CF FBS) and electric backstops (Elec BS) are not able to substitute conventional fossil fuels.

The decline of carbon emissions is mainly due to reduced total energy consumption. Obviously GREEN puts a large burden on adjustments on the energy demand side. The annual consumption of oil and gas is nearly the same in each scenario. Only in 2050 is the consumption of oil increasing, whereas the consumption of natural gas has been decreased by half in the carbon reduction scenario. The main burden falls on the gas production in the EEX region, which is only 23% compared to the baseline case (Figure 9.8).

ERB

Like GREEN, the ERB model shows a similar structure. Total energy consumption drops as a consequence of the imposition of a carbon tax. The consumption of oil is nearly unchanged until the year 2035; in 2050 oil is

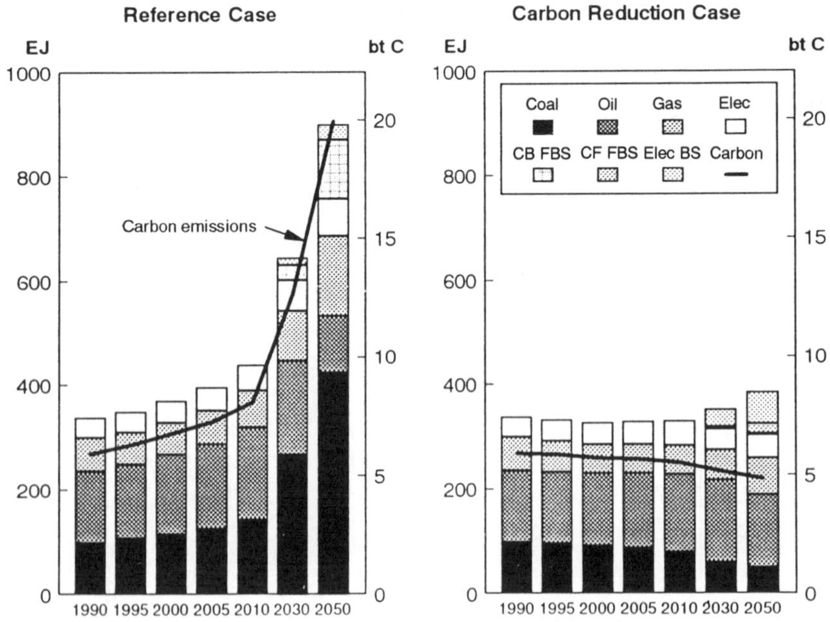

Figure 9.7 Composition of global energy consumption and carbon emissions in GREEN

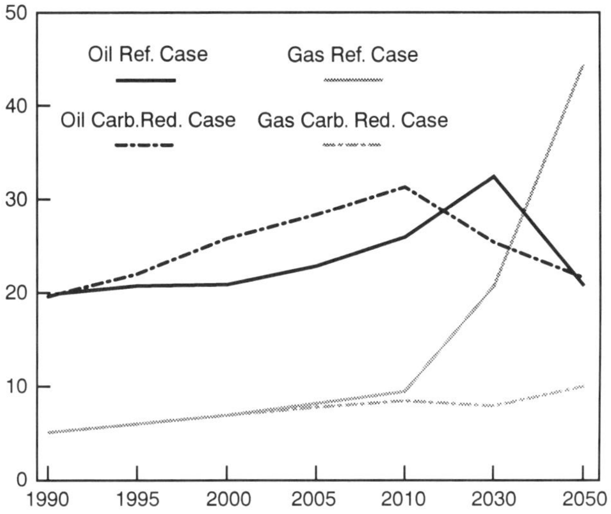

Figure 9.8 Oil and gas production (EJ) in the EEX region in GREEN

consumed more in the reduction scenario compared to the reference case. Interestingly, as in GREEN, natural gas consumption declines drastically as an effect of a carbon tax (Figure 9.9).

The results with respect to the supply of oil in the Middle East region, which is the region with more than 75% of proven oil reserves, seem questionable. In both scenarios, oil production is rising nearly identically until the year 2035. After that year, oil production falls sharply towards nearly zero production in the baseline case and only very small production in the carbon reduction scenario. In 2050 oil from the Middle East will be replaced by regional supply of synthetic oil from coal and biomass. Again, we doubt whether this is a good picture of what we may expect (Figure 9.10).

Comparison of the Energy Price Paths

As a consequence of the imposition of a carbon tax, one has to expect that the international producer oil price will go down under competitive conditions (see the section on "Resource pricing with an environmental constraint", p. 145). This is not only a basic concept of the theory of natural resources but sound economics in general.

The effects of a carbon tax on the international *supply price of oil* in the three models can be seen in Figure 9.11. The reference oil price path is set to the index number 100 throughout. In the ERB model, there is nearly no reaction of the oil price until the year 2035. Thereafter, oil prices rise only slightly until the year 2050. In GREEN the international oil price is always

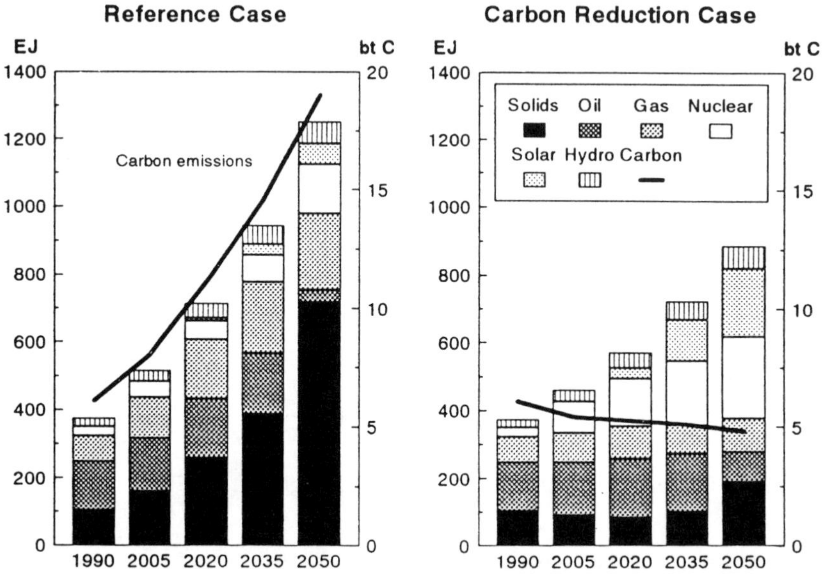

Figure 9.9 Composition of global energy consumption and carbon emissions in ERB

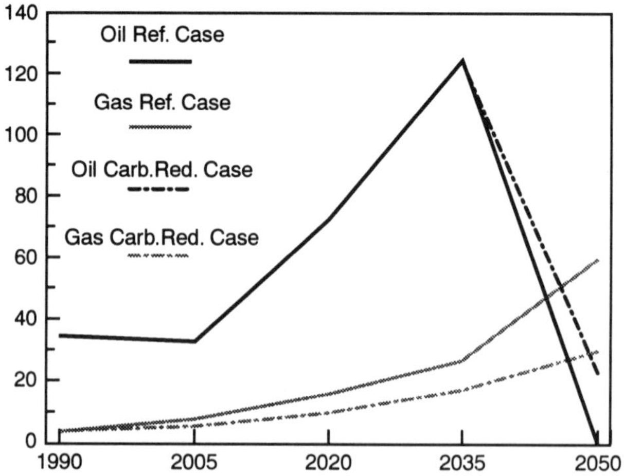

Figure 9.10 Oil and gas production (EJ) in the Middle East region in ERB

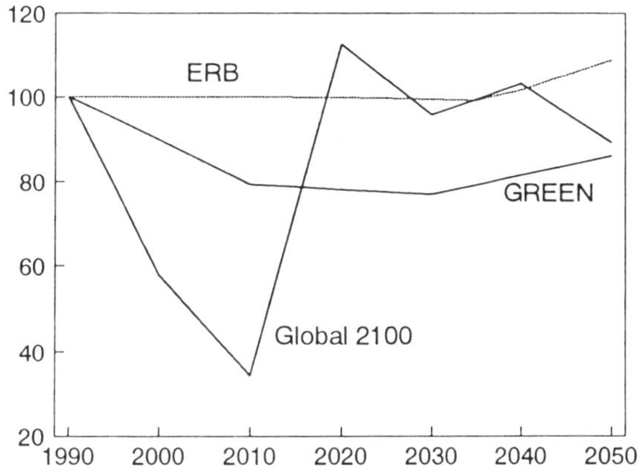

Figure 9.11 Oil price indices after following a CO_2 reduction strategy

lower after the imposition of a carbon tax. The price declines to 80% compared to the reference price by the year 2010. The rising demand for oil, as a substitute for the carbon-intensive use of coal, by the year 2050 leads to a recovery of the oil price, but the reference level will not be reached. Global 2100 shows a drastic decline in the oil price down to 37% compared to the baseline level by the year 2010, followed by a much more dramatic increase in the oil price, overshooting the reference oil price by 2020. After the year 2050 the oil price will be stabilized at the price of the non-electric backstop costs minus the (constant) carbon tax. The volatile oil price in Global 2100 can be explained by the fact that in this model oil supply is a simple linear function of proven reserves. Only the (price-dependent) resource additions are under the control of the producers. Since the resource addition might be very small in the case of an oil price drop, current production is to a large extent independent of current oil prices. While the UN planner (the user of the model) in fact optimizes over the total time horizon, there is no optimization behaviour of suppliers over time: otherwise they could discover the chances for extra profits via arbitrage of oil along the time axis, which would alter the whole price and supply path.

Comparison of the Main Results

In Table 9.2 we look at some key variables of the three models. As a fourth column we introduced what we learned from practitioners, i.e. people with special knowledge in energy markets and energy policy: this type of evaluation does not stem from a well-defined model, but is “fuzzy”. Nevertheless, it may represent “good knowledge”. Obviously, there are significant differences with

Table 9.2 Main results from the three models

	Global 2100	GREEN	ERB	Energy economist's expectations
Demand reaction	Weakly negative –	Very strongly negative – – –	Strongly negative – –	– –
Shift within fossil fuels:				
oil	– –	–	+	±
natural gas	– – –	– –	– –	+
coal	–	– – –	– – –	– – –
Backstop:				
nuclear	+ +	+	+ +	+
solar	+ + +	+	+ +	+
others	+ +	+	+	+
Oil supply price	Chaotic fall and increase	Weak depression (–20%)	Constant	Depressing on market forces

respect to demand reaction: while Global 2100 seems to be too pessimistic, an energy economist would judge GREEN to be too optimistic.

While in the very long run there may be a common belief in the role of backstops, there are severe differences in expectations with respect to oil and natural gas. Energy economists believe that the high times of oil and natural gas are still to come (especially for natural gas in the case of a CO_2 policy), but this is not reflected in any of the models! The price paths in Global 2100 (which is a "forward looking" model!) and ERB do not seem plausible. At today's market structure most energy economists would probably share GREEN's point of view. An OPEC expert would point to a change of the game in the coming decades, which would eventually prevent the pressure on suppliers' prices from a CO_2 policy.

Strategic Behaviour and Incomplete Competition on Fossil Fuel Markets

The global models leave no room for strategic behaviour of fossil fuel producers. Since each region simulates a (perfect) market or a planned economy, fossil fuel suppliers are perfectly competitive. Only Global 2100 allows the RoW region, which includes OPEC, to set the oil price in such a way that supply meets demand. None of the three global models incorporates any kind of incomplete competition or cooperation among producers. Consequently, the introduction of a carbon tax has a rent extracting effect for the benefit of oil-consuming countries or the tax-raising international "energy fund".

Actually, the market for fossil fuels is neither a perfect competitive nor a monopolistic market. Typically the supply side of fossil fuel markets is oligopolistic (see Figure 9.12).[20] This allows producers to react strategically if confronted with any kind of greenhouse gas policy. The ability to react strategically depends on the degree of possible collusion among the global players. Cooperation among producer countries may increase their ability to lessen the rent extracting effect of any carbon reduction scenario. The global players on the fossil fuel markets are the OPEC members and a few other oligopolistic energy exporting countries. The ability to reach and maintain collusive behaviour among the global players on the market for fossil fuels depends on the (expected) stability of such an agreement.

The strong position of some OPEC members in oil and gas reserves allows OPEC to focus its policy on both oil and natural gas. OPEC's share in proven world oil reserves today is 77%. Over the next decades, the significance of non-OPEC countries will decline. Since non-OPEC oligopolists are hit by the rent extracting effect of a carbon or energy tax, their willingness to behave cooperatively may be greater than without a tax threat. Furthermore, nearly 40% of the proven natural gas reserves are held by OPEC member countries (another 40% share is held by the GUS countries in the former Soviet Union).

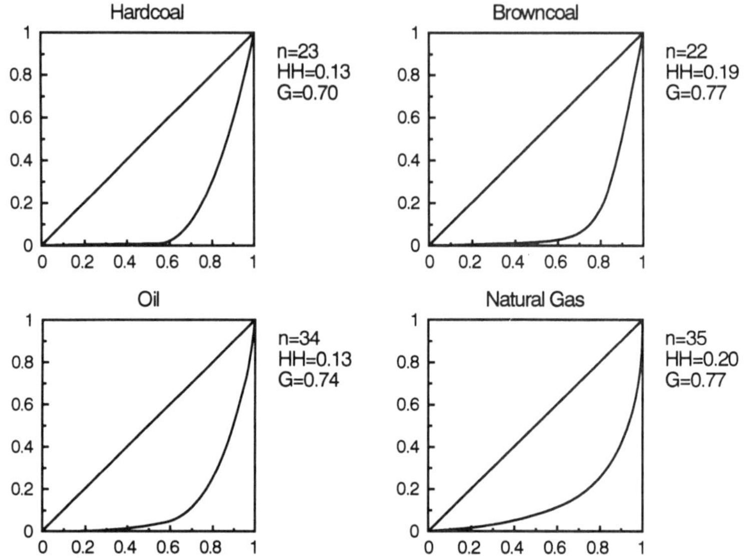

Figure 9.12 Concentration measure on fossil fuel markets. *n*, Number of countries representing 97% of proven reserves; HH, Herfindahl–Hirschman index; G, Gini coefficient. The Herfindahl–Hirschman index is the sum of squared market share for each country. When the market is occupied by only one supplier, the index equals the maximum value of 1. It becomes smaller with increases in the number of suppliers and increases with rising inequality among any given number of suppliers. The Gini coefficient is a measure for the market share distribution. For $G = 0$, all agents have equal market shares. The Gini coefficient rises to $G = 1$ with increasing uneven market shares. It represents the area under the diagonal line and above the Lorenz curve

Rising demand for natural gas will allow OPEC to reduce oil supply to maintain supply price but to export more gas to reach desired target revenues.

OPEC's future policy must be a sophisticated strategy to find the balance between oil and gas supply to maintain oil and gas prices high enough to prevent rent extracting taxes. OPEC members differ in their endowment of natural gas. OPEC members with relatively low natural gas reserves compared to their oil reserves, like Iraq and Kuwait, can produce a greater share of oil, while OPEC members with a relative high endowment of natural gas can focus on gas production and shorten their oil quota in favour of the non-gas oil producers. In the language of game-theory: a coordinated oil and gas policy will enlarge OPEC members' internal bargaining set.[21]

Hence, the impact of a carbon tax on the demand for fossil fuels is crucial to the market structure on the supply side.

CONCLUSION

Playing with alternative computer scenarios reflecting important issues of the energy demand and supply conditions over the next 100 years is the only method to find out how to cope with the problem of CO_2. Looking at the basic ingredients that one would like to find in a complex world energy model, one must conclude that the present state of the art of modelling looks rather well with respect to the demand side and still rather good with respect to the worldwide trade of energy. But the modelling of the (strategic) supply side of the most important energy markets has severe shortcomings in all three models looked at. The "youngest" model (i.e. GREEN) seems to be the most convincing.

The above discussion has shown why energy economists from the modelling scene meet certain reservations when proposing policy measures based on their model analysis. Such reservations are not only due to a lack of understanding and therefore distrust in general towards computer modelling, but merely to the reluctance of policy makers and practitioners in the energy markets to accept results that are too counterintuitive. Counterintuitive results may stem from such complex structures, that at first sight the uninformed cannot follow the possibly complex substitution processes and structural changes. But on the other hand, this type of result may also be a consequence of some immanent modelling shortcomings and assumed structural rigidities in the model at those points where the mechanisms in real-life problems are flexible.

Coping with the CO_2 problem may be similar to the effort necessary for the flight to the moon. People started with small and intermediate rockets and successively improved their knowledge. It may be that those of us engaged in world energy modelling have just developed some small-scale rockets that are really not intended to fly to the moon. Therefore, we should clearly point out where we stand and where we still have a lot of shortcomings.

NOTES

(1) Recent research results of natural scientists cannot preclude such sudden events, which makes an economic evaluation in terms of a cost–benefit analysis rather meaningless.

(2) Since there is a uniform damage caused by each unit of CO_2 (and no "hotspots"), all possible disadvantages of the Pigou tax or of tradable permits are not given here.

(3) This means that in all regions that have some functioning control with respect to energy balances, there is no real monitoring problem, which often renders taxes or market mechanisms in environmental policy rather useless.

(4) A theoretical fiction to find out an optimal system of reference.

(5) In more general models of resource economics one distinguishes between a pure rate of time preference in the utility function and the rate of interest. For our purpose we do not make this distinction, since it would not help to clarify our arguments.

(6) Among many of these factors, we have that capital markets are not perfect, that the discount rates of different suppliers may differ (high versus low absorber countries), that future demand cannot be contracted today, that backstop cost and potential are uncertain, that there are several important fuels (coal, oil, gas) competing in different segmented markets, political factors, imperfect markets and coalitions of suppliers. For a discussion of the Hotelling principle and its limitations, see Ströbele (1987), Watkins (1992) or McDonald (1994).

(7) Of course, one might argue that the relatively high rents and thus the high value of stocks of fossil fuel depended on a historical error: the true capacity of the atmosphere for CO_2 enforces a changed evaluation.

(8) These models have been published at several places. The basis sources are: GREEN (OECD, 1994); Global 2100 (Manne & Richels, 1992); ERB (Edmonds & Reilly, 1985).

(9) From our point of view the assumption with respect to coal is reasonable: at least for the next decades, the price of coal will be determined mainly by extraction costs and not by some significant rent component.

(10) Again, one observes the contrast in modelling: the building of the required capital stock for nuclear or solar energy until the end of the next century would be the main concern for an energy economist.

(11) Effectively, the reaction of domestic oil production due to a carbon tax is only marginal in all but the EEX regions.

(12) Alternatively, GREEN allows the world oil price to be determined exogenously. In this case, the EEX region is the residual supplier, guaranteeing equilibrium on the crude oil market.

(13) How these taxes are determined in a dynamic context is not spelled out in ERB. Of course, they should correspond with our λ of Equation (9.4).

(14) Even if this may seem rather low, it leads to an improvement in energy productivity of 30% from 1990 to 2050.

(15) Alternatively, ERB allows the imposition of a production tax or a combination of a production and a consumption tax.

(16) Many energy experts expect the high times of oil still to come in the next decades: so why should there be a transition to a rather expensive coal-based technology, if there is plenty of oil left in the Middle East supplier countries?

(17) This is simple but good economics of natural resources: see, for example, Ströbele (1987).

(18) After the year 2050 (high-cost) oil production is produced much more in the carbon reduction case than in the baseline case. Low-cost oil production is the same in both scenarios.

(19) The same is valid in the natural gas supply case.

(20) Due to the large resource endowment and market imperfections (subsidies etc.), the world coal market shows competitive outcome. Supply price equals marginal costs of extraction plus transportation costs.

(21) For a game theoretic approach to the world oil market, see Blank (1994).

REFERENCES

Barrett, S. (1995) How to introduce stricter environmental standards under the constraint of international competition: the case of CO_2, discussion paper.

Blank, J.E. (1994) *Marktstrukturen und Strategien auf dem Weltölmarkt–Spieltheoretische Betrachtungen*, Lit-Verlag, Münster.

Edmonds, J. and Reilly, J.M. (1985) *Global Energy*, Oxford University Press, New York.
Fankhauser, S. (1994) The social costs of greenhouse gas emissions: an expected value approach, *Energy Journal* **15**(2), 157–184.
McDonald, S.L. (1994) The Hotelling principle and in-ground values of oil reserves: why the principle over-predicts actual values, *Energy Journal* **15**(3), 1–17.
Manne, A.S. and Richels, R.G. (1992) *Buying Greenhouse Insurance*, MIT Press, Cambridge, MA.
Nordhaus, W. (1991) The cost of slowing climate change: a survey, *Energy Journal* **12**(1), 37–65.
OECD (1994) *GREEN Reference Manual*, OECD, Paris.
Ströbele, W. (1987) *Rohstoffökonomik* [*Economics of Natural Resources*], Vahlen-Verlag, Munich, (in German).
Watkins, G.C. (1992) The Hotelling principle: autobahn or cul de sac? *Energy Journal* **13**(1), 1–24.

Chapter 10

Exploring Oil Market Dynamics: A System Dynamics Model and Microworld of the Oil Producers

John D.W. Morecroft and Brian Marsh*

INTRODUCTION

Planners at Royal Dutch/Shell launched a project to develop a simulation model of global oil markets. This project was part of a broad-ranging scenario exercise to explore the strategic implications of changes in the structure of the energy industry. In Shell, scenario planning is viewed as a way to discover new concepts and language that enable the organization to become more agile in recognizing significant industry trends, defining emerging business problems and preparing the minds of senior managers to deal with such problems (de Geus, 1988; Wack, 1985a, 1985b). Scenario planning is *not* a way of forecasting or predicting the future (as many outsiders seem to think). It works by the development of consistent stories about *alternative futures*, as the basis for "what-if" thinking in the organization. A consistent story traces a time path into the future that forces managers and planners to think "what would I do,

* Before joining St Andrews Management Institute, Brian Marsh was a senior member of Royal Dutch/Shell's Group Planning department, Shell Centre, London SE1 7NA, UK.

Systems Modelling for Energy Policy. Edited by D.W. Bunn and E.R. Larsen

within my area of business responsibility, if this future were to unfold?" The internal consistency of the stories, creating credibility and persuasiveness, is an important factor in evaluating the usefulness of scenarios—much more important than the ex-post accuracy of the time paths. Scenarios are selected on their ability to make the organization a skilful observer of the business environment. As such they do not have to describe the most likely future, but the reasoning behind the scenarios must be plausible.

Models that fit the needs of scenario planning do not have to be accurate predictive models. But they should have the capability to stimulate novel thinking about future business options. Moreover, models that are used to construct consistent stories need to be understood by the story writers (often senior planners) in order to be communicated effectively to the story readers (corporate executives and business unit managers). A black-box predictor will not lead to the desired result, even if it has a good record of predictive power.

These criteria of scenario planning and scenario models help to explain the style of modelling adopted—the need for a comparatively simple model, the relatively closed process (the project did not make direct use of other world oil market models or enlist the aid of world oil market experts as consultants) and the intense participation of senior planners in model conceptualization. The project was not intended as an exercise in developing another general model of the oil trade to forecast better. The project team wanted to model *their understanding* of the oil market. Within the group there was an enormous amount of experience, reflecting knowledge about the actors in the oil market and observations about market behaviour. But the knowledge was scattered and anecdotal, and therefore not very operational. The group also recognized that the interlinkages in the system were complex. The desire of the group was to engage in a joint process through which their knowledge could be pooled and put into a framework through which real events could be interpreted. This framework could then be used to develop internally consistent scenarios. Most of the group members were not professional modellers. Other existing energy models to them were non-transparent black boxes, useful as a reflection of other's views, but quite unsuitable for framing their own knowledge.

THE MODEL DEVELOPMENT PROCESS

The first step was to assemble an experienced team of planners and managers (ten people in all) and to use their knowledge of the oil industry as the basis for a conceptual model. The team met three times for working sessions lasting three hours each. The meetings were facilitated by an experienced system dynamics modeller. One member of the team kept detailed minutes of the meetings (including copies of flip-chart notes and diagrams) in order to preserve a permanent trace of the model's conceptualization. A subgroup of the project team met separately to develop and test a full-blown algebraic model.

The team members agreed to represent the global oil system in terms of three categories of producers and a market. The components of their conceptual model are shown in Figure 10.1. The swing producer and the opportunists make up the oil producers' organization OPEC. This powerful group of producers has access to very large reserves of low cost oil. Their production decisions are motivated principally by social and political pressures, in contrast to the commercial logic of the independents. They coordinate production through quota setting. The opportunists agree to abide by quota, but will sometimes cheat by producing above quota in order to secure more oil revenues. The combined production of all three producer groups feeds the market where both price and demand are set. Price responds to supply and demand imbalances and then feeds back to influence both demand and the behaviour of producers.

Why three producer categories? Of course, there is no one correct or best way to represent producers. In reality there are hundreds of companies and dozens of nations engaged in oil exploration and production. But very often, experienced people evolve insightful ways of categorizing complexity. This team came up with a unique conceptualization that focuses attention on the

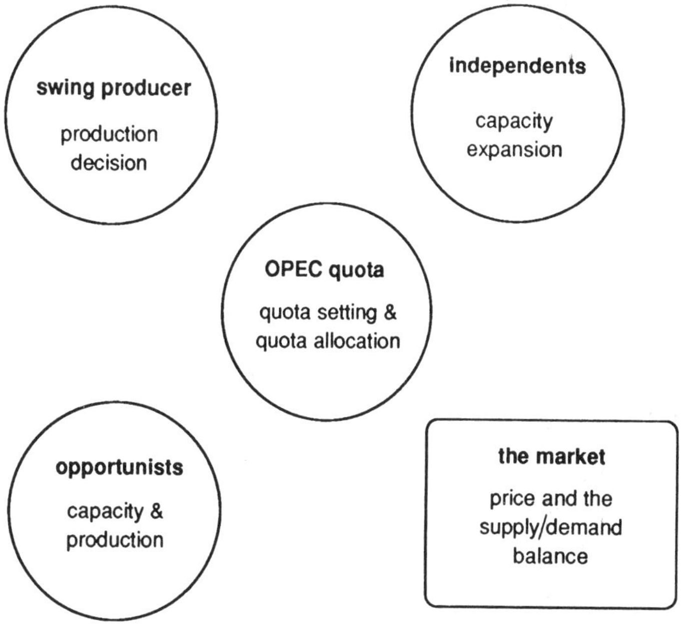

Figure 10.1 Producers and the market: components of a conceptual model

internal structure and political pressures that govern a producers' organization such as OPEC.

MAPPING THE DECISION MAKING LOGIC OF THE OIL PRODUCERS

The team spent half of its meeting time ($4\frac{1}{2}$ hours) on qualitative mapping of the producers' decision making, exploring questions such as: How much oil does each player produce? How (if at all) do they coordinate production? Which factors influence their decision making processes? How important is cost or price? Where and how do political factors enter? An experienced team has answers to all of these questions.

No computers were present in the meeting room, and the maps that were used to guide and capture team discussion were all drawn on flip-charts. By avoiding the early introduction into the meeting of detailed structural and algebraic modelling it was possible to generate a free flow of discussion and therefore to sustain active participation by the project team.

The meeting yielded about 20 flip-chart pages showing graphically the information flows entering the producers' and consumers' behavioural decision functions, with text notation to indicate the team's opinions on how the information is processed. Additional text pages captured the team's judgements on factual information such as production rates, operating capacity, surplus capacity, producers' market share, current price, price profiles and demand.

The diagrams that follow are similar to the flip-chart drawings from team meetings. A large circle with horizontal bars represents a decision making process. Labelled, curved lines with an arrow head represent flows of information or pressures that influence the decision making process in question. Straight lines in bold represent the output of a decision making process, such as the swing producer's production. Labels in bold represent important policy levers that shape the way in which information is processed in each decision function. Rectangles in bold represent "stocks" that accumulate the action flows arising from decision making processes.

THE SWING PRODUCER

The role of the swing producer is to produce just enough to defend the intended price, known in the industry as the marker price. A producer taking on this role must have both the physical and the economic capacity to increase or decrease production quickly, by as much as 2 million barrels per day or more, in order to absorb unexpected variations in demand (due, say, to an unusually mild winter) or to compensate for cuts in the output of other producers. (Few producers are capable of handling such variability of output and revenue.) But how does the swing producer decide how much to produce? Figure 10.2 shows the range of factors entering the decision making process.

The swing producer operates in two modes: normal swing mode and punitive mode. In swing mode, the swing producer sets a production rate that is equal to the swing quota, unless the oil price deviates from the intended price—meaning posted price or marker price for long-term contracts. When a price

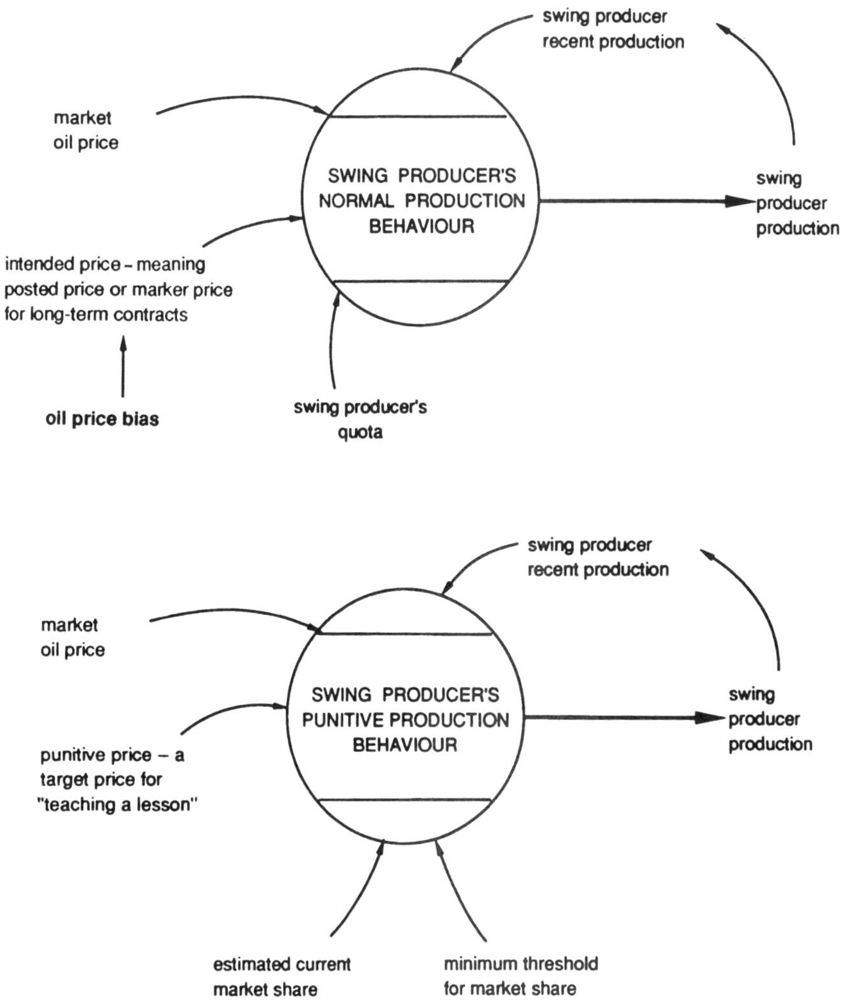

Figure 10.2 The logic of the swing producer. Two modes are shown: normal swing mode at the top and punitive mode at the bottom

deviation is detected, the swing producer quickly increases or decreases production of crude in order to regulate the price. In normal swing mode, the swing producer can influence market price by making production adjustments, the size of which depends on the price he is trying to defend. In the model, price control is exercised through the policy lever "oil price bias", shown on the left of the diagram. A value for oil price bias of 0 is neutral, meaning that the swing producer is defending a price equal to the average market oil price. A positive value, of say, 0.1 means that the swing producer is defending a price that is 10% higher than the average market oil price and so trying to edge market prices up. Conversely, a negative value for the bias means that the swing producer is trying to edge prices down.

In punitive mode the swing producer feels that his production is inadequate—he is not getting a fair share of the market—and so decides to re-establish his position by punishing the other producers.[1] In the model, the swing producer has a minimum threshold below which he is unwilling to allow market share to fall. When market share reaches the threshold, the swing producer increases the volume of production in order to flood the market and quickly lower the price. The team spent some time discussing the detail of punitive behaviour. For example, how does the swing producer decide on the volume of punitive production, and when does he switch back to swing mode? The proposal was to include a punitive price, a very low target price, for teaching a lesson to the other producers. Punitive production continues to expand until market oil price reaches the punitive price, or until the swing producer regains an acceptable market share (which is the signal to return to swing mode). The switch to punitive mode can send a powerful price signal to discipline the other producers, but it is an act of last resort, because in this mode the swing producer has abandoned the role of price regulator—essentially, the market is no longer managed.

THE INDEPENDENTS

The independents are all those producers—state-owned oil companies, the majors and other private producers—that are not part of OPEC. The independents are assumed to produce at economic capacity all of the time. So, unlike the OPEC members, they do not operate with economic capacity surplus. Their production rate is therefore dictated by their capacity expansion decisions, and the speed at which they can change production is limited by the long time-lag in constructing new capacity and in depleting existing fields.

The rationale for capacity expansion is dominated by commercial factors, as shown in Figure 10.3. The independents will expand capacity (and therefore production) when they judge that it is profitable to do so. If the investment climate is unfavourable (say, during a period of low oil prices) then no new

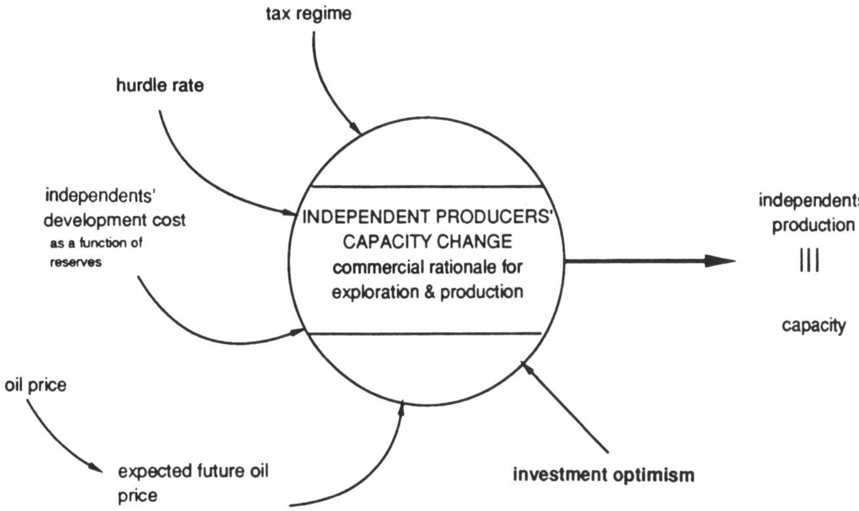

Figure 10.3 The logic of the independents

upstream capacity is added, resulting in a net loss of output as the production of established fields peaks and then declines.

The model replicates the major inputs to upstream investment decisions in order to calculate the average profitability of potential projects. The independents obviously have to take a view of the development costs of new fields and the expected future oil price over the lifetime of the field. Knowing future cost, price, the likely size of a new field and the tax regime, one can calculate the future profit stream and apply a hurdle rate to identify acceptable projects. In reality, each project undergoes a thorough and detailed screening, using well-tried upstream investment appraisal methods. The model treats project appraisal in the aggregate and computes a "recommended fractional increase in capacity". For example, if, at a specific hurdle rate, average industry project profitability is, say, 24%, then the recommended fractional increase of capacity as computed by the model is 20% per year. Executive control over the recommended expansion is exercised through the policy lever "investment optimism". A value of 1 is neutral, meaning that approved expansion is equal to recommended expansion. A value greater than 1 means that oil company executives are bullish about the investment climate and approve faster expansion than recommended—for example, 50% faster than the recommended rate if optimism is set at 1.5. A value less than 1 represents pessimism.

Development costs in the model are estimates of industry marginal costs starting in 1988. The independents begin with a large pool of undeveloped reserves.[2] The development cost profile for these reserves (assuming no cost improvements from technology) is, broadly speaking, a curve that rises as reserves are depleted. There is assumed to be a small quantity of low-cost reserves. Once low-cost reserves are exhausted the cost profile rises quite sharply and then levels out to a gently rising plateau. Finally, the cost profile rises steeply as known industry reserves approach zero—in other words, there is a finite supply of commercially viable oil.

Technology can undoubtedly be expected to lower costs as more efficient production and recovery methods are devised. The model assumes that technology improves significantly and steadily over the ten-year period from 1988 to 1998, thereby putting downward pressure on costs. Thereafter the effect of technology on development cost is assumed to remain constant.

QUOTA SETTING

Quota setting takes place in two stages, as shown in Figure 10.4. First, the cartel members agree on a quota for the cartel as a whole, then quotas for individual producers are negotiated by allocating the total quota among cartel members.

How much should the cartel as a whole agree to produce? The members need to form a view of the likely call on the cartel. To do this they need to estimate oil demand and the production of the independents over the quota period. The difference is the estimated call, which is the cartel's best guess of the quantity of oil (measured in millions of barrels per day) it needs to produce to just balance supply and demand. But the estimate need not be "spot-on". It could differ (and probably often does) from the actual call by as much as 1 or 2 million barrels a day. The swing producer will compensate for any mis-estimation that shows up in between cartel meetings through swing changes to production. More important than spot-on estimation of the call is the need for the cartel to agree on whether to set an overall quota that is *deliberately* less than the estimated call, or deliberately more. By setting a quota that is less than the call, the cartel is pursuing a policy aimed at increasing market prices. The cartel's decision to overproduce or underproduce is politically and economically motivated and is represented by the policy lever "cartel quota bias".

Quota negotiation, shown in the top half of the figure, allocates the cartel's agreed quota among the members. In reality, the negotiation is a political process that undoubtedly involves a lot of bargaining and side-deals—although the cartel uses formulae that include oil reserves, production capacity and population to provide a benchmark for the allocation. In the model, quota is allocated in proportion to each member's share of cartel operating capacity. Although this formulation is a simplification, it does capture the flavour of political bargaining by making the members' bargaining strength proportional to capacity share.

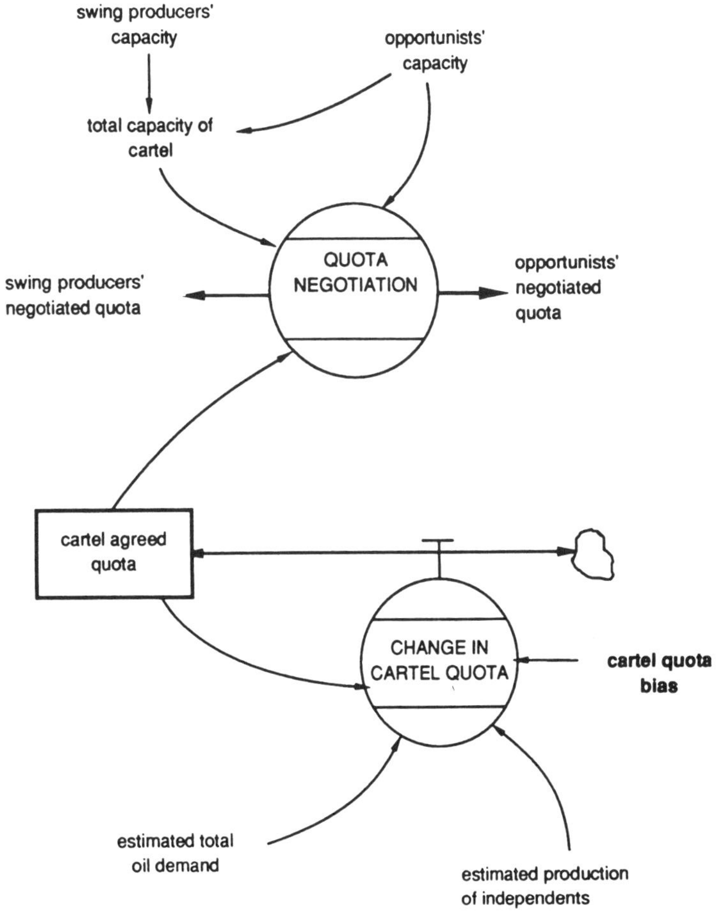

Figure 10.4 The logic of quota setting

THE OPPORTUNISTS

The opportunists are all of the other members of the cartel besides the swing producer. Some of the opportunists—countries like Venezuela—are known to adhere strictly to quota, so their production decision is straightforward: it is simply equal to negotiated quota. But other countries have a huge appetite for oil revenue to support their growing populations and developing economies. This huge revenue appetite, coupled with underutilized production capacity, provides the motivation to exceed quota and to strengthen their quota negotiating position by expanding capacity.

When viewed as a whole, the opportunists' production decisions are quite complex. Figure 10.5 shows some of the factors to consider. At the bottom left is the opportunists' negotiated quota. When playing by the rules, they produce at quota. But there is always the temptation to utilize capacity fully—if only they can get away with it. So they look at their capacity, and compare it to quota. Generally speaking, they aim to have some surplus capacity, at least 2–3%, partly to provide flexibility, but also to improve their bargaining position in quota negotiations. The policy lever "capacity bias", shown entering the production capacity of opportunists, allows the size of the surplus to be altered.

How do opportunists know whether conditions are right for utilizing surplus capacity and producing above quota? In reality, this utilization decision is likely to be a subjective judgement. But the project team felt that the apparent cohesion of the cartel would be an important factor in the minds of opportunists. Cohesion is high when the market oil price is close to the intended marker price—in other words, when the cartel is successfully defending its intended price. Under these price conditions, quota busting may go unnoticed. Cohesion is low when the intended marker price is greater than the market price. Under these price conditions, quota busting is likely to be visible and to incur the wrath of the other cartel members, especially the swing producer. Figure 10.5 shows graphically that intended marker price and oil price influence opportunists' production, and the inset at the upper-right shows the shape of the relationship between the apparent cohesion gap and surplus utilization.

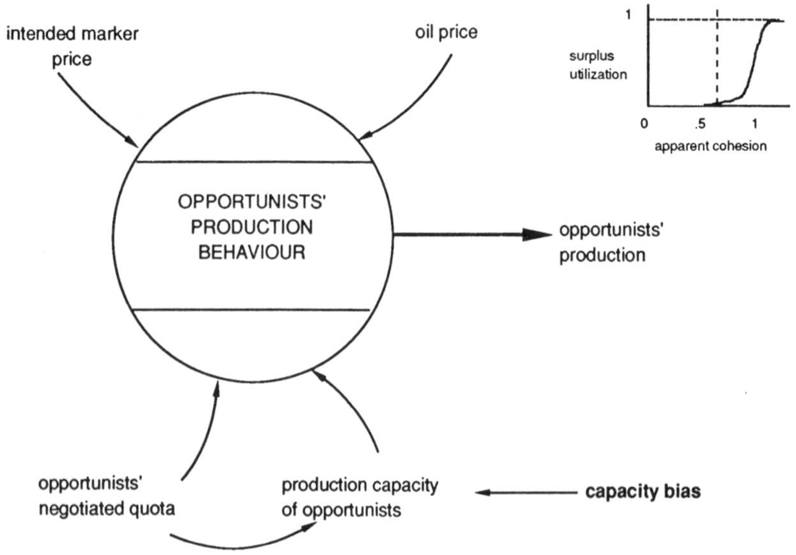

Figure 10.5 The logic of the opportunists

PRICE AND DEMAND

Market oil price responds to differences between demand and total production (the combined output of the independents, the opportunists and the swing producer). If demand exceeds production, say by 1 million barrels a day, then there is continual pressure for the price to rise, as long as the demand surplus persists. The price will move steadily upwards. Conversely, if production exceeds demand, there is continual pressure for the price to fall. When demand and production are exactly equal, then the price settles at whatever level it has just reached.

It is difficult to estimate accurately the sensitivity of oil price to production and demand imbalances. In the model, price increases at a rate of 4% per month when there is a shortfall of 2 million barrels per day in world markets. The rate of increase rises to 7.5% per month if the shortfall reaches 4 million barrels per day. There are corresponding rates of price decline resulting from a glut of production. These numbers produce plausible price profiles when tested in the simulator. Demand responds to market oil price as shown in Figure 10.6. The basic logic of demand changes is straightforward. When price goes up, demand goes down, and vice versa. But, in reality, there are more subtle dynamics of demand. If, for example, there is a sudden shock increase in price

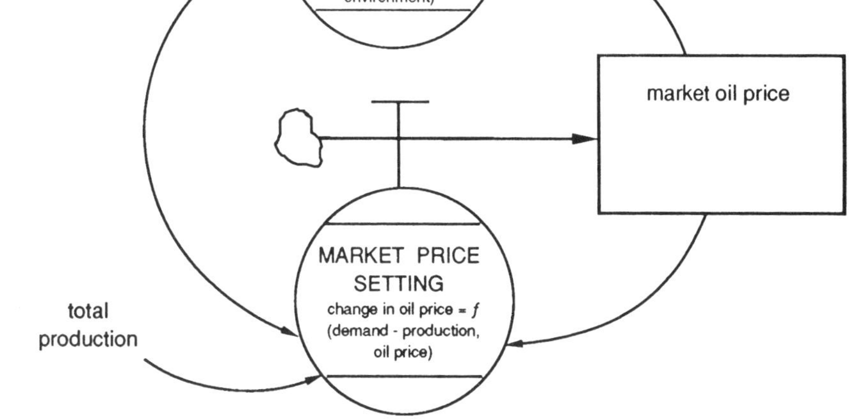

Figure 10.6 Price and demand

of, say, $10 per barrel, the resulting reduction in demand will be greater than if the same price increase were spread over a few years. In the model demand can change quickly in the short run due to conservation and easy substitution, but can also continue adjusting in the long run due to improvements in energy efficiency—for example, more fuel-efficient cars, or more heat-efficient homes.

It is impossible to state succinctly the price sensitivity assumptions of the model because of the mixture of dynamic long-run and short-run effects. However, to give an idea, suppose that price has been steady for a long time at, say, $20 per barrel, then rises suddenly to $30 per barrel, and settles at this new and higher value—a 50% increase. Then, in the model, consumers set out with the *intention* of reducing their demand by 20%. In fact, if the new price conditions prevail for many years (a highly unlikely event in reality, but nevertheless useful for illustration) demand will fall, but by less than 20%—perhaps by as little as 5%—as consumers get used to the new higher price level and abandon their good intentions to conserve energy.

Demand also responds to macroeconomic conditions through the policy lever "effect of GDP technology and environment".

DETAILED STRUCTURAL MAP REPRESENTING THE SWING PRODUCER

Figure 10.7 is a detailed structural map composed of system dynamics modelling symbols, representing the swing producer and implemented in the STELLA/Ithink simulation language (Richmond and Peterson, 1992). The full model contains 99 equations (including constants and graph functions). Separate pages of the structural map are devoted to the swing producer, the independents, OPEC quota setting, the opportunists and the market. Here only the swing producer page is shown to provide the reader with a feel for the model's complexity and degree of detail.[3]

The map was constructed by a four-person modelling team, which was a subgroup of the full project team, starting from the flip-chart drawings and meeting notes. The modelling team included two experienced managers, an external consultant acting as facilitator and model designer, and a model builder.

The figure traces how the swing producer's production responds to quota and to market oil price. The principal logic of changes in production is shown in the top left-hand branch of the figure. Indicated swing production (towards which production adjusts) depends on the swing quota and production pressure from the marker price. In the absence of price pressure, the swing producer produces at quota. But when the market oil price falls below the intended marker price (the price that OPEC is defending) the swing producer reduces production to

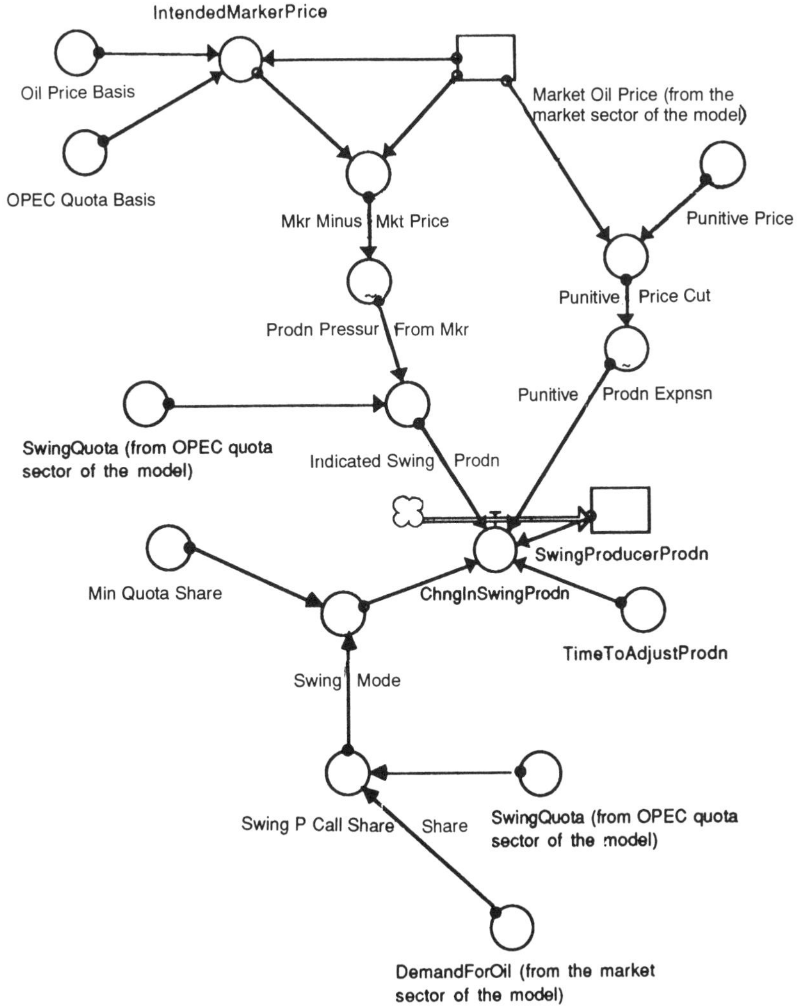

Figure 10.7 Structural map of the swing producer

compensate, and vice versa. The intended marker price is shown as a function—actually a two-year average—of the market oil price, modified by the oil price bias (an exogenous parameter representing the swing producer's tendency to edge oil price up or down) and the OPEC quota bias (another exogenous parameter, which represents the degree to which OPEC intends to underproduce or overproduce).

The right-hand branch of the figure shows the logic of punitive production. When the swing producer switches into punitive mode he ignores quota and price control and instead engages in punitive production expansion dictated by a punitive price cut. The size of the price cut depends on the difference between market oil price and a punitive price representing the low target price for disciplining the cartel and also the rock-bottom price that the swing producer is prepared to tolerate.

The switch between swing mode and punitive mode is shown in the bottom section of the diagram. Swing mode is switched on whenever the swing producer's call share (SwingPCallShare) exceeds the minimum quota share (MinQuotaShare). But when call share falls below the minimum quota share, swing mode is switched off and punitive production begins. The swing producer's call share is modelled as the ratio of swing quota (a variable whose value is computed in the OPEC quota sector of the model) to demand for oil (which is computed in the market sector of the model). Note that in drawing the detailed map, the modellers refined the original concept of market share to the more specific concept of call share, which measures the swing producer's market influence as set by quota negotiations.

USING THE MODEL TO GENERATE SCENARIOS

Having described the conceptual building blocks of the oil producers' model, we are now ready to simulate scenarios. The key to effective scenario modelling is to unfold several alternative futures (each an internally consistent story, but with a different plot and a different ending) in order to challenge conventional wisdom and to encourage users to think how they would act if this future were to unfold. It is important to remember that such simulated futures *are not* predictions of oil price, supply or demand, nor do they represent official company forecasts.

Two simulations are presented, each covering a 25-year period from 1988 to 2012.

Ten-Year Supply Squeeze Followed by Supply Glut

Imagine that, starting in 1988, OPEC had followed a policy in which quotas were set 10% lower than the call from the market (OPEC's estimate of the difference between total demand for crude oil and the independents' production). The policy was successfully enforced for 10 years until 1998, and then reversed. So, in the remaining 15 years of the simulation, from 1998 to 2012, quotas are set 10% higher than the call from the market. The result is a 10-year supply squeeze followed by a supply glut. Let us assume that OPEC maintains quota discipline throughout the squeeze and glut, so there is no cheating on quota by any member of OPEC. Let us also assume that non-price pressures on crude oil demand from the economy, technology and the environment are neutral so that, in the absence of price changes, demand would stay constant at 50 million barrels per day.

Figure 10.8 shows profiles of demand and production for the squeeze-then-glut scenario. Demand begins at 50 million barrels per day in 1988 and falls gradually to about 45 million barrels per day during the decade of supply squeeze. As OPEC switches policy from squeeze to glut in 1998, demand

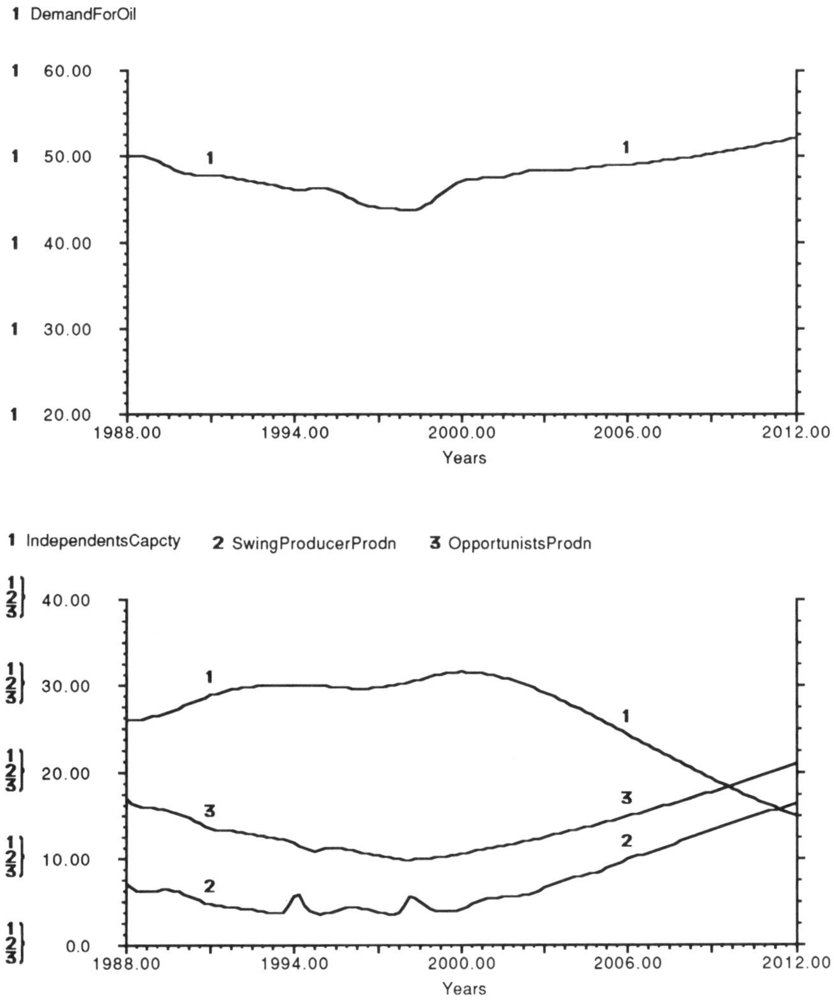

Figure 10.8 OPEC-induced squeeze then glut scenario, part 1. In this scenario, OPEC follows a policy from 1988 to 1998 in which quotas are set 10% lower than the call from the market. After 1998 the policy is reversed and quotas are set 10% higher than the call. The simulations in part 1 show profiles for demand and production

begins to rise and maintains a steady growth to about 52 million barrels per day by the end of the simulation in 2012. The lower half of the figure shows how demand is shared among the different producers. In 1988, the independents are producing 26 million barrels per day, the opportunists 17 million barrels per day, and the swing producer 7 million barrels per day. As the supply squeeze begins to bite, OPEC's output gradually falls. By 1993 the opportunists have cut production by 4 million barrels to 13 million barrels per day, while the swing producer's output has fallen to 4 million barrels per day. Meanwhile, the independents' production has risen to compensate.

In late 1993 and early 1994 there is a temporary surge in production as the swing producer switches to punitive mode to prevent further loss of market share. At this point the policy of supply squeeze is almost exhausted because the swing producer is unwilling to sacrifice further market share. The burden of any additional OPEC supply reductions is carried mainly by the opportunists. In late 1997 there is another burst of punitive production by the swing producer.

Then in 1998 OPEC reverses policy, changing a supply squeeze into a supply glut. Gradually OPEC's output begins to rise. Opportunists' production increases from a low of 10 million barrels per day in 1998 to 21 million barrels per day in 2012. Meanwhile the swing producer's output rises from 4 million barrels per day to almost 17 million barrels per day in 2012. It is interesting to note that OPEC's policy change has no immediate impact on independents' production. At the time of the policy change in 1998 the independents are producing 30 million barrels per day. Their output then rises to almost 32 million barrels per day by the year 2000. By 2003, five years after the policy change, independents' output is still 30 million barrels per day. Thereafter it declines steadily to 15 million barrels per day by the end of the simulation in 2012.

The impact of OPEC's policy on market oil price is shown in the top half of Figure 10.9. Oil price begins at $15 per barrel in 1988. Broadly speaking, price rises during the supply squeeze to a peak of $30 per barrel in 1998, then falls during the supply glut back to about $15 per barrel. But within the rising and falling trend there is much interesting detail in the movement of price that arises from the interplay of the feedback loops that balance supply and demand. The reader can view the state of supply and demand by comparing line 3 (DemandMinusProdn) with line 4 (RefZero). When line 3 is above line 4, demand exceeds supply. When line 3 is below line 4 there is a supply surplus and supply exceeds demand.

From 1988 to 1990 price rises sharply to $20 per barrel as the supply squeeze takes hold. Demand exceeds supply, as shown by line 3. Price then appears to decline slightly between 1990 and 1991 as falling demand creates a temporary supply surplus (despite OPEC's policy). Then, as the supply squeeze bites once more, price rises steadily to the next peak of $24 per barrel at the start of 1994.

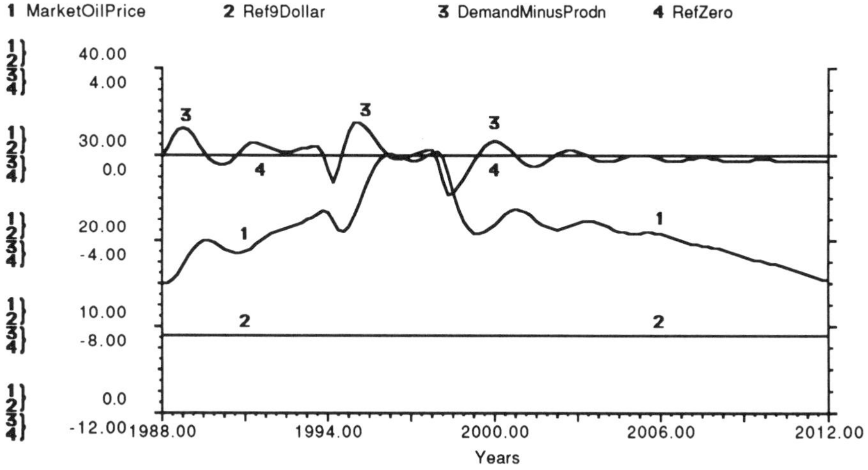

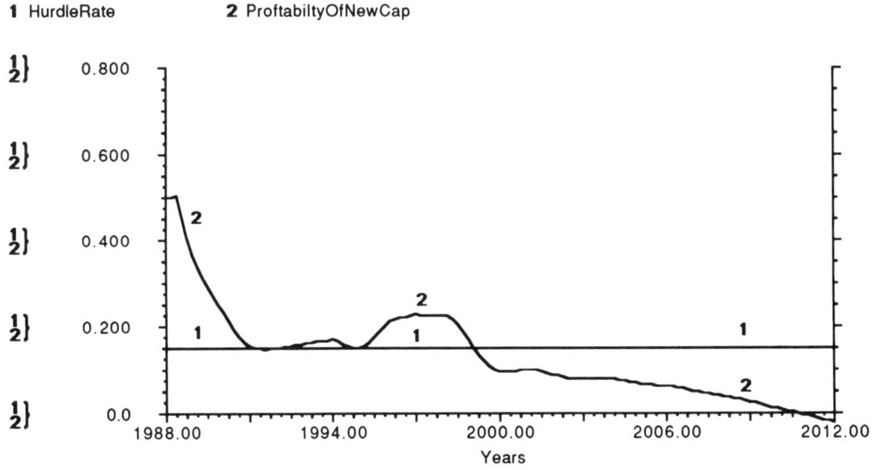

Figure 10.9 OPEC-induced squeeze then glut scenario, part 2. The simulations in part 2 show profiles for oil price and upstream investment profitability

Price dips during 1994 as the swing producer switches into punitive mode and rapidly expands production, creating a supply surplus. After the pulse of punitive production OPEC sharply curtails production, which causes price to rise sharply during 1995 to a plateau of $30 per barrel. The plateau continues until the start of 1998, and for a period of three years supply exactly balances demand. The era of price stability comes to a dramatic end during 1998 as the swing producer once more switches into punitive mode and price falls rapidly to $20 per barrel. Over the next two years the price level recovers slightly to a peak of about $23 per barrel in 2001 before settling into a slow but steady decline to its final value of $15 per barrel as OPEC's new policy of excess supply takes effect.

The lower half of Figure 10.9 shows the changing investment incentives for independent producers that help to explain the rise and fall of independents' production. The hurdle rate for upstream investment projects is assumed to remain constant at 15% (0.15). At the start of the simulation, when the oil price is $15 per barrel, the profitability of new capacity (shown as line 2, ProftabiltyOfNewCap) is 50%, reflecting the initial low cost of reserves. It is financially very attractive to add new capacity. But over the three-year interval to 1991 profitability falls sharply to equal the hurdle rate. This fall in profitability occurs despite the rise in oil price from $15 to $20 per barrel, because the development cost of new reserves is rising even faster (the independents are operating on the steep shoulder of the cost profile). As price rises to its second peak of $24 per barrel in 1994, the profitability of new capacity increases slightly above the hurdle rate. The price plateau of $30 per barrel between 1995 and 1998 creates an attractive era for investment, in which the profitability of new capacity rises to 25%, well above the hurdle rate. Then, as the swing producer generates a pulse of punitive production, and as OPEC changes policy, profitability falls quickly to only 10% in the year 2000. The investment climate continues to deteriorate throughout the remainder of the simulation. By the year 2011 the profitability of upstream investment has fallen to zero, meaning that the price of oil is now equal to independents' development costs. In the final year of the simulation, profitability actually goes negative.

Changing profitability drives independents' capacity as shown in Figure 10.8, but only after a lag that represents the time to develop a new field. The lag effect is most obvious following the attractive investment era between 1995 and 1998. Independents' capacity is static at 30 million barrels per day (actually declining slightly) during the three-year price plateau. Then, as profitability declines sharply between 1998 and 2000, independents' capacity rises gradually to 32 million barrels per day as new fields, approved earlier, begin to come on stream.

Quota Busting in a Green World

Imagine a world in which there are environmental pressures to reduce oil consumption—a green world. In the absence of price changes, these pressures are assumed to be sufficient to cut consumption by 10% per year. Imagine, too, that the opportunist producers in OPEC are inclined to cheat on quota. They deliberately plan to build capacity 10% greater than quota, and secretly use the excess capacity to produce more than quota whenever the opportunity arises.

Let us assume that OPEC adopts a neutral quota policy, setting quotas that exactly match its estimate of the market call on OPEC.

Figure 10.10 shows profiles of demand and production for this scenario of quota busting in a green world. In the top half of the figure, demand begins at 50 million barrels per day in 1988 and falls gradually to a final value of 43 million barrels per day by 2012. The decline is due to the assumption of a green

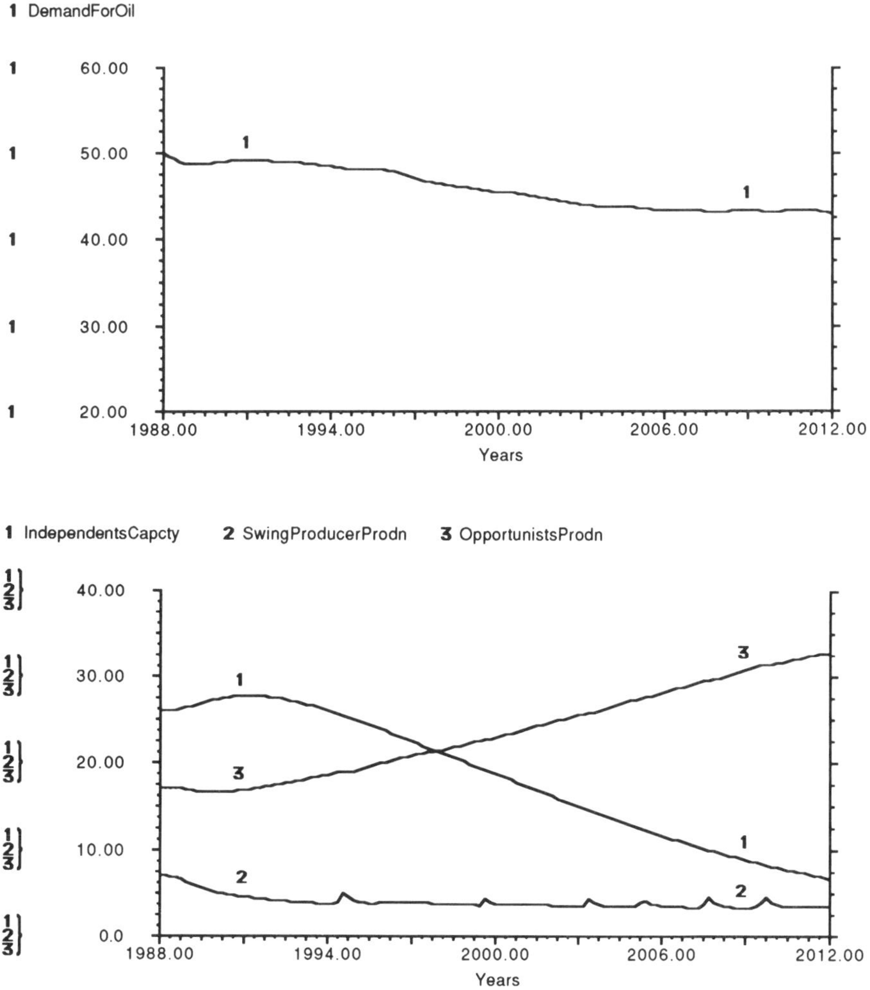

Figure 10.10 Quota busting in a green world, part 1. In this scenario the opportunists are assumed to adopt a covert policy of quota busting while at the same time consumers try to cut oil consumption in response to pressure for a greener environment. The simulations in part 1 show profiles for demand and production

world. But despite the *ceteris paribus* green assumption of a 10% cut in consumption per year, demand falls by only 7 million barrels per day over 25 years due (as we will see) to the compensating effect of lower price resulting from conservation and quota busting.

In the bottom half of the figure the opportunists' propensity to cheat pays off as they expand capacity and increase their bargaining strength in quota negotiations. Their production rises steadily from a starting value of 17 million barrels per day to a final value of 33 million barrels per day. Most of this expansion comes at the expense of the independent producers, who see their output fall from a starting value of 26 million barrels per day to only 7 million barrels per day at the end of the simulation. The swing producer, too, is squeezed by the opportunists at the start of the simulation, as output falls from 7 million barrels per day in 1988 to 4 million barrels per day in 1994. However, beyond 1994, and for the rest of the simulation, the swing producer uses mild pulses of punitive production to maintain market share and hold production at 4 million barrels per day. The first pulse occurs in 1994. The next pulse of punitive production occurs in 1999. Then, starting in 2003, there are four pulses spread evenly across the remaining years to the end of the simulation.

The top half of Figure 10.11 shows market oil price. For almost all of the simulation there is a slight surplus of production (line 3 below line 4) due to the combined effects of conservation and cheating. Consequently, price falls and stays at a low value. Price drops quickly from its starting value of $15 per barrel in 1988 to $11 per barrel in 1989 as demand falls in response to green conservation pressures. After this initial sharp drop, price falls very gradually during the rest of the simulation (with minor fluctuations caused by the swing producer's bursts of punitive production) to a final value of $9 per barrel by the end of the simulation. Needless to say, the sustained run of low oil price generates a bleak environment for the independents' upstream investment. In the lower half of the figure the profitability of new capacity quickly falls below the hurdle rate of 15%. By 1991 profitability is at about 11%, and continues to fall throughout the simulation to a final value of about 2% in 2012.

With no financial incentive for new investment, the independents' capacity (lower half of Figure 10.10) gradually declines as existing fields are depleted. Quota busting in a green world leads to a rapid demise of the independent producers, who end the simulation with no new capacity in the pipeline, a gloomy investment climate, and a market share of only 16% (down from 52% at the start of the simulation, and still declining).

Published Scenarios

A more elaborate version of the model (containing extra detail about independent producers and their reserves) generated simulations that were published in the company's scenario planning book, Group Planning (1989).[4] This modified

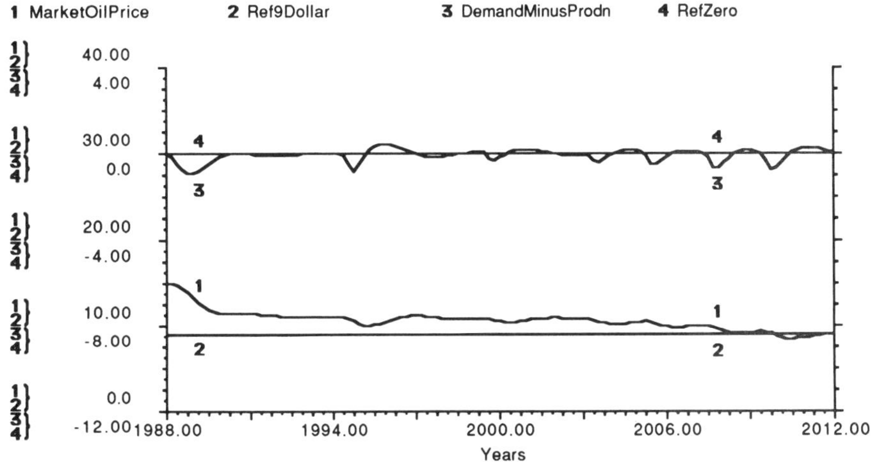

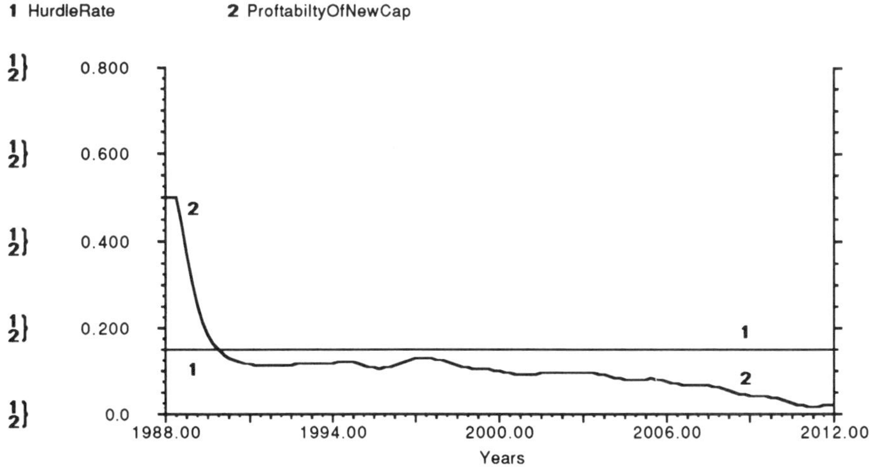

Figure 10.11 Quota busting in a green world, part 2. The simulations in part 2 show profiles for oil price and upstream investment profitability

model was simulated repeatedly to explore the effect on oil price and production of alternative OPEC strategies. For example, scenarios similar to the supply squeeze were created by simulating the effect of quota restrictions, ranging from 10% down to 0%. The supply scenario was extended by simulating the effect of OPEC quota surpluses, also ranging from 0% to 10%. New scenarios were created by varying other parameters that control producer behaviour, such as the degree of quota cheating in the cartel or the optimism of independents' investment decisions. A graphical compilation of the simulations revealed a "geography" of future price profiles. The geography showed three distinct areas, which were labelled mountains, plains and plateaux, depending on whether oil price fluctuates, or is stable at a low or higher price level.

DEVELOPMENT OF THE MODEL AS A GAMING SIMULATOR FOR TRAINING

The decision to develop a gaming simulator was motivated by Shell Group Planning's general interest in systems thinking and the specific desire to understand more about the packaging of models for dissemination. The resulting Oil Producers' Microworld has been used in several internal management training programmes.

Gaming simulation opens the possibility of widespread dissemination of model-based insights. A model built by a small team can be packaged in an attractive and user-friendly format for use throughout an organization. In principle, the insights derived by the original modelling team can be replicated for any group that works with the gaming software and accompanying courseware. In practice, there is a danger that the gaming simulator is treated as a black box whose structure is poorly understood. Simulations become an exercise in outwitting the computer and discovering weak points in the model—the so-called video game syndrome—rather than understanding the modelled business or social system.

Overcoming the video game syndrome is recognized as a major challenge in gaming simulation, which requires thoughtful design of both the pre-game briefing and the user interface.[5]

Design of the Pre-game Briefing: Recreating the Modelling Process

Pre-game briefing is intended to help participants to recreate the problem structuring and discovery process of the original project team. To set the scene, the participants read recent press cuttings about oil markets and a short booklet, *The Oil and Gas Industry in the 1990s* (Jennings, 1988). The articles provide participants with shared knowledge about the industry. Detail-rich industry information appears to be important for engaging participants with a gaming simulator which, by itself, is inevitably an abstract characterization of the industry.

The participants spend two or three hours in a guided tour of the model's feedback structure. The starting point is the balancing loop archetype from *The Fifth Discipline* (Senge, 1990). (If the participants are new to systems thinking and system dynamics, it is helpful for them to read chapter 5, "A Shift of Mind", beforehand.) The "balancing loop with delay" is a template to begin to organize the complex web of connections, pressures and forces that orchestrate the balance of supply and demand in global oil markets.

The workshop leader shows participants the basic archetype in the top half of Figure 10.12 and gives examples that link the structure with roller-coaster behaviour. Discussion then moves to the specific example in the lower half of Figure 10.12, which shows a balancing loop with delay outfitted to represent a simple oil supply system. A "what-if" story to accompany the picture might be as follows: suppose there is a production shortfall of 5 million barrels per day.

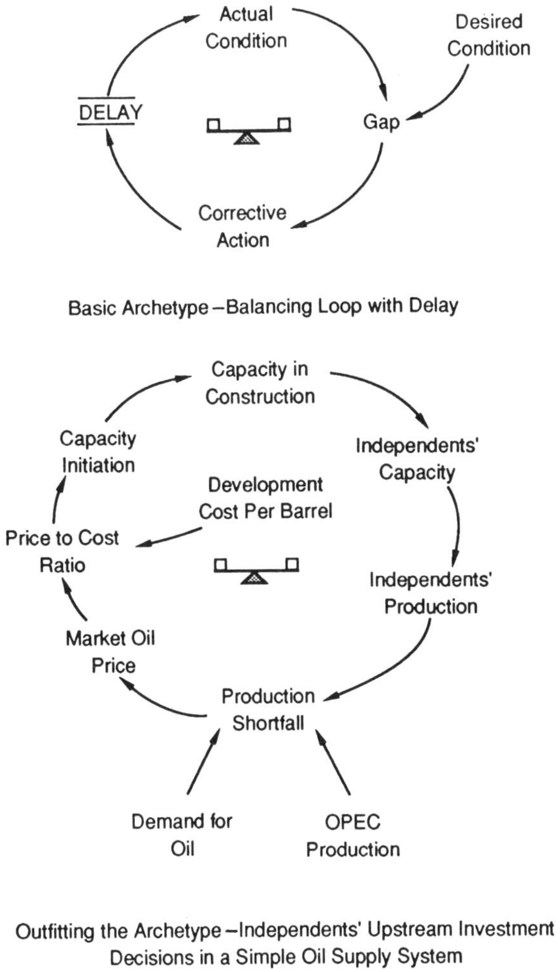

Figure 10.12 Template to begin organizing the complex web of connections in the global oil market (adapted from Senge, 1990)

What happens? The picture suggests that the shortfall causes the market price of crude to rise. A higher crude price improves the economic incentives for independent producers to expand capacity. After a long construction delay (several years) new capacity comes on line, allowing the independent producers to increase production and so reduce the production shortfall. The invisible hand of the market is at work.

A simulation of this partial model enables participants to see the story played out. They quickly realize that this feedback loop alone is not sufficient to explain real oil market behaviour. Simulated price rockets to a peak of 160 dollars per barrel, and production varies in a 20-year roller-coaster. Obviously the behaviour is implausible, but it is sufficiently engaging to draw participants into a closer look at the complete oil producers' model.

The next stage of the workshop recreates for participants the mapping exercise of the original project team. The oil world as conceived by the project team comprises not only the independent producers, but also a swing producer and other OPEC producers. These producers between them must, somehow, coordinate global oil supply and demand. Therefore the system's behaviour as a whole resembles a balancing loop with delay, but within the simple archetype there is much added complexity from the political, social and power factors that shape the capacity and production decisions of OPEC.

The workshop leader facilitates a discussion of the decision making logic of the independents, the swing producer, the opportunist producers (the other members of OPEC) and OPEC's quota setting. The objective is to map the key influences and information flows that make up the feedback structure of the full model, and to involve the participants in the process.

Imagine a mapping exercise for the independent producers. The workshop leader asks participants to name the factors that they think the independents consider when making upstream capital investment decisions. The comments are recorded on a whiteboard or flip-chart. Typically, participants mention many influences on the investment decision, but the workshop leader homes-in on the factors identified by the original project team. The result is a picture like the one shown in Figure 10.3. The independent producers' capacity change depends on commercial criteria, which include expected future oil price, independents' development cost, the hurdle rate (the rate of return required from capital investment projects) and the tax regime.

The exercise is repeated for the other producers. How does the swing producer set production? How do the opportunists set production? The discussion turns to quota, desired price (or marker price), market share, market power, revenue needs, social and political objectives, and a host of other related factors.

Participants have by now shared in the process that defined the architecture of connections in the oil producers' model. It just remains for them to discover the feedback loops that are formed as the system of producers is put together.

This final step is covered in pencil-and-paper exercises. Participants are presented with a set of feedback diagrams. But the diagrams are incomplete. There are phrases and words scattered over the page, but no connecting arrows. The task is to draw the connections, identify the archetype, and construct the corresponding story of time-dependent behaviour. For example, one diagram shows swing producer production, production shortfall, market price of crude and the price gap (market price minus marker price). When connected, it reveals a fast-acting balancing loop through which the swing producer exercises price control. Other diagrams reveal a reinforcing loop and a success-to-the-succesful archetype in OPEC quota bargaining and opportunists' capacity expansion.

By the end of the pre-game briefing, the participants have a basic grasp of the key feedback loops included in the model and gaming simulator.

DESIGN OF THE USER INTERFACE

The user interface was built using Microworld Creator (Diehl, 1990) starting with equations imported from the STELLA model mentioned earlier. Microworld Creator provides a screen template that allows a game builder to design a tailored, graphic and user-friendly interface to the simulation engine of the modelled system.

Microworld Creator sets aside an area on the right of the computer screen for menus of decision variables, reports, graphs and tables. The remainder of the screen is used for window displays of reports and graphs. Within this predefined format, the game builder has great scope and flexibility to design the visual layout of information and reports in a way that will appeal to the user.[6]

The Oil Producers' Microworld opens with a picture showing oil platforms, oil rigs, sea and sand dunes, intended to stimulate thoughts of the global oil system. The decision variables are in the upper right of the screen, as shown in Figure 10.13. There are five variables that give the user access to parameters of the simulation engine. EffGDPTcnolgyEnv (effect on demand of GDP technology and environment) allows the user to create alternative demand scenarios. The other parameters allow the user to specify policy characteristics of the producers that represent aggressive, defensive, optimistic and pessimistic behaviour.

The choice of decision variables presents many design options. Which of the many variables in the underlying model should be accessible to user teams? What role should the users play? Should users be an integral part of the feedback structure? Or should the users' role be to sit on the edge of the feedback structure, manipulating the policy levers that control loop gain and dominance, leaving the loop structure of the simulation engine intact.

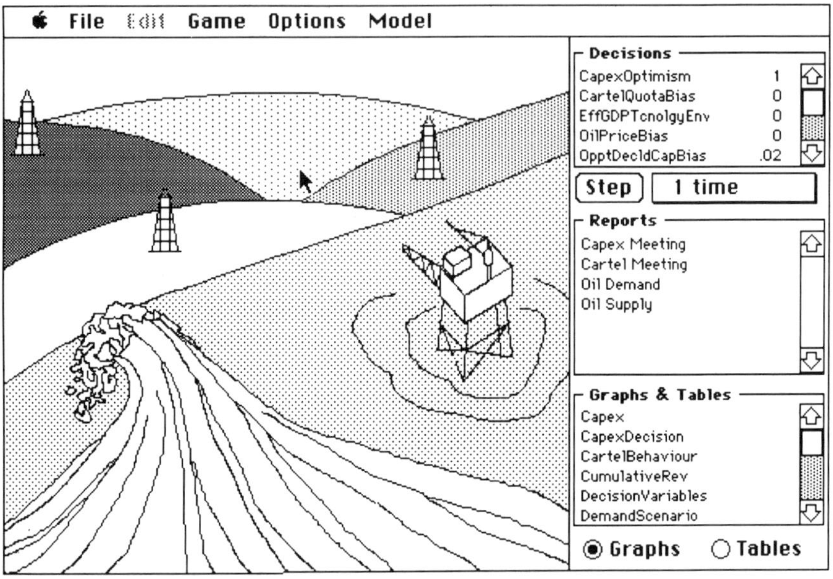

Figure 10.13 The opening window of the Oil Producers' Microworld

The Oil Producers' Microworld puts users in the role of the independents, swing producer and opportunists, manipulating policy levers. For example, the decision variable "cartel quota bias" allows users to define an upward or downward bias in quota relative to the computed call on OPEC. A value of 0 is neutral, meaning that the simulated cartel agrees a quota that is equal to the call, with the intention of balancing supply and demand. A value of −0.1 means that the cartel is setting a quota that is intentionally 10% less than the call. By design, users are not allowed to set quota or production rates directly.

Figure 10.14 shows two reports that were designed around the producers' decision logic. The Capex Meeting report shows a boardroom table seating a top management team from an independent producer. The team is deciding on capacity expansion in the first quarter of 1988 (the simulated date is shown in the top left of the picture). On the table is the information relevant to the independents: expected oil price, development cost, profitability and the hurdle rate. Out of the information comes the recommended expansion as calculated by standard project appraisal methods. The users are in control of management attitudes through the policy lever "capex optimism", which is displayed in the lower right of the report. When "capex optimism" is set to a value greater than 1, the independents are assumed to be bullish and approve more expansion than recommended. A value less than 1 represents pessimism. The final outcome of the capex deliberations for first quarter 1988 is shown as capacity initiation in the top right of the picture.

The Cartel Meeting report shows the cartel members sitting round a conference table. On their table is an estimate of the call on the cartel, opportunists' capacity and swing producer capacity. The "cartel quota bias"

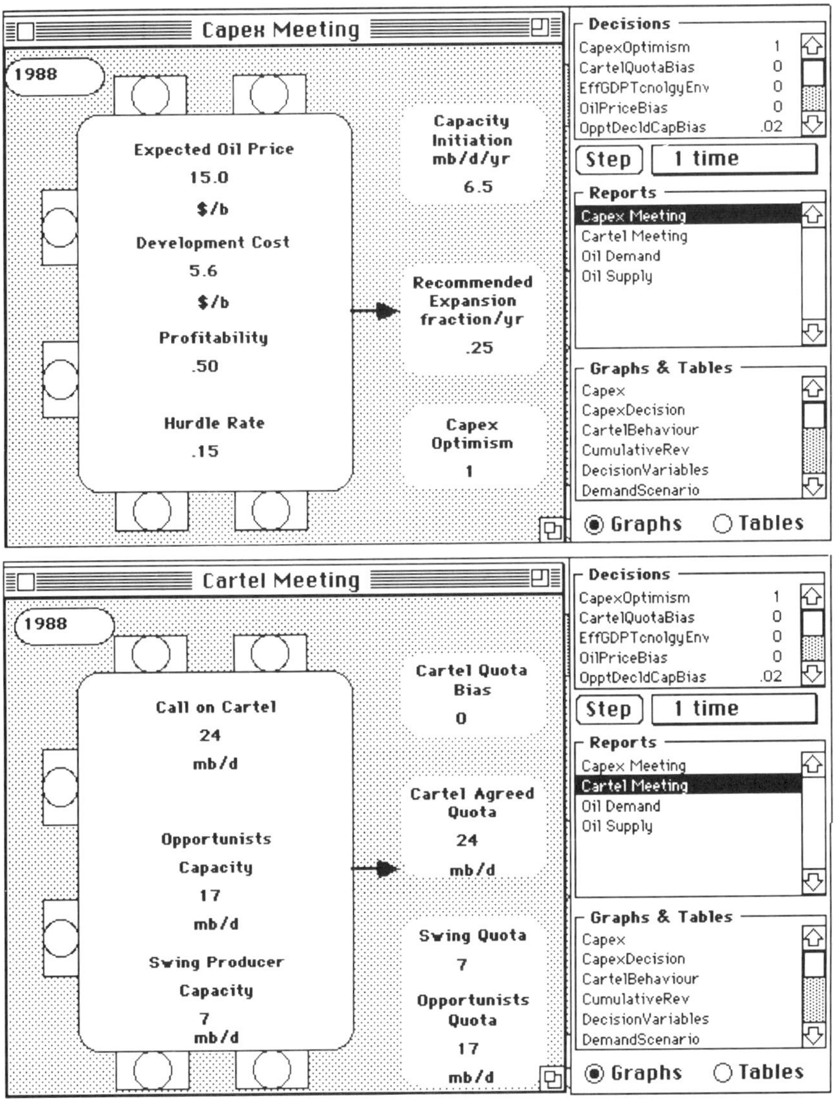

Figure 10.14 Reports generated by the Oil Producers' Microworld

shown in the top right of the report is set by the users and determines whether the agreed quota is greater than, equal to or less than the call on OPEC, as described above. A political bargaining process ensues within the model, which shares out the agreed quota between the swing producer and the opportunists.

The graphs and tables are designed with scenario-making in mind. The game

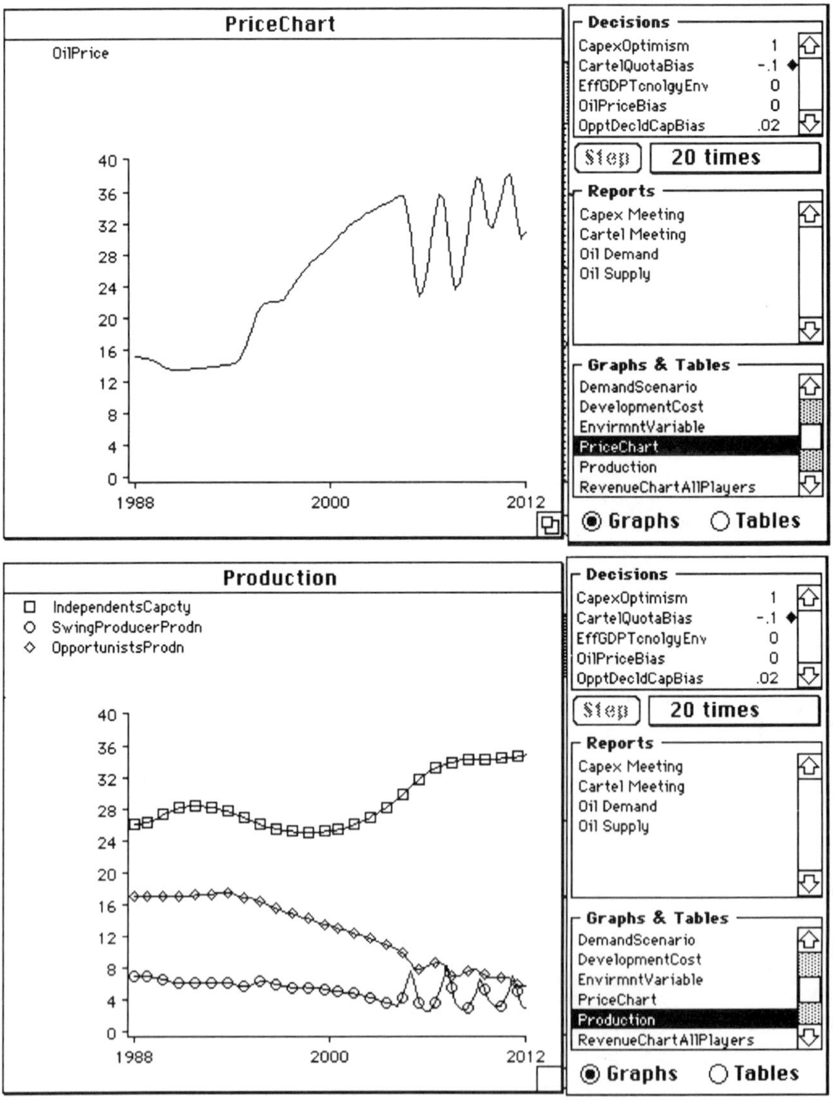

Figure 10.15 Sustained supply squeeze scenario

builder organizes and displays variables, and sets the scale of axes in such a way as to facilitate the construction of scenarios. A satisfactory arrangement of graphs takes time and is achieved through trial and error, with many simulations.

Figure 10.15 shows typical output from the Microworld. There are two graphs taken from a supply squeeze scenario, similar to the one presented earlier, but here the squeeze is maintained for the full 25 years of the simulation. The supply squeeze is created starting in 1994 by causing OPEC to set quotas that are 10% lower than the call from the market (cartel quota bias set to −0.1 as indicated in the decision window). The quota restriction is enforced to the end of the simulation in 2012.

A training workshop built around the Oil Producers' Microworld lasts for up to six hours (including the pre-game briefing) and provides an intense learning experience on long-term oil market dynamics. After the pre-game briefing, workshop participants are divided into small teams of three or four, each team sharing a computer. They first familiarize themselves with the Microworld interface and controls by playing through a pre-constructed scenario, following instructions in a user guide. Then they generate their own scenarios. The workshop is designed to encourage reflection, experimentation and team learning. Many protocols are possible, but one that has worked well requires teams to generate two contrasting scenarios: (i) prosperous, powerful OPEC in a green world and (ii) prosperous independents in a green world. For each scenario, teams must first think about how to manipulate the simulator controls in order to bring about the future envisaged by the scenario. This is a pencil-and-paper planning exercise intended to provoke discussion among the team members. Teams then use the simulator to enact their two scenarios, resetting the controls every five years in an effort to manage the outcome. The computer stores the simulated behaviour of key industry variables such as oil price, production rates, OPEC quota and revenues for use in team discussion and debriefing.

REFLECTIONS ON THE VALUE OF THE MODEL AND THE MICROWORLD

The scenario development team found new insights into oil market dynamics from the model's characterization of OPEC in terms of three behavioural decision making components: the swing producer having the ability to adopt punitive measures; opportunists who had the ability to cheat, up to a point, by supplying more than agreed quota levels; and the setting of those quotas. This contrasts with the unitary views of OPEC in other models: see Energy Modeling Forum (1982) and Powell (1990) for a review of published models.

Furthermore, the team owned this model and consequently felt confident of the conclusions they drew from the simulations and the influence that these had on the final scenario. The task of writing global scenarios that are relevant, challenging and internally consistent and which cover a score of years, or more, is not trivial. The team needs to make sense of the vast arrays of data, hard and soft, that a scenario exercise generates, to choose between the competing scenario themes, and to find ways of disseminating the scenarios through the organization. The exploration of oil market dynamics described in this chapter were of great value to the team on all three counts.

There are two other aspects of the exercise on which is it useful to reflect here in some detail. First, can the insights gained by the scenario team be used in a subsequent global scenario round, and, second, how successfully can the "oil producers' world" be exploited as a more general learning tool?

Although there is no doubt that the modelling experience was of value to the scenario team itself, the model was criticized by a number of people. Mostly the criticism was based on the accuracy and the modelling of the data embedded in the model. Some critics attempted to address this issue and developed other models for other applications.

There was, however, a more important aspect of the accusation: that the real world is far more complicated than the model supposes. Consequently, some lacked confidence in the ability of the model to reflect oil price patterns realistically. This criticism loses some force in the context of scenarios where fit and point prediction are not the focus of attention. But unless there is confidence that the model makes a genuine contribution to understanding of oil market behaviour, then the model will at best be ignored by those at whom the scenarios are directed and, at worst, be derided as a distracting video game. If a modelling team wishes to hand over its model to others to use, then it is not enough that the team knows that the model is sound: they need also to impart that knowledge.

In the next global scenario round (they run in roughly three-year cycles) a new team was formed, which included none of those who built the previous model. As always, the senior members of the new team effectively determine the way in which the scenario exercise will be conducted, and none had any real experience of system dynamics. Indeed, some senior members of the new team were critical of system dynamics on theoretical grounds. They also felt that the mental models of seasoned and experienced professionals could be exploited in other ways. Furthermore, the structure and dynamics of the oil market were not central to the themes that the new scenarios wished to address.

It is therefore interesting to note that some months later, when the new scenarios were beginning to gel, feedback loops were being used to map the team's understanding of the fundamental scenario dynamics and to facilitate dissemination of the scenarios. Furthermore, the scenario study had prompted some members of the team to develop a deeper understanding of systems thinking and of system dynamics!

The conclusion we would tentatively draw from this aspect of the story is that to extract the value offered by a formal systems thinking approach, key members of the team need to be familiar with the concepts and techniques, and have a positive desire to experiment with modelling of the kind described here. Time needs to be taken to explore the process, in the "playful learning" sense. And if learning is to be accelerated in the organization, then it may be a requirement that one or two influential individuals with successful experience of previous exercises join the new team. This is an observation not limited to scenario teams, or to oil market modelling.

Let us return now to the criticism that the oil producers' model did not adequately reflect reality. No one expects any model, even a sophisticated mental model, to fully reflect reality. However, there is a more profound aspect of this problem on which to dwell. There will always be some relationships that are hardwired into the structure of a model. Although "what-ifs" are played out by changing parameters at the start of a run, there is always something "static" in a system dynamics model! In the oil producers' model, simulations were typically run over a quarter of a century and seemed, to some observers at least, to ignore the possibility that the oil producers' world, as modelled, would change structurally in response to experience.

Thus, in practice, real players might change their relationships, their biases and their ambitions to such an extent that over the period of the run, the model ceases to adequately represent the underlying structure of the oil industry. Some critics argued that what is missing is a meta-model that steers the evolution of the base model to allow for deep structural change.

This concern about deep structural change touches at the heart of modelling, particularly when used for long-term scenario thinking. Modelling is the art of leaving out detail, of identifying the core enduring features of a system that will not change quickly or easily over the time horizon explored by the model.

The desire for a meta-model is an understandable response for overcoming perceived rigidities in the feedback structure of a dynamic model and for creating a model that can itself learn and adapt. It was the critics' need to see explicit adaptation of players' behaviour and relationships that for them cast doubt on the value of the oil producers' model.

This criticism carries two challenges. The first is for modellers to justify and communicate those relationships that are assumed to be resistant to adaptation. Not all players in a system do learn and adapt, even when they are performing poorly or acting at cross-purposes. Sterman's work on misperceptions of feedback in dynamic decision making documents experimental work on barriers to learning in complex systems (Sterman, 1989). The second and more subtle challenge is to use scenario models to examine the resilience of assumed policy and structure—in other words, to use model simulations to provoke exploration and dialogue on when and with what force the underlying structures themselves could begin to change and what would need to transpire for that change to be rapid, for example, or occur without warning. In this way, the core model becomes the vehicle for discovering radically different ways of conceptualizing the system.

Thus, the three-party oil producers' model might exist only for as long as: (i) there is a swing producer with low cost reserves; (ii) there is a single set of participants who are motivated by commercial needs; and (iii) there is a cartel motivated by economic development needs with an ability to cheat without destroying the harness of the quota system. Indeed, the scenario development team accepted this mix of players and motives as enduring structural features of the oil market. The suggestion here is that model simulations should be used to gain insights into what would need to occur for this three-party structure to break down or, at least, to change materially. This implies using the model in interaction with seasoned oil professionals in a game playing manner: what are the circumstances that would provoke behaviour outside the bounds predetermined by the model? Thus, in a very real sense this model, and any like it, would become a transitional object that would allow managers to address their more profound perceptions about their environment.

Finally, we will reflect on the experience we have had with this modelling of the oil producers' world as a training device in management development programmes.

There is the story, often told by de Geus and others, of the man from the jungle who had never been to a modern city before and whose only reaction to the many wonders with which he was confronted was to comment on the use of carts to carry bananas. Transporting bananas was the only thing to which his experience allowed him to relate.

In the same way, microworlds are likely to be better training vehicles if they relate to the experience of the trainees. The well-known People Express simulator (Sterman, 1988) represents a fast growing airline business, and is likely to work better with executives in start-ups or service firms than with farmers. For farmers, a gaming simulation such as Fishbanks (Meadows et al., 1989), that deals with the management and harvesting of fish stocks, might be more appropriate. People who have some experience of the oil business or of commodity markets may find the Oil Producers' Microworld a more rewarding learning medium. The challenge for the trainer is thus to find and present a modelling situation to which participants can relate.

The persuasive power of model-based insights lies in the opportunity that simulation provides for people to experience the future and to role-play a microcosm of their world, as Meadow's pioneering work has demonstrated (Meadows, 1984, 1989). Microworlds are the tools of the learning organisation and the learning society, giving us the means to experiment with the design of business and social policy in the safety of simulated reality.

Our experience with the Oil Producers' Microworld in a training situation suggests that its effectiveness depends on both the pre-game briefing and the task set for teams as they interact with the simulator. As noted earlier, the overriding need in the pre-game briefing is for a properly designed sequence of steps in the training situation. For example, it is necessary to excite interest in the way a given system actually seems to work before introducing mapping, and for trainees to discover the value of quantifying relationships before introducing system dynamics structures, and to identify issues that the model will illuminate before introducing trainees to the model itself.

The choice of simulation task seems to affect the amount of time trainees spend thinking and talking about the system, rather than mechanically operating the simulator, although our evidence is based only on casual observation of four training programmes. When the task was simply to maximize total industry revenue, we noted that trainees were easily trapped into the video game syndrome, manipulating parameters and simulating without reflecting on the reasons for observed changes in system performance. One problem with simple revenue maximizing is that it places teams in the unrealistic role of managing industry production as a whole. When the task was to generate contrasting scenarios, such as "prosperous, powerful OPEC" versus "prosperous, powerful independents", we noted more team dialogue. Teams now have to place themselves in realistic producer roles and reflect on the industry conditions that would favour either OPEC or the independents. Another approach we have considered (but not yet tried) is to set each team the task of explaining how a meta-model of the industry might react to specific simulations generated by the base model. Teams then interacting in plenary debate may well reflect more deeply about the structure and dynamic behaviour of global oil markets and also expose their own mental models of these markets. If so, trainees would have a valuable glimpse of the process of model conceptualization and team learning.

In summary, therefore, there are three questions that this work seems to answer. First, was the modelling experience of value to the original modellers? The answer was a clear "yes". Second, could other scenario building teams benefit? While blind adoption would be misleading, lessons will be learnt if time is allocated to doing so. Third, is there a benefit for management development? Again, the answer is "yes", provided the learning experience is carefully engineered and delivered.

ACKNOWLEDGEMENTS

The authors would like to acknowledge the assistance of Andrew Davis, formerly in Shell's Business Consultancy Division, for his contributions to building and calibrating the model described in this chapter. Also, Ged Davis, a senior member of Group Planning, made many thoughtful suggestions during the model's conceptualization and adapted the model creatively for use in the 1989 scenario publication (Group Planning, 1989).

This chapter builds on two previous papers from the Oil Producers' project: Morecroft (1992) and Morecroft and van der Heijden (1992).

NOTES

(1) The team members debated whether market share or revenue should be the principal variable driving the swing producer in and out of punitive mode. In the end they selected market share. The reason is that share is a power element. A swing producer can only operate from a position of power. His production level must be large enough to allow swings to take place, without affecting too much his internal economic conditions. A swing producer requires a minimum share of the market in order to be effective. The swing volume required to stabilize the market is proportional to the size of the market, just as required stock levels in manufacturing are related to days throughput. This requirement for swing volume is more important to the swing producer than his revenue position. As soon as he becomes driven by revenue he has to give up the role of swing producer. For further background, the reader is referred to Hargreaves (1991).

(2) The estimation of fossil fuel reserves is a major topic in its own right that has received the attention of system dynamics modellers (Sterman & Richardson, 1985; Sterman, Richardson & Davidsen, 1988). Their work has shown that reserve estimation is essentially a behavioural, judgemental decision making process, despite the seeming "hard", scientific and geological inputs to the process. The search for oil, its discovery and ultimate production take place in a feedback system that supplies the economic incentives for exploration and production through oil price. When commonly-used estimation algorithms are embedded in such a feedback system they show a tendency to exaggerate the size of recoverable reserves, especially during the early-to-mid stages of the depletion life cycle. The 1988 starting value of independents' crude oil reserves used in the oil producers' model is a global aggregate (excluding reserves of Eastern bloc countries) compiled from the best estimates of corporate planners.

(3) A full description of the Oil Producers' model is beyond the scope of this chapter. However, energy modellers might be interested in a few comments on the model's relationship to the literature. Powell (1990) has provided a useful summary of existing world oil market models which helps to position the oil producers' model. In his view, most models (such as those described in the Energy Modeling Forum summary report (1982) and by Griffin & Teece (1982)) can be classified as either intertemporal optimization models or behavioural simulation models. The optimization models usually assume economically rational behaviour on the part of the actors in the market, whereas behavioural models portray decision makers as highly constrained by limited information, limited computational ability, rules of thumb and limited decision making sophistication.

The oil producers' model clearly belongs in the class of behavioural simulation models even though it was created by a team without detailed structural knowledge of existing oil models. However, it differs in significant ways from what Powell describes as the dominant behavioural model—the so-called Target Capacity Utilization model, variants of which appear in the Energy Modeling Forum's 1982 study and, more recently, in Gately (1983, 1986) and Sweeney & Boskin (1985). The Target Capacity Utilization model portrays OPEC as a unitary swing producer, supplying the difference between world oil demand and non-OPEC supply (the call on OPEC), and also setting oil prices. The assumed decision rule for pricing is a behavioural rule of thumb based on the gap between OPEC's current capacity utilization and a target level of capacity utilization. By contrast, the oil producers' model disaggregates OPEC into a swing producer and a group of opportunists. The model explicitly represents a behavioural decision rule for OPEC quota setting (which allows OPEC the option to set quotas either above or below the call on OPEC) and a political bargaining process of quota allocation based on the ratio of swing producer to opportunists' production capacity (as a surrogate for bargaining power). Price is set in the market by the balance of supply and demand and cannot be set directly by OPEC, although the swing producer can influence price by withholding production when the market price is lower than OPEC's intended marker price (or vice versa). The oil producers' model also represents the political and socioeconomic pressures that cause opportunist producers to exceed quota. Capacity is determined endogenously through capacity expansion policies for both opportunists and independents.

(4) The elaborated model, which was known as the "Industry Structure Model", represented OPEC in terms of a swing producer, opportunists and quota setting, the same as the smaller oil producers' model on which it was based. Extra detail was added to the portrayal of independents' reserves, exploration decisions and upstream capacity expansion decisions. Some of these more detailed formulations were borrowed from a much larger upstream investment model. In addition, the independents were disaggregated to show explicit producing countries.

(5) Recent gaming simulations such as Fishbanks (Meadows et al. 1989), the People Express Management Flight Simulator (Sterman, 1988), the Claims Game (Kim, 1989; Bergin & Prusko, 1990) and OwnerTransformLE (Naumienko & Dlugosz, 1990) show a variety of constructive approaches to game design that focus on gameware and workshop protocols. Gameware includes the materials handed to participants before and during the gaming simulation. It also includes the user interface—the layout of symbols and information on the computer screen or gameboard. Workshop protocols include the pre-game briefing, the steps of play, the guidelines for teamwork and team decision making, and the post-game debriefing.

(6) Langley (1995) has completely redesigned and rewritten the Oil Producers' Microworld using Visual Basic. The new Microworld runs under the Windows operating system. The revised user interface makes use of high quality colour graphics and digitized photographs to improve visual appeal. Langley used the revised Microworld in a PhD dissertation to evaluate game-players' learning about feedback structure and dynamic behaviour.

REFERENCES

Bergin, R. and Prusko, G.P. (1990) The learning laboratory, *The HealthCare Forum Journal* **33**(2), 32–36.

Diehl, E.W. (1990) Microworld creator, Microworlds Inc., PO Box 1400, Cambridge, MA 02142, USA.

Energy Modeling Forum (1982) *World Oil: Summary Report*, 406 Terman Engineering Center, Stanford University, Stanford, CA 94305-4022, USA.

Gately, D. (1986) Lessons from the 1986 oil price collapse, *Brookings Papers on Economic Activity* **2**, 237–284.

Gately, D. (1983) OPEC: retrospective and prospects 1973–1990, *European Economic Review* **21**, 313–331.

de Geus A.P. (1988) Planning as Learning, *Harvard Business Review* March–April, 70–74.

Griffin, J.M. and Teece, D.J. (1982) *OPEC Behavior and World Oil Prices*, Allen & Unwin, London.

Group Planning (1989) 1989 Scenarios, challenge and response, Group Planning Publication (confidential), PL 89 S12, Shell International Petroleum Company, London.

Hargreaves, D. (1991) Saudis back in control, *The Financial Times*, 14 March, p. 18.

Jennings, J.S. (1988) The oil and gas industry in the 1990s, an address to the Third Annual North Sea Conference of the British Institute of Energy Economics, London, 27 June, Shell Publications, Public Affairs, Shell International Petroleum Company, London.

Kim, D. (1989) Learning laboratories, designing reflective learning environments, in P.M. Milling and E.O.K. Zahn (eds) *Computer Based Management of Complex Systems*, Springer-Verlag, Berlin, pp. 327–334.

Langley, P. (1995) An experimental study of the impact of online cognitive feedback on performance and learning in an Oil Producers' Microworld, PhD thesis, University of London.

Meadows, D.L. (1984) Stratagem 1, a management training exercise on long-term economic–environment interactions, monograph, Institute for Policy and Social Science Research, Hood House, University of New Hampshire, Durham, NH 03824, USA.

Meadows D.L. (1989) Gaming to implement system dynamics models, in P.M. Milling and E.O.K. Zahn (eds) *Computer Based Management of Complex Systems*, Springer-Verlag, Berlin, pp. 635–640.

Meadows, B.L., Fiddaman, T. and Shannon, D. (1989) Fishbanks Ltd, a microcomputer assisted simulation that teaches principles for sustainable management of renewable resources, monograph, Institute for Policy and Social Science Research, Hood House, University of New Hampshire, Durham, NH 03824, USA.

Morecroft, J.D.W. (1992) Design of a learning environment—the Oil Producers' Microworld, in *Proceedings of the 1992 International System Dynamics Conference*, pp. 465–474.

Morecroft, J.D.W. and van der Heijden, K.A.J.M. (1992) Modelling the oil producers—capturing oil industry knowledge in a behavioural simulation model, *European Journal of Operational Research* **59**(1), 102–122.

Naumienko, E. and Dlugosz, M. (1990) Gaming simulation and the challenges of Eastern Europe, presented at the XXI ISAGA Conference, Durham, New Hampshire, July 1990 (available from E. Naumienko, School of Management, Warsaw University, Szturmowa Str. 3, 02-678 Warsaw, Poland).

Powell, S.G. (1990) An evaluation of behavioral simulation models of OPEC, in *Proceedings of the 1990 International System Dynamics Conference*, Chestnut Hill, Massachusetts, July, pp. 878–892 (available from the System Dynamics Society, Massachusetts Institute of Technology, E40–294, Cambridge, MA 02139, USA).

Richmond, B.M. and Peterson, S. (1992) I think technical reference, High Performance Systems Publications, 45 Lyme Road, Hanover NH 03755, USA.

Senge, P.M. (1990) *The Fifth Discipline*, Doubleday Currency, New York.

Sterman, J.D. (1988) People Express Management Flight Simulator (software and documentation available from author, Sloan School of Management, MIT, E52–562, Cambridge MA, 02139, USA).

Sterman, J.D. (1989) Misperceptions of feedback in dynamic decision making, *Organizational Behavior and Human Decision Processes* **43**, 301–355.
Sterman, J.D., Richardson, G.P. and Davidsen, P.I. (1988) Modeling the estimation of petroleum resources in the United States, *Technological Forecasting and Social Change* **33**, 219–249.
Sterman, J.D. and Richardson, G.P. (1985) An experiment to evaluate methods for estimating fossil fuel resources, *Journal of Forecasting* **4**, 197–226.
Sweeney, J.L. and Boskin, J.M. (1985) Analysing impacts of potential tax policy changes on US oil security, *Energy Journal* **6**, Special Task Issue, 89–108.
Wack, P. (1985a) Scenarios: shooting the rapids, *Harvard Business Review* November/December, 139–148.
Wack, P. (1985b) Scenarios: uncharted waters ahead, *Harvard Business Review* September/October, 72–89.

FURTHER READING

Forrester, J.W. (1968) Market growth as influenced by capital investment, *Sloan Management Review* **9**(2), 83–105 (also in *Collected Papers of Jay W. Forrester*, Wright-Allen Press, Cambridge, MA, pp. 111–132, 1975).
Fossli, K. and Wilkinson, M. (1986) Subroto warns of oil price collapse, *The Financial Times* 4 May.
Goldstein, W. (1990) The supply and demand for oil in the 1990s, *Energy Policy* September, 593–595.
Kalff, D. (1989) Strategic decision making and simulation in Shell companies, in *Strategic Planning in Shell* ("Blue Guide to Planning"), no. 8, Shell International Petroleum Company, Group Planning, Shell Centre, London.
Morecroft, J.D.W. (1982) A critical review of diagramming tools for conceptualizing feedback system models, *Dynamica* **8**(I), 20–29.
Morecroft, J.D.W. (1983) System dynamics: portraying bounded rationality, *Omega* **11**(2), 131–142.
Morecroft, J.D.W. (1985) The feedback view of business policy and strategy, *System Dynamics Review* **1**(1), 4–19.
Morecroft, J.D.W. (1988) System dynamics and microworlds for policymakers, *European Journal of Operational Research* **35**, 301–320.
Morecroft J.D.W. (1992) Executive knowledge, models and learning, *European Journal of Operational Research* **59**(1), 9–27.
Richmond, B.M. (1987) The strategic forum: from vision to operating policies and back again, High Performance Systems Publications, 45 Lyme Road, Hanover, NH 03755, USA.
Sterman, J.D. (1987) Testing behavioral simulation models by direct experiment, *Management Science* **33**(12), 1572–1592.
Vennix, J.A.M. and Geurts, J.L.A. (1987) Communicating insights from complex simulation models, *Simulation and Games* **18**(3), 321–343.
Wilkinson, M. (1988) Opec's dangerous market game, *The Financial Times* 27 September.
Wolstenholme, E.F. (1992) The definition and application of a modular approach to model conceptualisation and analysis, *European Journal of Operational Research* **59**(1), 123–136.

Chapter 11

Diary of an Oil Market Model: How a System Dynamics Modelling Process Was Used with Managers to Resolve Conflict and to Generate Insight

David C. Lane

INTRODUCTION

This chapter describes the construction of a series of system dynamics (SD) simulation models by a group of managers and consultants working in the Royal Dutch/Shell group of companies. This section therefore briefly introduces the general experiences of system dynamics work in Shell[1] that led up to this project and the particular circumstances of the project itself. It also describes and justifies the particular style that I have chosen to employ in this chapter.

Shell and System Dynamics: The "Planning as Learning" Initiative

The project described in this chapter was sponsored by the Group Planning Department of Shell. This department has had an extended, and significant,

Systems Modelling for Energy Policy. Edited by D.W. Bunn and E.R. Larsen

involvement in the development of new approaches to strategic planning, perhaps most notably the work done on "scenarios" (see Wack, 1985). Such developments tended increasingly towards a new role for Group Planning: rather than attempting to act as a dispenser of wisdom to all of Shell, that department began examining different ways in which it might coax into being a continual process of organizational learning. The development of this line of argument is described in a paper by the then coordinator of Group Planning, Arie de Geus (1988), and the department's "Planning as Learning" initiative had this article as its intellectual starting-block. During the mid-1980s various activities that constituted the Planning as Learning initiative had involved the use of different modelling processes as forums to aid learning in management teams in Shell. Attention increasingly turned to the idea of providing "transitional objects" in the style of Papert (1980)—artefacts which somehow represented parts of reality and which could be experimented with to promote learning. In his quest for a "business LOGO"—a reference to the computer language developed by Seymour Papert to help children learn about geometry by interacting with a small programmable robot—Arie become interested in the contribution of system dynamics models.

Aided by the System Dynamics Group at MIT, members of Group Planning began to explore and experiment with these ideas. The purpose of interactive model building sessions is not intended to be the solution of a problem but rather to allow the participants to learn together about a system. Forrester (1971) emphasizes the superior importance of the modelling *process* compared with the final model itself and, with Shell, the term "Modelling as Learning" was coined for the approach (Lane, 1992).

To bring these ideas into Shell, Group Planning initiated a series of modelling projects. Some took place entirely within the department. This led to projects in other parts of Shell, with Group Planning offering support by providing appropriate staff. The aim here was to demonstrate SD in many parts of Shell and to accumulate experience and, it was hoped, success stories. An example of such a project is described in detail by Morecroft, Lane & Viita (1991). Although the project described here was, in general structure, of this type it also had important links back to the work within Group Planning. The specific backdrop of this work is described below.

Specific Project Background

When the opportunity arose to do this project, there was already a well-established Upstream Model project, "upstream" referring to the exploration and production end of the oil business. This work had been done within Group Planning over a period of two to three years and had ultimately yielded a large model that was intended to study the dynamics of investment, production and technological innovation within the global oil and gas industry under different oil price scenarios. More recently, a "Downstream Challenge" had attempted to emulate this work as it sought to understand the refining, supply and marketing elements of the industry. Concerning, as it did, a less homogeneous and technologically influenced part of the industry, the Downstream Challenge employed a number of different approaches in its early stages. The project was led by Heenk van der Boor (a pseudonym) and had a core team that was based in Group Planning, but had contacted many other members of Shell staff in order to call on their experience and support. This had been done using a NOTEPAD conference, an electronic mail based conferencing facility in which senior Shell staff had "posted" ideas onto the network's bulletin board and then been invited to comment and build on them. Over a period of weeks, this Delphi process had led to various hypotheses about the future of the Downstream industry.

In what became known as Phase I of the Downstream Challenge, members of Group Planning and seconded members of the Business Consultancy department worked on these hypotheses during March and April of the year in question to extract from them a sense of the driving forces in the Downstream and the common concerns of the contributors. For Phase II, conducted during late May into June, causal loop diagrams and stock and flow diagrams were used with members of the Downstream core team to illustrate the various hypotheses. By mid-June a particular issue had been identified: the management of refining capacity. Although work on other hypotheses continued—in particular, an SD representation of product differentiation was fairly well advanced and some SSM techniques were employed—it was the refinery issue that was the object of increasing interest.

Heenk's objective was now to support a study on a current issue within the Downstream business. To add extra vigour to this proposed Phase III, he had arranged for the support of an MIT-based modeller. The hope was that success in such a situation would show how SD could contribute to the Downstream Challenge, that it would generate *wider* interest among various senior players in the Downstream business for SD model building as a learning process and so encourage more formal participation by senior staff. In July such an opportunity arose.

We learned that there was a long-standing debate within Shell's Cathay Region concerning the relationships in that region between price, marginal costs and marginal revenues. The debate had recently become more specific, evolving into a disagreement about the nature of the relationship between refinery output and total cash generation in circumstances of, for example, maintenance shut-downs. If the Shell volume went down slightly, what would happen to cash generation? Members of the Refining Strategy department of Cathay Region argued that cash generation might actually rise slightly. But in

Shell Khitomer, the Shell operating company in one of the major countries in Cathay Region, people in the trading department believed that cash generation would fall off in virtually the same proportion as the temporary volume reduction. This was the opportunity that Heenk had been waiting for: by proposing and supporting a project centred on this debate about marginal volumes and marginal revenues, he would address the dual aims of doing work in the Downstream part of the business and continuing to demonstrate the use of SD modelling as a learning approach.

The Approach Taken in this Chapter

Before embarking on the main body of this chapter, let me say two things about how I have chosen to present this case. Firstly—and the reader may very well have observed this already—I have changed the overwhelming majority of names, of both people and places. Additionally, the dates and, most importantly, the parameters of the models have been shifted, factored up, factored down or in some other way altered. However, for reasons of verisimilitude, I have tried to preserve the relative cultural standings of the pseudonyms employed for people. I have also tried to ensure that the sense of geographical configuration of the place names reflects something of the original setting of the work. The dates that I use certainly preserve the order of events and in most cases (but not all) preserve the actual durations. In the case of the model parameters, the values have been extensively reworked, with the product and crude prices moved to a new base. However, the logic behind the causal connections has been rigorously preserved, as have the resulting modes of behaviour. The explanations of the dynamic effects that result are similarly unaltered.

These changes were made for reasons of personal and commercial confidentiality. Some are almost invisible, while others are only too visible, but they are one of the constraints within which management researchers operate. This does of course raise questions regarding the veracity of the account given here. The account is not a complete one since I have chosen to exclude details made extraneous by the focus of this chapter. However, as the narrative will make clear, I was centrally involved in the whole duration of the project. I am able to give my professional assurance that, notwithstanding the alterations described above, from the point of view of drawing out some interesting SD lessons, the essentials of the case are authentic, being accurately and truthfully represented.

My second point concerns the decision to present a description of how events unfolded and how the various SD models evolved. The SD logic of modelling has it that the process of model building should be kept close to the clients, with the aim being not solution but the articulation of ideas and the creation of insights and learning. Nevertheless, the overwhelming majority of published examples describe finished models and their features. This omission is of ever greater significance if a less objective, or functionalist, view is taken (Lane, 1994a). However, it is not necessary to choose between an objective and a subjective approach (Burrell & Morgan, 1979) to agree that the process of model building is a useful occurrence to document. Accounts of the technical elements of model creation and the elements of social interaction and cognitive understanding are valuable in showing how SD is actually done. To do this it is necessary to take a micro-sociological approach and to attempt, albeit post hoc, the role of "participant-as-observer" (Morrison, 1986).

To record some of this information, to try to give a sense of the evolution of the model and a glimpse of the changing thinking of those involved, I have therefore chosen to use the format of a diary. I hope that this will display the participants not as elements of an organizational chart but as characters, real people facing problems and uncertainties, who make contingent, boundedly rational decisions. It may also help to convey the frustrations and disappointments, the pleasures and the satisfactions of one example of SD modelling.

DIARY OF THE MODEL BUILDING PROCESS

Monday 25 July: Start-up (London)

The full team assembles. We are managed by Heenk van der Boor of Group Planning. The lead consultant is Martin Zandt from MIT, with Kate Williams and me, both members of Business Consultancy, who bring experience of previous Downstream Challenge work. We review a note of 22 July from Edward Hoskins, a strong and articulate advocate of the "increased cash generation view". In the note he describes his ideas using influence diagrams that employ engineering symbols. These are quite detailed and we are concerned that a very large model may be necessary. However, we are encouraged that he adds that, "we do not expect to be able to predict the effect[s] but more to give substance to the bounds". This indicates that he is not necessarily looking for a forecast from a huge model. Edward's ideas have support from the head of his department of Cathay Refining Strategy, Heenk Taube, and there is also interest at the most senior level in Cathay Refining.

Heenk explains that the goal for our team is to study the issue and make a presentation at the next quarterly Downstream meeting in London on 2 September. To this meeting come the most senior managers in the Downstream business; if we can show an interesting study of some aspect of refining, which has involved staff from Cathay Region, refining, trading and supply/ marketing then Heenk can get interest in a larger scale project on a strategic issue.

Tuesday 26 July: Flying out

Martin and I fly out to Xanadu, the capital city of Khitomer, for the first meeting. We rest in a hotel before the activities of the next day.

Wednesday 27 July: First Khitomer visit

Morning: Briefing by Cathay Region refining strategy (Xanadu). The next morning, we meet Edward and two of his colleagues, John Mill and "Gerry" Gérin. Edward spends about an hour with us giving a very detailed briefing on oil supply, demand and refining in Cathay Region. He divides the region into a Western and an Eastern Zone and describes how product supplies flow between them. He describes some basics of refineries to us (see Figure 11.1a). Refineries take in crude and split it into various products. Higher prices are obtained by the "lighter" products. Conversely, "heavy" products command relatively lower prices. The refineries that use a crude distillation unit (CDU) and a platformer, and so offer a basic hydroskimming process, are said to have "simple capacity". Other refineries have "complex capacity"; they can take *some* of the heavier fuel oil and split it down further in a catalytic cracker, or using some other upgrading process. This is important because of the "gross refiners margin" (GRM), the difference between the cost of the crude feedstock and the total value of the mix of products produced. To grasp the GRM idea, one might consider the fate of a single barrel of oil passing into a refinery. Some of its volume is lost as it is burned up to provide energy for the refining process. The majority is split into a number of products. Part of the barrel emerges as mogas, other parts as naphtha, gas oil, fuel oil, etc. If we give each of the fractions of a barrel of these products that are produced a label, W_i, and the prices available on the market for each of the products the designation Product Price$_i$, then:

$$\text{GRM} = \sum_i (W_i \times \text{Product Price}_i) - \text{Crude Price} \qquad (11.1)$$

The W_is do not sum to 1 because of the losses during the process. Now the product prices increase with the lightness of the product—Figure 11.1a shows indicative values. Complex capacity has high weights, W_is, for the lighter products. Complex capacities therefore generally have higher GRMs, sometimes called "complex refiners margin". Other refiners earn only "simple" or "hydroskimming" margins. Note that this is only a simplified description but it is useful as a framework for understanding and remembering the detailed information that Edward is able to give to us. His briefing has all the realistic detail used in managing refineries and employed previously by Business Consultancy in modelling work (Uhlmann, 1988). For this, and the description below, he is able to provide us with full documentation, some confidential,

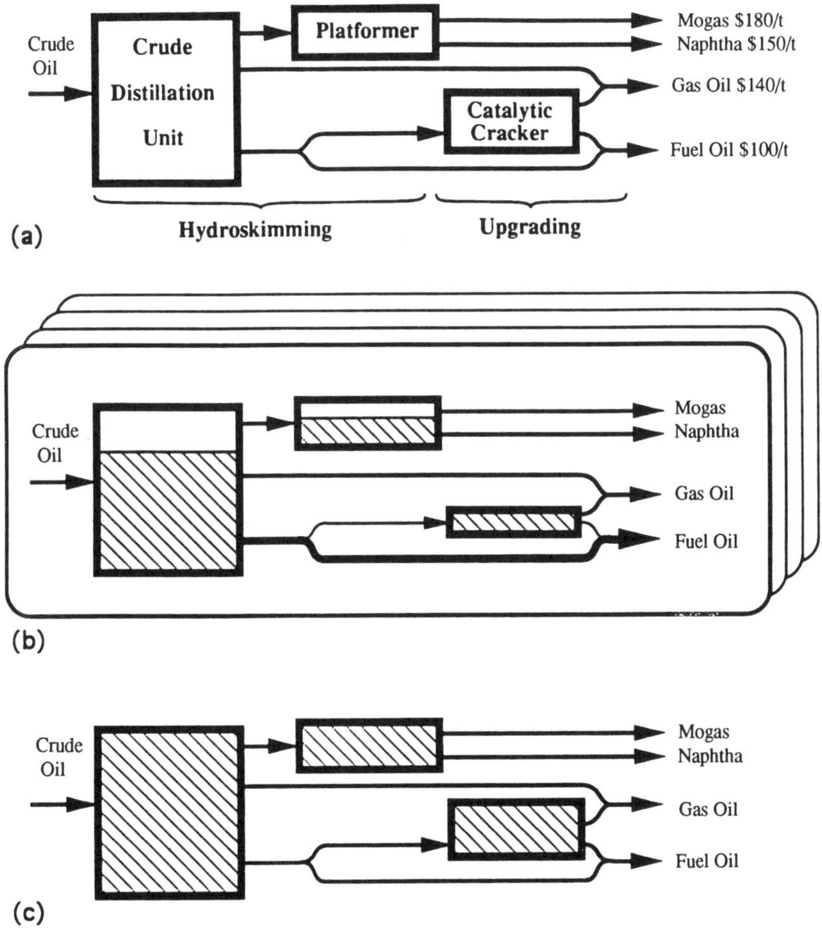

Figure 11.1 Representation of refining activities in Cathay Region, gleaned from Edward Hoskins' briefing on 27 July. (a) Highly simplified diagram of the refining process in a facility with both "simple" (hydroskimming) and "complex" (upgrading) capacity. Note that the product prices (quoted here in tonnes) are indicative only. (b,c) The basic configurations of the refineries in the region, utilized capacity being indicated by shading. In (b) the limited amount of complex capacity of the numerous competitor refineries is fully utilized while excess simple capacity is unused. In (c) Shell has proportionally more complex capacity, and this, as well as its simple capacity, is utilized to the maximum

some available publicly (see van Hilten, 1986; Shell Group Public Affairs, 1986).

Edward and his colleagues have done a detailed analysis of all the many dozens of competitors' refineries in the region and have discovered two things (see Figure 11.1b):

(1) Refineries always use all the complex capacity that they have at first. Eventually, the CDUs are producing more fuel oil than the complex capacity can upgrade so this passes straight out and the refinery is then producing a "darker barrel". All of the competitors are currently using all of their complex capacity.
(2) There is excess refining capacity in the region. Because margins are low, competitors are not using all of their simple/hydroskimming capacity.

Turning to Shell, Edward knows that Cathay Region has invested in a lot of complex capacity, more than the competitors (see Figure 11.1c). There is therefore more value added to a barrel of crude passing through a Shell refinery and Shell picks up a complex refiners margin.

Edward thinks that the issue is: "How will refiners' margins change in response to changes in the output volume of Shell's refineries?" His argument is that if Shell lost some of its output due to maintenance, it would temporarily subtract from the market a lighter barrel. However, the competitors cannot produce a complex barrel to replace that volume; they only have capacity to produce simple output that does not command sufficient margin. So, for the supply gap to be bridged, the margin on the extra—the marginal—barrels of the competitors will have to rise. So the hydroskimming margin will have to rise to encourage competitor production; that is, the product prices will rise. But then, Edward's argument continues, Shell's complex margin will have risen even further, because of its lighter barrel, so that Shell might actually make more money.

Martin now talks about SD modelling, how it is done and what it can achieve. Edward, John and Gerry are clearly interested; the approach seems to suit the background in natural science that they all share.

Afternoon: Briefing by Khitomer trading (Shakaree). We have lunch and then go with Edward by car to Shakaree, the area's largest port. This city is the focus for the traded product market and is therefore where Shell Khitomer is based. There we see the head of Shell Khitomer trading, Charlie Herbert, and spend about an hour and a half with him. Charlie listens good-naturedly to Edward's case, smiling and commenting at various points; this is clearly not the first time that they have had this discussion. In response, he gives us a detailed explanation of the world of trading, punctuating his exposition with laughter.

Charlie is rather dismissive of the refinery-by-refinery analysis, the details obviously being of little interest to him. He is quite emphatic that market responses are so overwhelmingly efficient and fast that any variation in Shell's output will be compensated for by a redirection of supply, effected by the activities of traders. Irrespective of any detailed analysis of competitor capacity, any rise in product prices caused by a supply gap will produce a rapid response from the traded market that will fill the supply gap. Any reduction in Shell volume caused by maintenance shut-downs will therefore only result in a reduction of cash generation. Charlie asserts that anybody from supply/ marketing, if asked, would support this view. He suggests a few names, staff now in London, whose opinion and experience he respects, among them Chris Durbridge, Carlos Martinez and Paul Earnest.

We drive back to Xanadu with Edward, Charlie grabbing the opportunity of a lift there. In the evening we fly back to London, to report to Heenk and Kate and to have a think.

Thursday 28 July: Clarifying the task (London)

Martin spends some time summarizing our progress and the current goal of our work in a note to Heenk. He is then able to use some previous analysis in Group Planning to produce a first STELLA model, initially called "sh_simple_cap". Martin is very keen that we avoid going by product prices. Meanwhile Kate is working from our notes of the meeting with Charlie Herbert to produce CLDs of the factors that he described.

Tuesday 2 August: Developing a model (London)

Because Heenk has asked Martin also to look briefly at other prospects for SD modelling raised by Phase I, over the last few days it is Kate and I who have worked on data collection and calibration of the model, producing sh_simple_cap_dcl/kw. Today Martin produced for Edward and others in Cathay Refining Strategy a note summarizing our ideas so far and containing a diagram only of the model.

Wednesday 3 August: First cut model, V0.00 (London)

We have moved to a first quantitative model, currently called Ref_Proto_V0.0, and written in STELLA (see Richmond, Vescuso & Peterson, 1987). This model deals with the interaction between Shell's capacity and the market (see Figure 11.2). It has 22 equations and 2 feedback loops. We have managed to go directly from supply to margins: the ratio of supply and demand is calculated and, via a nonlinear relationship, this sets the value of GRM. One loop involves changes in the utilization level of Shell's capacity. The other loop, acting over a much longer period, allows the amount of capacity to be changed. We have guessed the time scale of 90 days as being appropriate so there is effectively only one loop, the second having no effect in this time scale. If

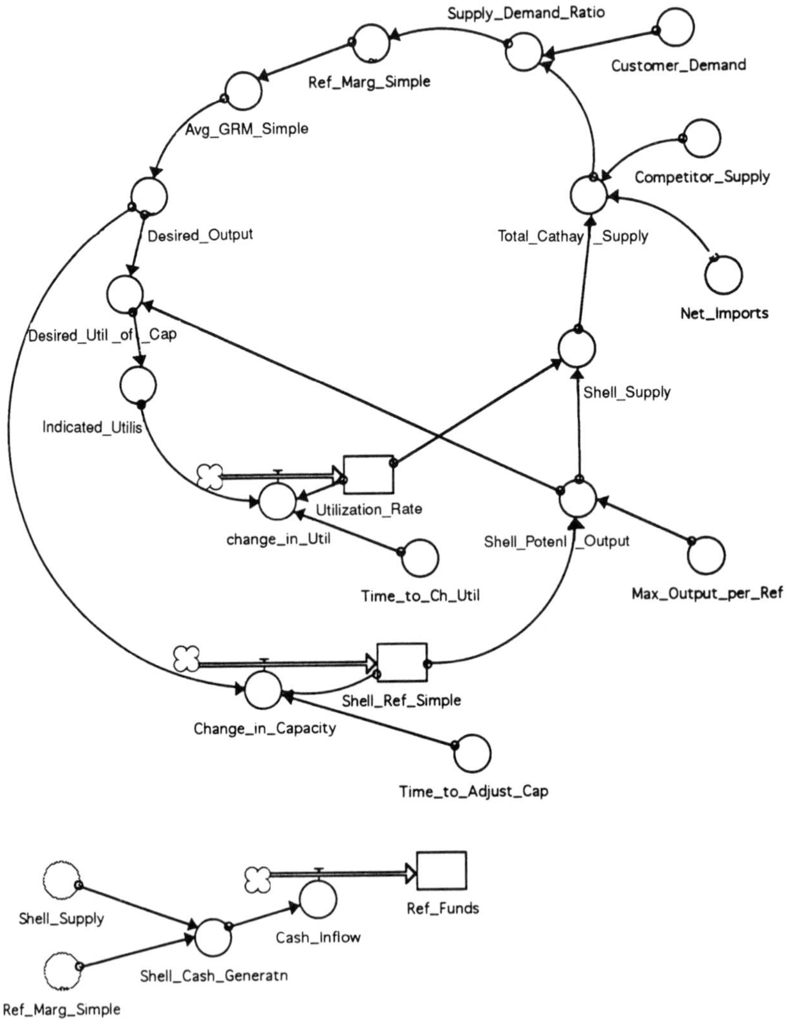

Figure 11.2 Stock and flow diagram of the first model, Ref_Proto_V0.0, dated 3 August. This deals with the interaction between Shell's capacity and the market. This model has 22 equations, 3 stocks and 2 feedback loops

Shell output is arbitrarily dropped, margins rise as the market is undersupplied and Shell cash generation falls. However, many of the parameters are guesses and there is no response from competitors or traders. We think of this model just as a useful framework for the next stage.

Thursday 4 August: Second Khitomer visit (Shakaree)

When we visit Shakaree again Edward is abroad at another meeting so that only Gerry and John are there. We knew that this was the case but all five of us wanted to keep the momentum of the process going. Martin and I present the modelling framework. Its explicitness is liked. However, our link between supply/demand ratio and indicated margin is received badly; Gerry and John do not like the way that it implicitly allows long-term shortages or excesses in the market to endure. They insist that supply and demand must be brought back into balance.

We see that they strongly support a structural model but that it has to involve product prices if it is to command any confidence. To start off this process we begin collecting some idea of the general shape of the relationships by asking them to work with us on the volume/margin response curves for the competitors (see Figures 11.3a and b). As we had previously learned, simple output could increase if the margins came available. However, complex output cannot increase from its current level, although it would drop if margins really fell. We also discuss how traded imports might be affected by changes in product price (see Figure 11.3c). We fly home, having agreed to put the ideas into a model and to send them a copy. We have also agreed to contact more staff in supply/marketing over the parameters, particularly concerning the volume of imports, and to return with a calibrated model in about a week.

Friday 5 August: Firming-up the model (London)

Martin and I debate the next steps. We need a model that equilibrates supply to demand. Martin knows another way of representing such effects with free-floating prices and he discusses this with me. He is concerned by the complexity that it would add and hopes that another approach might avoid this problem. He therefore introduces demand elasticity with respect to price as a feedback loop, thinking that this should clear the market. To make this work, of course, we will have to represent explicitly the price of some amalgamated product.

In the afternoon Martin, Kate and I meet with Chris Durbridge, a member of the supply/marketing division working on the analysis of oil demand. We wish to explore with him the issue of refining margins in general but we have also prepared some general layouts of the three relationships that we want to examine (see Figure 11.3) as well as the supply/demand ratio to price transfer function. Chris proves to be both an affable and very experienced man.

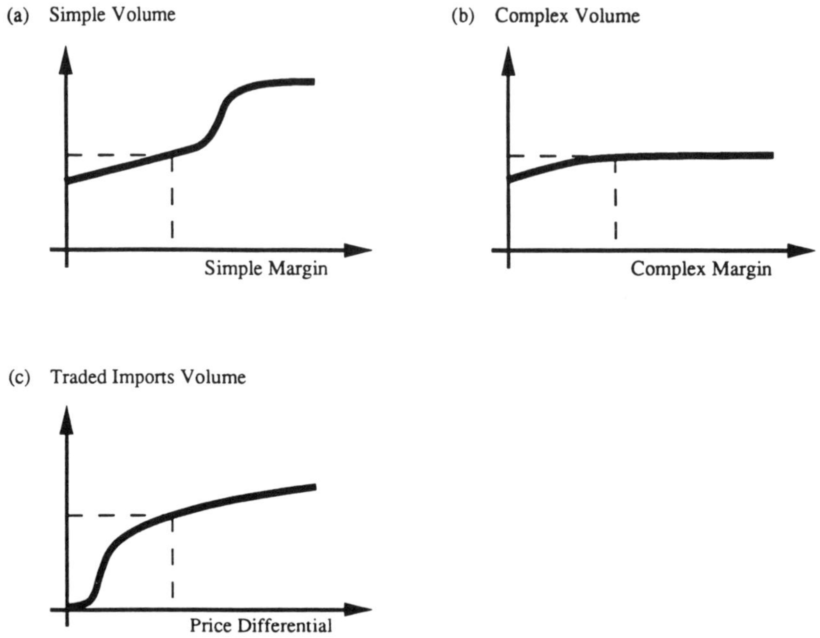

Figure 11.3 General view, as of 4 August, on the form of three key relationships in the model. The broken lines indicate the base case values of the variables. (a) Relationship between competitors' simple output volume and the simple, or hydroskimming GRM; the existence of excess capacity means that more volume could be produced if the GRM became available until all capacity is in use. Compare with (b): relationship between competitors' complex output volume and the complex GRM. The higher level of margin available from this capacity and the lack of excess capacity mean, respectively, that this curve is very flat on the left of the base value and completely flat on the right. Compare with (c): the volume of product imports traded into the area as a function of price differential. Note the smooth nature of this curve

Although he supports Charlie Herbert's views on margins and revenues, he is able to confirm Edward's analysis that the competitors' complex capacity is fully loaded. However, he informs us that in the Eastern Zone of Cathay Region there is excess simple capacity. By presenting him with some carefully labelled axes and offering some gentle prompts, we begin to calibrate the relationships concerning imports and the indicated price mechanism. After a very useful meeting, he leaves us with a cheery, "Drop by my office any time, there's an open door for you". Over the next days Kate maps out some influence diagrams of what Chris said in the meeting to be sure that we had understood his contribution and valued his time and expertise.

By the end of the day we are coming to believe that the relationships for simple, complex and imported volume are crucial (Figure 11.3). We must install them in the appropriate feedback structure but it looks as though the nature of these curves is *central* to understanding the relationship between volumes and revenues.

Thursday 11 August: Towards a calibrated, multi-loop model (London)

Martin has spent the week laying out the basic structure while I have been collecting parameter values. By Thursday we have put all of our ideas into Ref_Proto_V0.7 (see Figure 11.4). This now has six feedback loops: product price and demand; product price and competitor production (from simple and complex capacity); product price and traded imports; and Shell margin influencing Shell supply (via utilization short term and capacity long term). In fact, demand is not altered by price because we have no data on this. Similarly, Shell capacity cannot expand on the time scale being considered, Shell utilization has a fixed step downwards and the competitor complex production is calibrated to be incapable of increase from its base value. The model represents the situation in all of Cathay Region and is set to run for a simulated

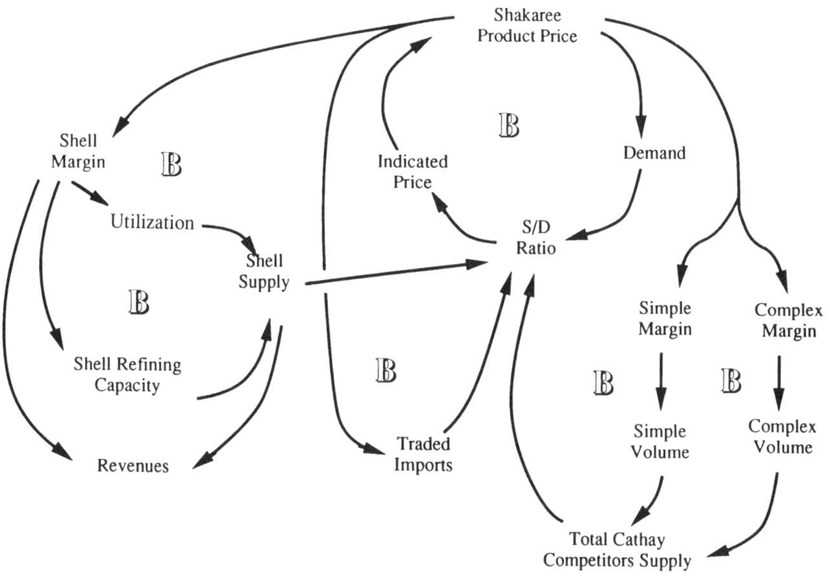

Figure 11.4 The major loops of the model at 11 August, called Ref_Proto_V0.7. Note the indicated pricing mechanism. The other five loops indicate: responsiveness of Shell to GRM changes via adjustments of utilization (short term) and capacity (long term); and responsiveness of competitor output from simple and complex output to respective GRMs and changes in traded imports. The relationships of Figure 11.3 are implicitly embedded within this structure. This model therefore has 36 equations, 4 stocks and 6 feedback loops

90-day period. If 10% of Shell capacity is lost due to maintenance then a combination of extra (simple) competitor production and import fills the resulting supply gap and Shell cash generation falls.

Friday 12 August: Working on the detail (London)

Preliminary alterations to the imported volume curve by Kate and I yield Ref_Proto_V0.8. This behaves in the same way as version 0.7. I have also been working on a clear statement of the hypotheses proposed by Cathay refining strategy and Khitomer trading as Heenk wants to circulate these for discussion.

Monday 15 August: Getting the detail right (London)

Kate has received data from Chris Durbridge and we further hone the imported volume curve. With the resulting Ref_Proto_V0.9, Shell cash generation falls, in line with the view of Charlie and others. As Martin gets drawn into other work for Heenk, I have had to work further on the model in order to get a clear layout and good variable names as well as installing switches to control the loops. He is not free to fly to Xanadu until Wednesday but we decide that Kate and I will visit tomorrow, as Edward will be back and we want to keep him interested.

During the day Kate and I work on an "agenda to depart from" for the visit, noting the questions that we would like to look at and the points of the model that we want to display. Four questions are of particular importance. Firstly, exactly which geographical area are we seeking to model? At present we have all of Cathay Region but there may be a case for dividing this into West and East. Secondly, how can we clarify what it means to use a single product price somehow to represent the whole set of products on the market? Thirdly, this simple formulation yields a value of refining margin from simple capacity but to calculate a complex margin, the margin earned by refineries with upgrading capacity, we are simply adding an arbitrary amount per barrel. This "fix" is used to calculate Shell's cash generation and is certainly worth further examination. Finally, what is the elasticity of demand with respect to price?

In parallel with this activity, we evolve the model into Ref_Proto_V0.911 as Chris now provides a view on the pricing mechanism. This is a model of 43 equations, 4 stocks and 6 loops, although only 3 of these are active (imports and competitor volume; simple and complex). Prices are changing in one day and refineries are altering their output in seven days. With all the model functioning—the "base case"—additional imports are triggered very fast. The supply gap is rapidly filled with virtually no contribution from the competitors. However, the interplay of parameters means that the system stabilizes with the market very slightly (0.5%) short. Shell cash generation increases, in line with Edward's hypothesis.

Tuesday 16 August: Third Khitomer visit, day 1 (Shakaree)

We meet with Gerry and Edward and they listen while Kate demonstrates the model. I then describe the outstanding issues. There is a big problem: Edward does not like the model: "It's much simpler than I expected," he comments. He has a number of difficulties with what we have so far. Firstly, he does not understand what the single product is. He can understand that, for the purposes of this study and this issue, we do not want to detail all of the products refined from oil, but he insists that one product hides what is going on. Edward also rejects the indicated pricing mechanism, firstly, because there is no natural price level at which supply and demand are in line and, secondly, because the market, in his view, must clear. Gerry agrees with this. They want to see market equilibration clearly occurring in the model. We knew that this was a problem waiting to happen and had hoped that the demand/price feedback might cope with it but it now looks as though we will have to go to free-floating prices. Edward is slightly apologetic at the extent of his criticisms but we try to assure him that this is not a problem. I suggest a break while Kate and I think about ways forward. Left to ourselves, Kate and I ponder the prospect of changing the model so that it has many products. Eventually we decide that the way that our clients think of the refining world simply demands some form of product split. What we decide to suggest is a division into two products, one being a "basket" of the light products and the other a basket of the darker ones. This will enable us to model the difference between simple and complex capacity.

We meet again with Edward and Gerry. They have been thinking also. They say that we should be working on a model of the Western Zone only of Cathay Region. This geographical area can be considered as a coherent market-place and should therefore be the focus of our work. They also agree that we have a time scale of a few months, so our run of 90 days is fine. They go on to say that this means that Shell capacity expansion can just be removed completely from the model. Edward adds that he likes the clear statement of the two hypotheses that we have brought. On this encouraging note, we describe the two-product idea and this is well received. We talk about it for a while and agree that we can just call them "distillate" for the light products (mogas, naphtha, etc.) and "fuel oil" for the heavier ones (fuel oil itself, bitumen, etc.).

After lunch we all spend the afternoon developing this more complex model. We split the products and have two prices. We also change the imported volume and divide that proportionally into the two products. At day's end we have Ref_Proto_V1.1. Curiously, the margins stabilize but the two product prices continue to change; there seems to be no stable solution but no sustained oscillations either. We will have to study this further.

In the early evening Martin arrives and we describe the day's events and the state of the model. He regrets the additional complexity that will result from the changes but accepts that the developments were necessary.

Wednesday 17 August: Third Khitomer visit, day 2 (Shakaree)

Today all three of us worked on the model with Edward, changing the structure and looking at possible parameter values that he had found. In the middle of this, an idea floats by: "of course, if imports changed then prices in the other markets could change". Looking back, I cannot remember who said this but it was an idea that was to return. At the end of a hard day we fly back with Ref_Proto_V1.20.

Friday 19 August: Setting up the two product model (London)

Martin and I are now dealing with a model that has "floating prices". In this formulation, the ratio of supply to demand drives a rate of change of product price (Meadows, 1970). When the ratio is greater than one, the price is factored downwards; when it is less than one, the price is factored upwards. Only when the ratio equals one does the price stabilize, although there is no natural level implicit in the model for where this might be.

The structure of the model is now more complex and can best be explained in subsystems terms (see Figure 11.5). We have 44 equations with 3 stocks (2 product prices and accumulated Shell funds). There are neither short- nor long-term Shell response loops. For every barrel (bbl) passing into the Shell refineries, 0.2 bbl of fuel oil and 0.74 bbl of distillates are produced. The utilization of this capacity is simply stepped downwards arbitrarily. There are

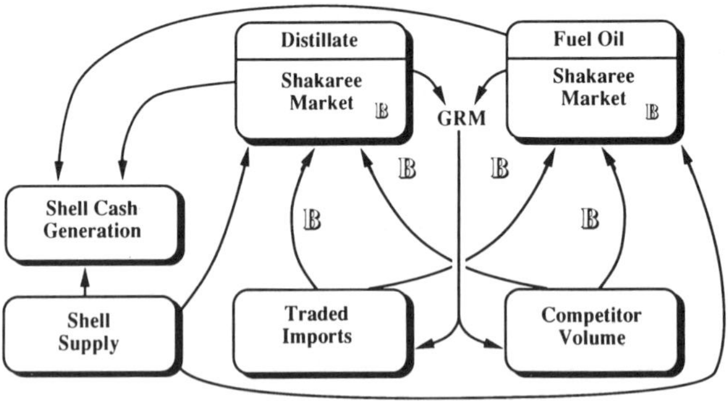

Figure 11.5 Subsystem overview of the model Ref_Proto_V1.20 on 19 August. Western Zone refinery output is now represented by two products, and the adjustment loops for Shell have been deleted. In addition, floating prices have been introduced for the products, as has the structure for price/demand feedback. Both traded imports and competitor volume respond to GRM. The relationships of Figures 11.3a and c are implicitly embedded within this structure, but the complex capacity of the competitors is modelled as producing at its maximum throughout. This model therefore has 44 equations, 3 stocks and 6 feedback loops

floating product prices and each product has a structure for price/demand effects. In fact, only fuel oil has parameters that lead to the stimulation of demand by lower prices, the logic being that there are some industrial sites with the capacity to switch from coal to fuel oil if the price is sufficiently low. The two product prices lead to the GRM which triggers competitor volume. As these have only simple capacity free to produce, they yield 0.4 bbl fuel oil and 0.56 bbl distillate. The complex capacity of the competitors is modelled as producing at its maximum throughout. The imported volume is divided into two products in the same proportion as the split for the simple refineries in the Western Zone, the reasoning being that imports must be coming from similar, simple, refining capacity outside that Western Zone. The model therefore has six feedback loops. What does it do? Shell Cash generation rises but it remains rather unstable. Our guess is that this is because of the floating prices. We are still scratching our heads about this one.

Monday 22 August: Handover and more analysis (London)

Martin is now heavily embroiled in Heenk's other Downstream Challenge activities. His time is now almost entirely spent pulling together a presentation on what SD has been used for in Group Planning over the past year. The model is passed over to Kate and me as Martin works on a presentation of this material for 2 September. Edward and I arrange to have telephone conversations and exchange long faxed notes. Throughout all of this, Kate works on Charlie's comments on trading and the information supplied by other people in Shell Centre to help calibrate the model.

Going back over previous notes, I see that an implication of Edward's first briefing is that the split of products emerging from a refinery changes with its utilization level. I send Edward a long note proposing that this should be built into the model and suggesting parameter values based on his previous figures.

I spend the rest of the day trying to understand the dynamics of the model, now cautiously nudged up to be Ref_Proto_V1.22. It seems clear that because the lost Shell volume is "lighter", in the current model it simply cannot be replaced by extra runs of simple capacity refineries, either in the Western Zone or in the zones from which trading is done. This reinforces my view that we need to think very carefully about the light/dark mix of the marginal barrels lost from Shell and added by the competitors and the refineries supplying the traded imports. I await Edward's answer to my note with considerable interest.

Tuesday 23 August: Understanding the dynamics (London/Khitomer)

Today Edward and I had a long telephone conversation about the way to model the product splits in the model and he likes the note that I faxed to him. The idea is that the first barrel of crude oil passing through a refinery goes through the complex capacity. This creates a relatively high loss fraction because the process is more extensive but it yields a lighter output barrel. However, as utilization (or output volume) rises, the complex capacity eventually becomes full. The next barrel passes through the simple capacity only and although this burns less oil it yields a darker product barrel. Subsequent barrels have this "history". As a result, the average proportions across all the output from a refinery are graphs against utilization. Hence the fraction of product that is distillate decreases as utilisation increases and more "simple" output is used. The loss fraction has the same form, reducing as proportionally fewer barrels are upgraded. With the fuel oil fraction the curve rises as less and less of the throughput is upgraded.

We decide to model the Shell output using curves from fuel oil and distillate output as a function of total output. With these changes we now have Ref_Proto_V1.25. This model is structurally the same the same as Ref_Proto_V1.20 but it now settles down to a new stable state in which the replacement volume is provided predominantly by traded imports, with a small proportion originating in the competitors' refineries. The explanation derives from the fact that previously I had Shell's lost volume consisting of too high a proportion of distillate. With the new formulation this figure is lower so that it can be filled more easily, and the fuel oil that is simultaneously produced by extra production no longer floods the market. Shell cash generation still rises. I describe all of this to Martin, who is very satisfied with progress. I mail a copy of the model to Edward.

Wednesday 24 August: Challenging our assumptions (London/Khitomer)

Another long discussion with Edward by telephone. He does not believe the results. He says that he wants to see the current, rather artificial, splitting of traded imports replaced with completely separate traded markets in the two products. He also wants to model the Eastern Zone markets. I ask him why. Edward is beginning to think that he may be wrong.

He now accepts that trading effects are faster and more powerful than he had assumed. But, in building the model, his thinking has evolved. He reasons that the obvious way for the distillate gap resulting from Shell's partial shut-down to be closed is for independent traders simply to purchase distillate *only* and to import it into the Western Zone. This is what happens in reality and would have the effect of reducing any distillate price rise. Shell cash generation might well fall in such circumstances. The current model does not allow us to experiment in this way because it implicitly unites import traders with the source refineries so that only a (simple) package of distillate and fuel oil can be imported. He therefore wants the traded imports subsystem reworked into two parts, one for each of the products. However, he is not yet convinced that his original hypothesis is false, for the following reason. It is now clear to him that most of the traded volume is coming from one place; the Eastern Zone of Cathay Region. But he thinks that if all the necessary distillate were being traded out of that area, the price would rise, which would pull it back. So, in addition, he wants me to build a model of the two product markets in the Eastern Zone. We can then model independent traders arranging imports between each of them.

After the call I rework the model. This is fairly easy in structural terms as I simply copy some of the previous structures and rename them. I then write a note describing the new layout and listing the parameter values to which Edward should give some thought, in particular the response of prices in the Eastern Zone to supply/demand imbalances and price/demand effects. I fax him the note and mail the new model to him. This does not simulate yet since it lacks some parameters. I copy the note to Martin to keep him informed.

Reflecting on the events of the day, I can see that Edward is still inclined to believe his hypothesis regarding the behaviour of the system. However, he is now using more structural elements of his mental model of the system and is using SD to help him with this process.

Friday 26 August: A two-product, two-zone model (London)

We now have a model called Ref_Proto_V2.00. It has 74 equations and 10 loops (see Figures 11.6 and 11.7). It models the collected activities of the Shell refineries in the Western Zone of Cathay Region. Two products represent the

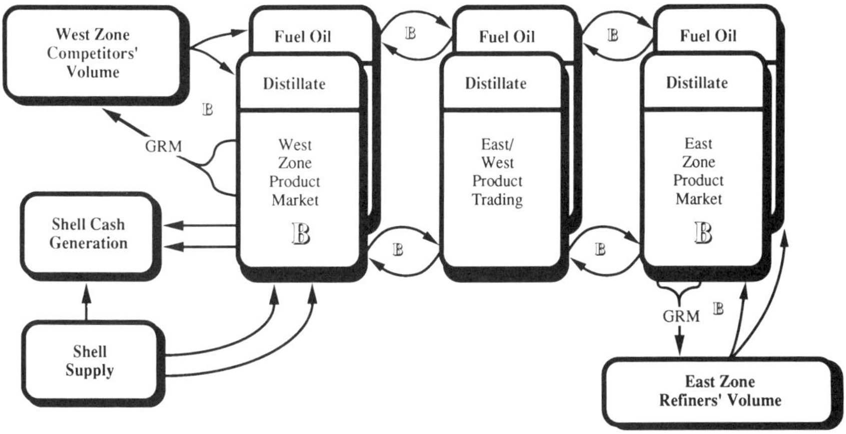

Figure 11.6 Subsystem overview of the model Ref_Proto_V2.00 on 26 August. This is a two-product, two-zone model with separate traders controlling imports between the two zones. The refineries in each zone respond to simple GRM changes. All four products have floating price mechanisms and the structure for price/demand feedback. Decisions on imports for each of the products are made by comparing the prices in each of the zones. This model therefore has 74 equations, 5 stocks and 10 feedback loops

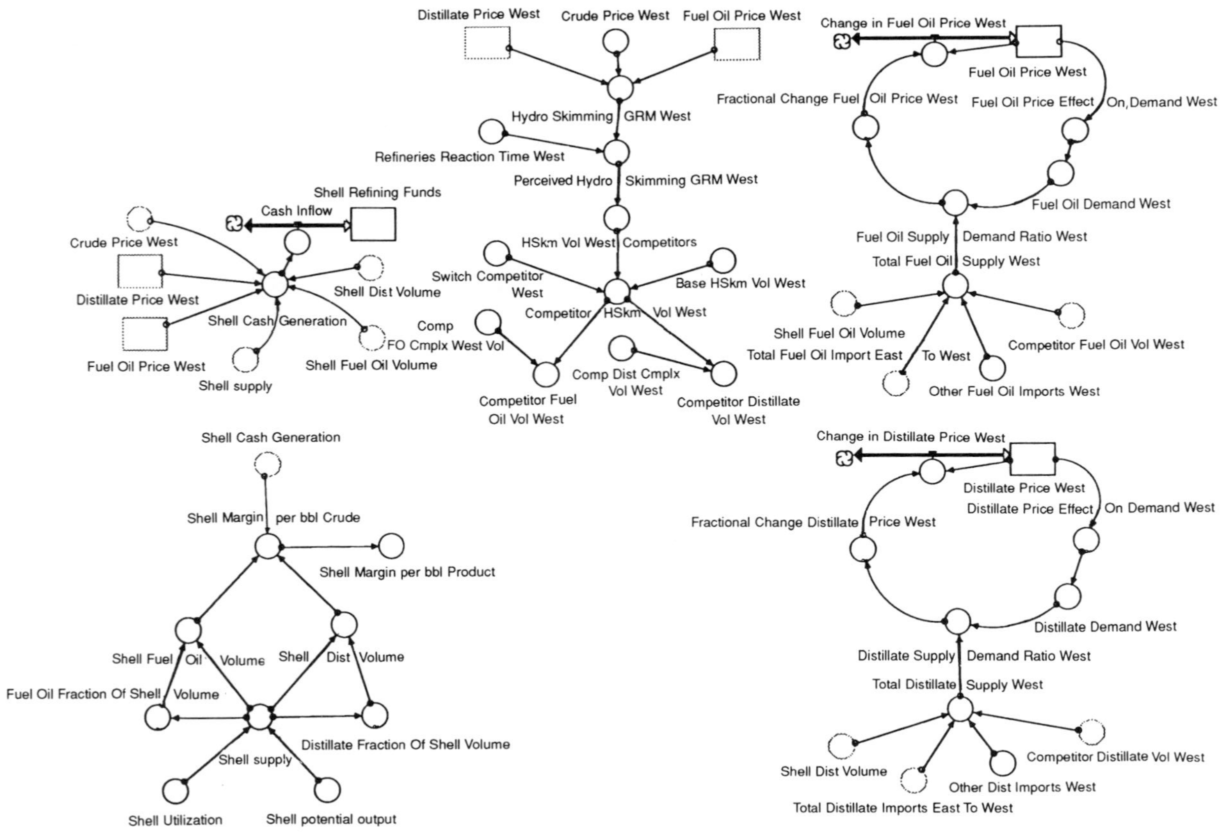
Distillate Price West
Crude Price West
Fuel Oil Price West
Hydro Skimming GRM West
Refineries Reaction Time West
Perceived Hydro Skimming GRM West
HSkm Vol West Competitors
Switch Competitor West
Base HSkm Vol West
Competitor HSkm Vol West
Comp FO Cmplx West Vol
Comp Dist Cmplx Vol West
Competitor Fuel Oil Vol West
Competitor Distillate Vol West
Change in Fuel Oil Price West
Fuel Oil Price West
Fractional Change Fuel Oil Price West
Fuel Oil Price Effect On,Demand West
Fuel Oil Demand West
Fuel Oil Supply Demand Ratio West
Total Fuel Oil Supply West
Shell Fuel Oil Volume
Total Fuel Oil Import East To West
Competitor Fuel Oil Vol West
Other Fuel Oil Imports West
Shell Refining Funds
Cash Inflow
Crude Price West
Distillate Price West
Shell Dist Volume
Shell Cash Generation
Fuel Oil Price West
Shell supply
Shell Fuel Oil Volume
Shell Cash Generation
Shell Margin per bbl Crude
Shell Margin per bbl Product
Shell Fuel Oil Volume
Shell Dist Volume
Fuel Oil Fraction Of Shell Volume
Distillate Fraction Of Shell Volume
Shell supply
Shell Utilization
Shell potential output
Change in Distillate Price West
Distillate Price West
Distillate Price Effect On Demand West
Fractional Change Distillate Price West
Distillate Demand West
Distillate Supply Demand Ratio West
Total Distillate Supply West
Shell Dist Volume
Competitor Distillate Vol West
Other Dist Imports West
Total Distillate Imports East To West

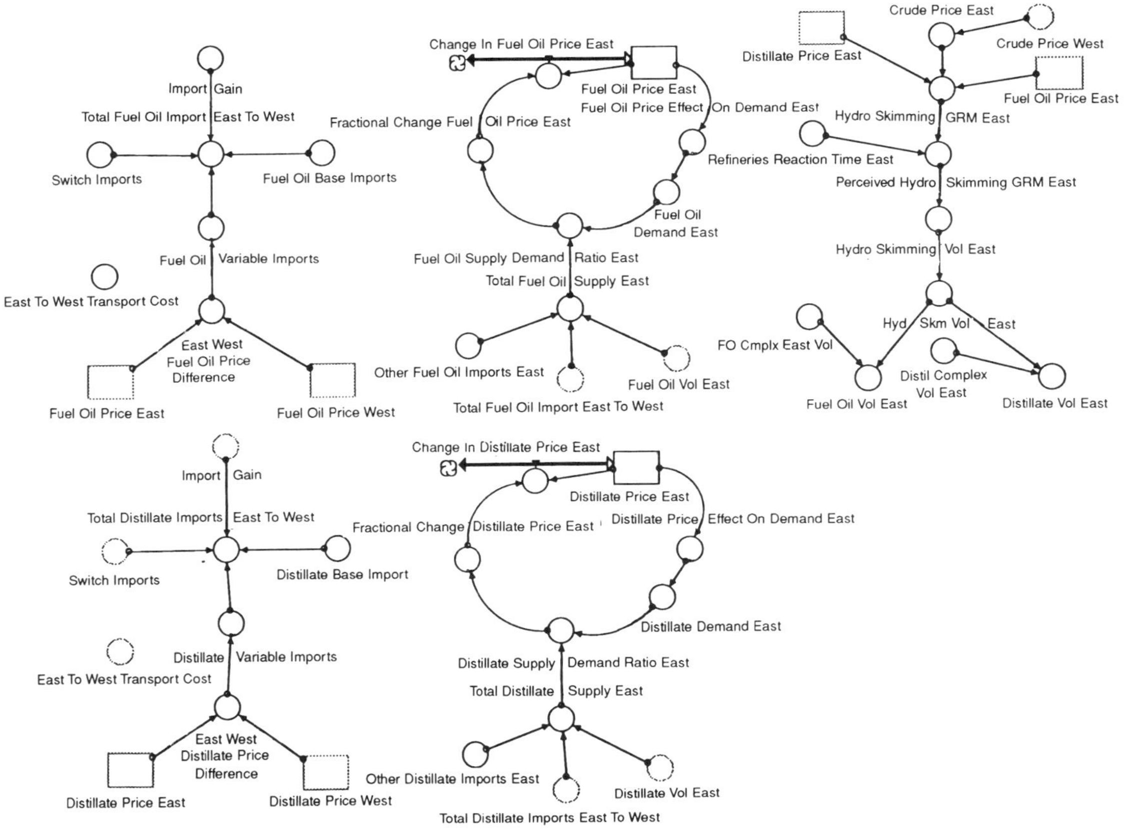

Figure 11.7 Complete stock and flow diagram of the two-product, two-zone model, Ref_Proto_V2.00. The layout is very similar to that in Figure 11.6. This model has 74 equations, 5 stocks and 10 feedback loops

market, each having price/demand feedback (two loops). The competitor refineries in the Western Zone are represented collectively. They monitor the product prices via the hydroskimming margin and alter their simple output in response (one loop). Their complex capacity is always fully in use. The Eastern Zone is modelled in the same way structurally (three loops). New subsystems representing traders for each of the products are introduced. Each monitors the product prices in the Eastern and Western Zones and adjusts imported volume in response (two times two loops). It should be noted, however, that the parameters of the model have distillate prices in either of the zones leaving demand unaffected; unlike fuel oil, these are not products that are rapidly substituted. The model therefore has only eight functioning feedback loops.

Some parameter values have been obtained and Edward, Kate and I are unearthing more. The model behaviour is quite complex so I have introduced many more graphical outputs to understand what is happening. Martin comes by in the afternoon and we spend an hour exploring. We end by realizing how useful this time has been for us in gaining insights into the basic dynamics of this system. In trying to understand the model, we begin to get a sense of how these interacting markets work and we experience again the power of expressing these ideas in a model. Our job is to ensure that this learning experience is delivered fully to Edward and is described effectively by Martin on 2 September. We send Edward the new model.

Monday 29 August: A requisite model (London)

After a working weekend by all, we have a model that Edward, Martin, Kate and I are satisfied with. We are content that we have the right variables mapped out in the structure and we have now calibrated the model. We are also beginning to understand the dynamic effects. This is Ref_Proto_V2.01.

With V2.01 Edward is right. The new requirement for distillates in the Western Zone produces traded imports from the Eastern Zone and initially Shell Cash generation *falls*. However, this shorts the Eastern Zone distillate market and limits trading. But the key effect is that the while the extra distillates are traded away from the Eastern Zone there is a glut of fuel oil that is left there. This has the effect of reducing the Eastern Zone fuel oil price, which then pulls down hydroskimming margins. This then limits production of both products, and prices are somewhat restored, reducing traded imports. Eventually Shell cash generation in the Western Zone is *higher* than at the beginning of the run.

But now the structure of the model challenges Edward. We have reproduced the price/demand balancing loop and he knows that the Eastern Zone has a considerable number of older power stations that can switch fuel back and forth from coal. He therefore has to include a significant demand/price elasticity effect for Eastern Zone fuel oil. *Without the model* he infers that *this* will allow extra production in the East, the distillate being traded into the Western Zone and the extra fuel oil sold locally as lower prices stimulate demand. In consequence, it is his belief that Shell cash generation will be *lower*. He feels that he can now see all of this without a functioning model, by just using mental simulation, the process having advanced his mental models to this extent. However, I encourage him to suggest suitable numbers, almost for my own gratification but also for use at the meeting of 2 September. After a little work he is able to do so and the resulting Ref_Proto_V2.02 behaves just as Edward expected it would (see Figure 11.8). Clearly we have produced a requisite model, one that is sufficient to resolve the issue that is at hand.

CONCLUDING THE STUDY

There were three different elements to the means by which this project was concluded. Firstly, in the case of Edward Hoskins—and, in fact, his colleagues, Gerry Gérin and John Mill—there was a significant increase in understanding of the dynamics of the refining business in Cathay Region. His understanding of the parameters and the interlinking factors had been altered. By giving him access, albeit vicarious, to the expertise of various staff on the supply/marketing and trading side of the business, the modelling process arguably improved his mental model. Two examples are his clarification of the relationships between volume and margin for traded imports and the speed with which such effects came into play. These we might count as improvements in his mental model. However, the provision of dynamic simulation modelling, in a form that he could readily understand, allowed him to "bring to life" his evolving mental model, to see how these interlinking factors played themselves out over time. It is also relevant to comment that the new understanding that Edward and his colleagues derived from participating in the process was in agreement with the view taken by Khitomer trading, a situation of no little advantage to Shell.

From the description given in the core section of this chapter it might be deduced that the divergence of views between Cathay Region refining strategy and Khitomer trading was resolved simply because the representative of the former was shown to be in the wrong. In fact, adjustments of views occurred on both sides. With Edward's permission, we reported our work and our findings and observations back to Charlie Herbert. What we found out in early September was that although the dominant feeling in Khitomer trading was that they had been correct, they had also noted that refining strategy's detailed analysis of the refineries in the region was significantly better than anything that they had compiled. They had not appreciated fully the amount of knowledge of such fundamentals of the business that refining strategy

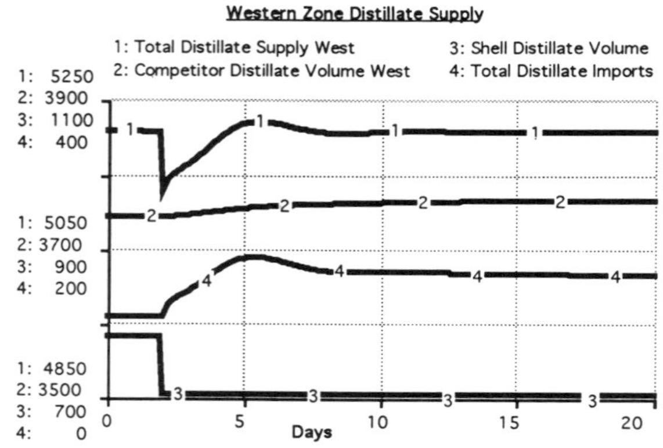
Western Zone Distillate Supply
1: Total Distillate Supply West
2: Competitor Distillate Volume West
3: Shell Distillate Volume
4: Total Distillate Imports
1: 5250
2: 3900
3: 1100
4: 400
1: 5050
2: 3700
3: 900
4: 200
1: 4850
2: 3500
3: 700
4: 0
0
5
10
15
20
Days

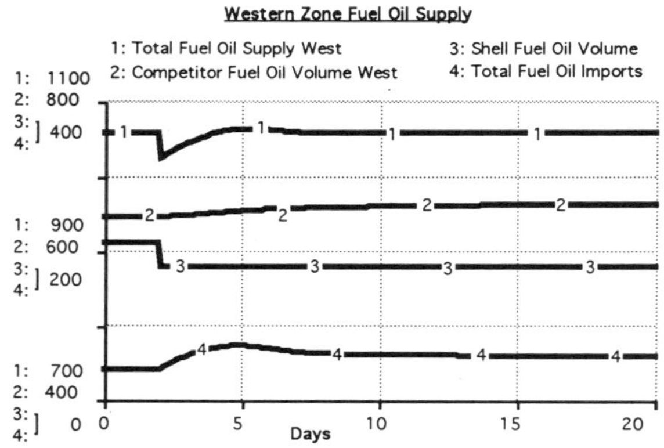
Western Zone Fuel Oil Supply
1: Total Fuel Oil Supply West
2: Competitor Fuel Oil Volume West
3: Shell Fuel Oil Volume
4: Total Fuel Oil Imports
1: 1100
2: 800
3:] 400
4:
1: 900
2: 600
3:] 200
4:
1: 700
2: 400
3:] 0
4:
0
5
10
15
20
Days

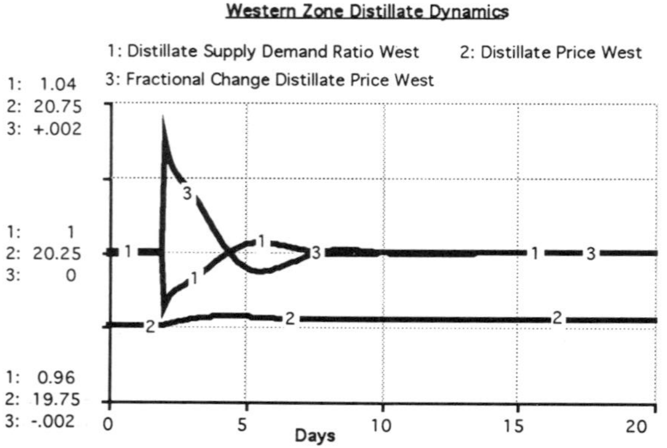
Western Zone Distillate Dynamics
1: Distillate Supply Demand Ratio West
2: Distillate Price West
3: Fractional Change Distillate Price West
1: 1.04
2: 20.75
3: +.002
1: 1
2: 20.25
3: 0
1: 0.96
2: 19.75
3: -.002
0
5
10
15
20
Days

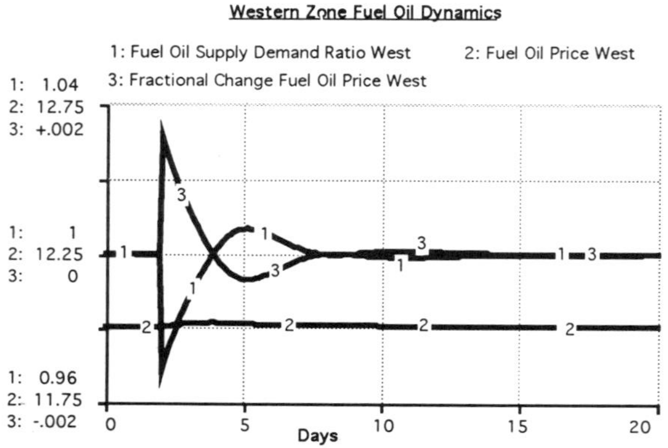
Western Zone Fuel Oil Dynamics
1: Fuel Oil Supply Demand Ratio West
2: Fuel Oil Price West
3: Fractional Change Fuel Oil Price West
1: 1.04
2: 12.75
3: +.002
1: 1
2: 12.25
3: 0
1: 0.96
2: 11.75
3: -.002
0
5
10
15
20
Days

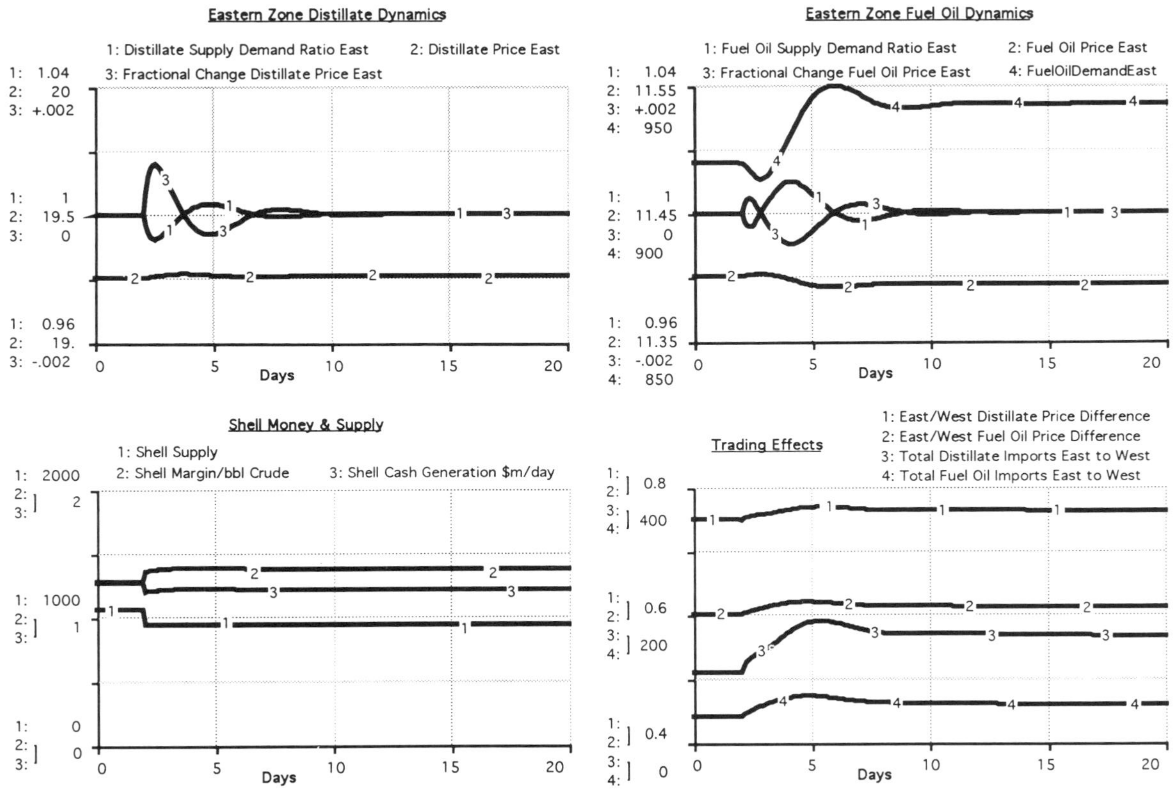

Figure 11.8 A selection of outputs from the final version of the model, Ref_Proto_V2.02, on 29 August

possessed. We might speculate, while noting that it is only a speculation, that the confidence that Khitomer trading derived from their role in the modelling work led to their request to meet more frequently with refining strategy to draw on their views. This interest in continued cooperation was an encouraging second element of the work's conclusion.

Finally, we must consider the role of the project in the success of the presentation of 2 September. The full details of this are part of a wider story, while the focus of this chapter is on a particular project, so that great detail is not appropriate. Martin gave a wide-ranging presentation on the benefits of SD, its previous successes in Shell and its potential for the future. Describing the structure and the behaviour of the Ref_Proto_V2.02 model (Figures 11.7 and 11.8), and the benefits of the work, he was able to use this as an example of the inter-divisional benefits that could be obtained from such a process. The feeling of the meeting—which included Heenk Taube and Carlos Martinez' representative—was that the approach was fruitful and that further support would be given to Heenk van der Boor's Downstream Challenge process.

REFLECTIONS ON THE STUDY

In reflecting on this particular study we address a number of questions. In the first subsection we remain with the detail of this specific case and try to answer the question: What was the longer term consequences of the work? The project consisted of the use of a series of icon-driven system dynamics simulation models, constructed to facilitate a process of negotiation between different groups. In the other subsections we therefore take a more general view of this approach to ask what elements of the approach made the project a success, why this was the case, and how the approach compares with other methods. In reflecting on the method that was used we may view the approach from three perspectives: technical, information processing and social process.

The Longer Term Consequences of the Work

This project had a short-term goal of displaying the usefulness of the approach. This was achieved and was then used to create longer term benefits. As part of the Downstream Challenge, Heenk van der Boor harnessed support for a study of the interaction of the whole portfolio of business activities in Shell. Regrettably, the needs of commercial confidentiality do not allow this work to be described further. However, alongside this, other SD-based work continued both in Group Planning and in other parts of Shell. For example, assisted by staff from Business Consultancy, John Morecroft worked with senior members of Group Planning to create a model of the dynamics of oil quota negotiations in OPEC (Morecroft & van der Heijden, 1992) and SD modelling contributed to a pair of famous Shell scenarios (see Kahane, 1991). Within Business Consultancy, the success of the oil market model and the accumulation of experience that resulted from this and other work resulted in the adoption of SD modelling as one of that Department's core skills. Projects in a wide range of areas resulted (see Lane, 1992, 1993; Davis, 1994) and the activities continue today. We might also comment on the later experiences of those who took part in the study. As a result of this and other projects, I chose to move in academia to research, teach and consult in the area of interactive SD modelling full-time. As stated above, Heenk van der Boor chose to use the approach further. Perhaps the most striking effect may be seen in the experience of Jeremy Bentham. Having worked on a project similar to that described here, he chose to become a Sloan Fellow at MIT, writing his Masters thesis on the use of systems thinking in organizations, before returning to Shell as a line manager to put these ideas into practice (Bentham & de Visscher, 1994).

Observations on the Method

The technical elements

The technical elements of the method refer to the ability of the approach to provide analytical capability for dealing with a complex problem. In this case the situation that was studied was conceptualized as a system of elements interconnected by information feedback loops. The use of such models within the social sciences has been studied by Richardson (1991), who discerns two "threads" of tradition and current practice. The "cybernetics thread" focuses interest on the role that feedback plays in processes of communication and control within systems. Areas of contemporary operational research lying in this thread would be the "viable system model" of Stafford Beer (see Beer, 1979, 1981) and the "soft systems methodology" (Checkland, 1981; Checkland & Scholes, 1990). In contrast, the "servomechanisms thread" attaches prime importance to the dynamic behaviour of feedback systems. System dynamics lies within this second thread. This perspective led to the system variables being conceptualized either as rates or as levels to create a mathematical formulation, the behaviour of which could be studied (Forrester, 1968a, 1968b). From a mathematical point of view, the models therefore consisted of a series of simultaneous nonlinear differential equations.

Although this purely analytical view is seldom employed when dealing with clients (it would be a hurdle to the "process" elements of the work described below), the need rigorously to formulate feedback systems as simulation models is central to the definition of SD. The view is that it is not possible to infer the behaviour of such complex mathematical systems without the deductive powers of modelling (Forrester, 1968b). Of equal importance to this notion is the view that social systems can be represented using such formulations and therefore produce behaviour that is "counterintuitive" (Forrester, 1975, 1987). This dynamic complexity, combined with time delays, distortions and gaps in information and the ability to see only one example of behaviour, makes learning about such social systems virtually impossible. In contrast, the technical elements of SD modelling allow the creation of an effective learning cycle, this being achieved by closing the time gap between strategy implementation and outcome, by excluding "confounding factors" and by providing unambiguous information (Sterman, 1994).

The obvious competitor regarding the mathematical analysis of social systems is econometrics, and there has been a regrettably long-standing lack of understanding between that field and the discipline of SD. This is not the place to revisit in detail this particular area of friction, partly because it has been addressed extensively elsewhere (see Meadows, 1980) and partly because the area of SD labelled "non-conformist economics" (Lane, 1994a) has achieved a more constructive engagement between the two. We shall limit ourselves to commenting that the detailed quantitative analysis for precise prediction and the use of statistical validation techniques in econometrics are quite different from the validation processes and the goal of policy design to which SD aspires. There are other technical differences, including the persistent use in econometrics of linear, open models with many exogenous variables. This contrasts sharply with the nonlinear, closed (in an information sense) and endogenous perspective adopted by SD practitioners.

The provision of the technical capability of SD to the study of oil markets is not unique to this case. For example, Meadows (1970) studied a variety of such situations in the case of generic commodities and Rahn (1989) took the study of such systems further. Specific examples of oil market modelling have been reported in the SD field (for examples, see Choucri, 1981; Karsky, 1988; Aizhu & Zheng, 1992). In each case, the technical element of SD—the ability to probe the dynamic consequences of a system structure—is central to the nature of the works.

Considering the nature of the case reported here—the experiments performed and the insights conveyed—it is difficult to find an approach that might substitute for SD. From its foundation, the field of SD has been critical of the focus in operational research on individual decisions, preferring instead to study the repeated application of a policy on a stream of incoming information (Forrester, 1968c). In the modelling work done here, Figure 11.3 is an example of such policies; rather than linking one input to one output to form a decision, these transfer functions show the relationships between information streams so that the consequences of the rates within the model—the prices—can be deduced as new streams. Recently encouraged to compare and contrast the SD approach with modern developments in operational research (Lane, 1994b), Forrester re-asserted the value of being able to "bring to life" the variables in a situation so as to perform meaningful experiments (Forrester,

1994). This powerful advantage of the SD approach is one that I readily acknowledge, and the contribution that this technical capability brought to the case detailed here is clear.

However, convincing as this technical view may be, it can make SD appear little more than a subset of simulation modelling, and this is a view that has been taken by some observers of the field who have not understood, or who have significantly underestimated, the importance attached to the other elements of model building (see, respectively, Pidd, 1992; Flood & Jackson, 1991).

The information processing elements

In taking an information processing perspective on SD modelling, we make reference to the cognitive sciences. This is not a standard or well-explored view to bring to bear but initial forays indicate that it holds some promise, and it is these works that have been drawn on in making the following observations (see Robey & Taggart, 1981; Klein, 1994). The simplest facts emerging from the cognitive sciences are that information processing consists of two elements. The short-term—or working—memory has a low capacity, being able to work consciously with only seven or eight pieces of information or some four "propositions". Linked to this is long-term memory, containing vast amounts of information. Information passes from long-term to working memory as associative links are activated. Hence, concepts that are linked to many other concepts are more easily recalled. Following Klein (1994), we may then consider SD using two cognitive factors: (i) the contribution of the approach to altering the ability consciously to process information in working memory, and (ii) the effect of the approach on the nature and quantity of the information that is recalled from long-term memory.

The visual and interactive aspects of SD modelling provide information processing advantages to users. The visual representation of variables and relationships using causal loop diagrams, subsystem overviews and computer models with iconic interfaces all allowed the participants to deal with more information than unaided working memory normally allows. This means that the range of concepts included in any considerations is wider and the consequent thinking more comprehensive. This is true when we are simply considering the various judgements made by participants as they considered these different forms of diagram. What also seems to be the case is that symbols that are familiar to users allow a form of perceptual processing, as opposed to logical processing. In the case of SD the distinction between stocks and flows has proven very powerful; for example, Lane (1993) describes the ease with which he was able to introduce these to clients and have them work with them. The software employed in this study was founded on the familiarity of these symbols (see Richmond, 1985). In this study we can particularly note the ease with which the request for floating prices on 16 August was accepted as being correctly represented using a symbol that explicitly recalls the notion of material flowing in and out of a tank.

Modelling processes that show the interconnectedness of variables are more likely to allow information flow between long-term and working memory—cognitive mapping (Eden, 1988) employs this fact to considerable effect. One consequence of this is that information representations that put variables in context are more likely to trigger links to related pieces of information held in long-term memory and thus are valuable in facilitating the elicitation of mental models. A second consequence is that processes that encourage working memory to construct interrelationships are more likely to result in the new structures being stored in long-term memory. This point is perhaps the most readily observed aspect of this whole study in that the diary section of this chapter is an account of an increasingly complex story being recalled, worked on and reinstalled in long-term memory. We might very cautiously suggest that a demonstration of this is perhaps Edward's description of 29 August, in which he seems to access the entire model and then change a relationship and perform mental simulation in order to generate a new piece of information.

Many of the comments made in this section overlap with the aspects of the SD method described below. This reflects the considerable overlap between the information processing and the social process perspective. The contents of this subsection seem most readily to fit with the former approach, but below we turn to the latter.

The social process elements

Most SD models are not intended to be stand-alone artefacts. Rather, their value is intimately related to the individuals who created them. Such a comment might be seen to be true of any model in operational research but the field of SD attempts in most cases to practise "interactive modelling", in which the model is seen to be a representation of the assumptions of the participants. Such aspirations can be related to the information processing arguments described above but in this subsection we will consider the social process advantages of SD.

The process of building a model with a group of individuals has always been seen as providing benefits above and beyond technical analytical capability. Although this strand of thought was less clear in early SD practice, hampered as it was by limits on computer hardware and software, the suggestion that this is a new development in the field is false (Keys, 1988; Dash, 1994). Lane (1994a) offers the alternative view that the process benefits that are obtained from model building were described in the field's first book (Forrester, 1961) and that over time, this view became increasingly operationalized and, in consequence, elaborated as other publications added to this line of thought (see

Morecroft, 1988; Senge, 1990; Lane, 1992). With SD something rather more than the functionality of applied mathematics is therefore being employed. These "social process" elements of SD may be divided into three: mental model access, intuition creation and social negotiation. We deal with each of these in turn.

Mental model access. The process of building an SD model causes the participants to articulate their own mental models of a situation (Forrester, 1961). In so doing, they access, as a group, what is probably the most important database for the organization of which they are members. Although the paper and electronically based data of an organization are valuable and influential, when decision makers act, some of the most important facts, ideas, parameters and policies of that organization will nevertheless reside solely in the mental models, in the heads, of actors within that system. This view was proposed at the creation of the field and is held today (see, for example, Forrester, 1992). These mental models may, of course, contain falsehoods. What is certain, however, is that they are in a form that renders them closed to discussion and challenge by colleagues. SD offers a framework that provokes model builders to consider parameter values and to see how variables relate to each other (see Eden, 1988). The information processing advantages of SD in accessing mental models have already been mentioned but there is an additional advantage. Mental models, be they pieces of structure, behaviour modes or parameters, are articulated in a form that is explicit, open and challengeable. The distinction offered by Forrester is that although these articulated ideas may not be accurate, they are precise: "by 'accuracy' we mean the degree of correspondence to the real world", whilst, by "'precise' is meant 'specific', 'sharply defined', and 'not vague'" (Forrester, 1961, p. 57). It is from this precision, this sharp definition, of the articulated mental models that informed debate is assisted and supported.

In the study here, causal loop diagrams and interfaces employing stock and flow symbols were used, and they proved to be very flexible and comprehensible as a means for representing ideas during discussions and sharing views on the major structures. The articulation and clear representation of the variables and loops perceived to be important to the problem allowed the participants to articulate their mental models. They were able to express the relationships between variables and hence explore the importance and subjective meaning that they attached to them.

Intuition creation. The outcome of a successful SD project is intended to be intuition creation for the participants, an idea that we have touched on above. The purpose of building an SD model is not to have a final model that can then be used for optimization and prediction. An SD model cannot be separated from the process or from the group that created it. It is not intended to stand alone. Rather, "[the] primary function of model building should be educational rather than predictive ... modelling [is] a process rather than ... a technology for producing an answer" (Sterman, 1988, p. 165). Similarly, Richardson & Pugh (1981) judge that, "A model's utility and effectiveness depend ... on the degree to which the model communicates, helps to generate insights, enhances understanding, and in general reaches and influences its audience" (pp. 312–313). This notion of experiential learning, imparted by the process of model building and running, links the field of SD to the "transitional object" concept (Morecroft, 1988). The style of presentation chosen in this project account is intended to convey the slow process of intuition creation that occurred in the case of the "client" group in Cathay Region and the "modelling" group from Shell in London.

Social negotiation. The majority of modelling processes involve some element of discussion and negotiation between the participants. Beyond the technical analysis is an interaction of human characters, a process by which a group negotiates a joint understanding of an issue and commits to a course of action in relation to it. The SD field recognizes these elements: Schein's ideas of facilitation and process consultation are an acknowledged contributor to projects (Morecroft, 1988; Lane, 1992).

In this specific case, the negotiation element is very clear. At the start there were two conflicting camps, with diametrically opposed views about the way that the market would respond. The consultants used SD tools in a process consulting mode to articulate the mental models of the two "camps". The consultants were thus able to move between the two groups and provide a common language that enabled those managers to share their views and to debate their differences. A more open dialogue resulted in the resolution of the conflict as one of the camps accepted that its view was incorrect. The two sides were thus able to negotiate an agreed understanding of the detailed structure and macro-behaviour of the system in question. The negotiation process used here has some linkages with "strategic assumption surfacing and testing" (SAST; Mason & Mitroff, 1981) in that the two groups were separate during the study. However, rather than being allowed separately to develop their respective ideas as SAST advocates, the groups interacted via the consultants.

To put this process element of SD into context, we may observe that an emphasis on the social negotiation aspects of operational research practice and theory has been one of the crucial elements of developments in that field over the past two decades. The "problem structuring techniques" of "soft" OR all share a belief in the importance of handling explicitly and well these parts of any project. For example, the "Strategic Choice" approach of Friend & Hickling (1987) consists of a stream of activities for managing the interaction between the technical and the social elements of any work involving team discussions. Similarly, a distinction between the "content" (our technical elements) and the "process" parts of an intervention lies at the heart of SODA (Eden, 1989). Finally, Checkland (1985) describes how the goal for analysis should not be solution by analysis but the aligning of the participants behind a shared understanding derived from the learning that has taken place. However, the field of SD has engaged only weakly with developments in OR so that comparisons and contrasts are only beginning to emerge. Morecroft's (1988) comparison of SD with OR fails, apparently, to recognize (and, hence, engage with) the profoundly different social theories underlying the methods that he studied. In a more detailed comparison, Lane (1994b) points out that while "soft" OR is located within an interpretive sociological paradigm, SD remains predominantly within a functionalist one (see Burrell & Morgan, 1979). Hence, while there are features in common, comparisons remain difficult and constructive engagement highly problematic until a clarification of the deeper assumptions of SD is available.

However, Senge advocates the use of SD tools to support "dialogue", a free sharing of meaning between people (Senge, 1989, 1990, after Bohm). He contrasts this term with "discussion" in which conflicting ideas are exchanged but not understood—the Latin root of discuss is *discutere*, to agitate. In this way, individual understanding can be expressed and debated. Both the knowledge about a situation and the meaning of that knowledge can then be created among participants so that they "align" behind a course of action, and Senge believes that organizational "visions" can be created and shared by such approaches and that continued organizational learning can result.

It is this last view that perhaps offers the most exciting grounds on which one might advocate the type of modelling described here. The concept of the learning organization, or simply of improving individual decision making processes, very quickly runs aground unless some tools are provided. These may be tools to represent knowledge, tools that allow experimentation and exploration or tools that allow knowledge to be conveyed to others. In the case described here, we see how a clear articulation of mental models allowed the participants to interrogate their own mental models, to learn from the knowledge of others and to manufacture more meaning from the accumulation of knowledge that had been created. If groups of individuals can at least express themselves in a clear language—perhaps SD, perhaps another—then they can take the first steps from discussion to dialogue. Then methods, such as interactive SD model building, which offer technical, information processing and social process advantages, may enable the ideas of individuals to contribute to the well-being of a group, an organization or a society (Senge & Sterman, 1992). It is this distinct perspective which—it is hoped—justifies the format of this chapter. And it is this perspective that inspires and supports the form of modelling that took place in the example of SD modelling described here.

NOTE

(1) In this chapter, for convenience, the collective expression "Shell" is used when referring to the companies of the Royal Dutch/Shell Group in general, or when no purpose is served by identifying a particular Shell company or companies.

REFERENCES

Aizhu, L. and Zheng, H. (1992) An oil field planning support system based on system dynamics, in J.A.M. Vennix, J. Faber, W.J. Scheper and C.A.Th. Takkenberg (eds) *System Dynamics 1992: Proceedings of the International System Dynamics Conference*, Utrecht, The Netherlands, System Dynamics Society, Boston, MA, pp. 11–20.

Beer, S. (1979) *The Heart of the Enterprise*, Wiley, Chichester.

Beer, S. (1981) *The Brain of the Firm*, Wiley, Chichester.

Bentham, J. and de Visscher, A.G. (1994) Systems thinking and its influence on operational culture, in E. Wolstenholme and C. Monaghan (eds) *Proceedings of the 1994 International System Dynamics Conference: Organisational Learning*, University of Stirling, pp. 1–12.

Burrell, G. and Morgan, G. (1979) *Sociological Paradigms and Organisational Analysis: Elements of the Sociology of Corporate Life*, Heinemann, London.

Checkland, P.B. (1981) *Systems Thinking, Systems Practice*, Wiley, Chichester.

Checkland, P.B. (1985) From optimization to learning: a development of systems thinking for the 1990s, *Journal of the Operational Research Society* **36**, 757–767.

Checkland, P.B. and Scholes, J. (1990) *Soft Systems Methodology in Action*, Wiley, Chichester.

Choucri, N. (1981) Dynamic adjustments and change in the world oil market, in D.F. Andersen and J.D.W. Morecroft (eds) *System Dynamics 1981: Proceedings of the International System Dynamics Conference*, Rensselaerville, NY, System Dynamics Society, Boston, MA, pp. 317–319.

Dash, D.P. (1994) System dynamics: changing perspectives, *Systems Practice* **7**(1), 87–98.

Davis, A. (1994) The application of system dynamics to re-engineering career plans, in E. Wolstenholme and C. Monaghan (eds) *Proceedings of the 1994 International System Dynamics Conference: Business Decision-Making*, University of Stirling, pp. 31–39.

de Geus, A. (1988) Planning as learning, *Harvard Business Review* **66**(2), 70–74.

Eden, C. (1988) Cognitive mapping, *European Journal of Operational Research* **36**, 1–13.

Eden, C. (1989) Using cognitive mapping for strategic options development and analysis (SODA), in J.V. Rosenhead (ed.) *Rational Analysis for a Problematic World: Problem Structuring Methods for Complexity, Uncertainty and Conflict*, Wiley, Chichester, pp. 21–42.

Flood, R.L. and Jackson, M.C. (1991) *Creative Problem Solving: Total Systems Intervention*, Wiley, Chichester.

Forrester, J.W. (1961) *Industrial Dynamics*, MIT Press, Cambridge, MA.

Forrester, J.W. (1968a) Market growth as influenced by capital investment, *Industrial Management Review* (now the *Sloan Management Review*) **9**(2), 83–105.

Forrester, J.W. (1968b) *Principles of Systems*, MIT Press, Cambridge, MA.

Forrester, J.W. (1968c) Industrial dynamics—a response to Ansoff and Slevin, *Management Science* **14**(9), 601–618.
Forrester, J.W. (1971) "The" model versus a modeling "process", *System Dynamics Review* **1**(2), 133–134.
Forrester, J.W. (1975) Counterintuitive behavior of social systems, in *Collected Papers of Jay W. Forrester*, Wright-Allen Press, Cambridge, MA, pp. 211–244 (first published in 1970).
Forrester, J.W. (1987) Nonlinearity in high-order models of social systems, *European Journal of Operational Research* **30**(2), 104–109.
Forrester, J.W. (1992) Policies, decisions and information sources for modeling, *European Journal of Operational Research* **59**(1), 42–63.
Forrester, J.W. (1994) System dynamics, systems thinking, and soft OR, *System Dynamics Review* **10**(2/3), 245–256.
Friend, J. and Hickling, A. (1987) *Planning under Pressure: The Strategic Choice Approach*, Pergamon, Oxford.
Kahane, A. (1991) Global scenarios for the energy industry: challenge and response, Shell selected papers series, Shell Group Public Affairs, London.
Karsky, M. (1988) The world oil market: a system dynamics approach, in N.B. Forrester, J.B. Homer, A. Ford and D. Nemanich (eds) *Proceedings of the 1988 International Conference of the System Dynamics Society*, La Jolla, CA, System Dynamics Society, Boston, MA, pp. 188–200.
Keys, P. (1988) System dynamics: a methodological perspective, *Transactions of the Institute of Measurement and Control* **10**(4), 218–224.
Klein, J.H. (1994) Cognitive processes and operational research: a human information processing perspective, *Journal of the Operational Research Society* **45**(8), 855–866.
Lane, D.C. (1992) Modelling as learning: a consultancy methodology for enhancing learning in management teams, *European Journal of Operational Research* **59**(1), 64–84.
Lane, D.C. (1993) The road not taken: observing a process of issue selection and model conceptualization, *System Dynamics Review* **9**(3), 239–264.
Lane, D.C. (1994a) Social theory and system dynamics practice, in E. Wolstenholme and C. Monaghan (eds) *Proceedings of the 1994 International System Dynamics Conference: System Dynamics: Methodological and Technical Issues*, University of Stirling, pp. 53–66 (expanded as LSE discussion paper LSEOR.95. 20).
Lane, D.C. (1994b) With a little help from our friends: how system dynamics and "soft" OR can learn from each other, *System Dynamics Review* **10**(2/3), 101–134.
Mason, R.O. and Mitroff, I.I. (1981) *Challenging Strategic Planning Assumptions*, Wiley, New York.
Meadows, D.L. (1970) *Dynamics of Commodity Production Cycles*, Wright-Allen Press, Cambridge, MA.
Meadows, D.L. (1980) The unavoidable *a priori*, in J. Randers (ed.) *Elements of the System Dynamics Method*, MIT Press, Cambridge, MA, pp. 23–57.
Morecroft, J.D.W. (1988) System dynamics and microworlds for policymakers, *European Journal of Operational Research* **35**(3), 301–320.
Morecroft, J.D.W. and van der Heijden, K.A.J.M. (1992) Modelling the oil producers—capturing oil industry knowledge in a behavioural simulation model, *European Journal of Operational Research* **59**(1), 102–122.
Morecroft, J.D.W., Lane, D.C. and Viita, P.S. (1991) Modelling growth strategy in a biotechnology start-up firm, *System Dynamics Review* **7**(2), 93–116.
Morrisen, M. (1986) *Methods in Sociology*, Longman, London.

Papert, S. (1980) *Mindstorms: Children, Computers and Powerful Ideas*, Harvester Press, Brighton.

Pidd, M. (1992) *Computer Simulation in Management Science*, 3rd edn, Wiley, Chichester.

Rahn, R.J. (1989) Aggregate dynamics of a population of commodity cycle models, in P.M. Milling and E.O.K. Zahn (eds) *Computer-Based Management of Complex Systems*, Springer-Verlag, Berlin, pp. 277–284.

Richardson, G.P. (1991) *Feedback thought in Social Science and Systems Theory*, Unversity of Pennsylvania Press, Philadelphia, PA.

Richardson, G.P. and Pugh, A.L. (1981) *Introduction to System Dynamics Modeling with DYNAMO*, MIT Press, Cambridge, MA.

Richmond, B. (1985) STELLA: software for bringing system dynamics to the other 98%, in D.F. Andersen, N.B. Forrester and M.E. Warkentin (eds) *Proceedings of the 1985 International Conference of the System Dynamics Society*, Vol. II, System Dynamics Society, Boston, MA, pp. 706–718.

Richmond, B.M., Vescuso, P. and Peterson, S. (1987) STELLA for business (distributed by High Performance Systems, 13 Dartmouth College Highway, Lyme, NH 03768, USA).

Robey, D. and Taggart, W. (1981) Measuring managers' minds: the assessment of style in human information processing, *Academy of Management Review* **6**(3), 375–383.

Senge, P. (1989) Organizational learning: a new challenge for system dynamics, in P.M. Milling and E.O.K. Zahn (eds) *Computer-Based Management of Complex Systems*, Springer-Verlag, Berlin, pp. 229–236.

Senge, P. (1990) *The Fifth Discipline: The Art and Practice of the Learning Organization*, Doubleday/Currency, New York.

Senge, P. and Sterman, J.D. (1992) Systems thinking and organizational learning: acting locally and thinking globally in the organization of the future, *European Journal of Operational Research* **59**(1), 137–150.

Shell Group Public Affairs (1986) Refining under review: refining technology, refinery location, regional refining, Shell Briefing Service, no. 1, Shell Group Public Affairs, London.

Sterman, J. (1988) A skeptic's guide to computer models, in L. Grant (ed.) *Foresight and National Decisions*, University Press of America, Lanham, MD, pp. 133–169.

Sterman, J. (1994) Learning in and about complex systems, *System Dynamics Review* **10**(2/3), 291–330.

Uhlmann, A. (1988) Linear programming on a micro computer: an application in refinery modelling, *European Journal of Operational Research* **35**(3), 321–327.

van Hilten, D. (1986) *Refining—Today's Changing Patterns and Tomorrow's Prospects*, Shell selected papers series, Shell Group Public Affairs, London.

Wack, P. (1985) Scenarios: uncharted waters ahead, *Harvard Business Review* September/October, 72–89.

Chapter 12

The Changing Role of Simulation Models: The Case of the Pacific Northwest Electric System

Andrew Ford

PURPOSE

This chapter describes the use of system dynamics (SD) modeling to better understand the electric system in the Pacific Northwest region of the United States. The model was developed during the early 1980s to help conservation planners at the Bonneville Power Administration. The model has been further developed during the remainder of the 1980s and into the early 1990s to meet the changing needs of resource planners and strategic planners at Bonneville. It begins by describing the historical context for the model's development and the main features that made it useful. The chapter concludes with a description of the changing role of modeling during the early 1990s.

BACKGROUND

During the 1980s, electric utility companies in both the USA and Europe became interested in programs to encourage their customers to invest in

Systems Modelling for Energy Policy. Edited by D.W. Bunn and E.R. Larsen

conservation (ACEEE, 1986; Commission of the European Communities, 1984). The programs included general information such as advertising, specific information such as audits, and direct financial incentives such as zero interest loans. The programs were needed to help customers to overcome market obstacles that limited their investment in measures that would improve the efficiency of electricity use. Utility conservation programs were viewed as a better use of company funds than investment in conventional coal or nuclear power plants.

Conservation was and still is important in the Pacific Northwest. The region encompasses the states of Washington, Oregon and Idaho and part of Montana, and is famous for its great rivers and mountainous terrain, which provide its huge hydroelectric potential. Development of this potential began in 1933 with the Bonneville and Grand Coulee dams. The region now has one of the world's largest hydroelectric systems and, historically, some of the lowest electric rates. Because of the low rates, the region's homes and businesses have not made the same level of investment in conservation as in other parts of the country, and the potential conservation saving is large.

The long period of low electric rates ended in the 1970s when events combined to increase utility spending on new generating capacity. Like many utilities across the country, those in the Pacific Northwest looked to nuclear power plants to meet anticipated growth in demand. And, like utilities elsewhere, the Pacific Northwest companies were hit hard by double digit escalation in construction costs and unanticipated reductions in the rate of growth in electric load. The combination of problems led to soaring electric rates, cancellation of several partly constructed plants, major defaults on bonds, and, finally, to the passage of the Pacific Northwest Electric Planning and Conservation Act of 1980 (Bonneville Power Administration, 1981).

The Act created the Northwest Power Planning Council to take responsibility for setting broad policies for the regional development of electricity resources. The Act also created major new responsibilities for Bonneville, the arm of the US Department of Energy (DOE) with authority to market power from federal resources in the region. The Act called for Bonneville to act as a "power broker" and to take a central role in implementing the best plan for the region. Three provisions of the act are especially relevant to conservation. First, Bonneville was obligated to provide for any future need for electric power in the region. Second, Bonneville now had the authority to acquire resources to meet loads received from other utilities in the region. Third, and perhaps most important, Bonneville was required to consider conservation (along with renewable and co-generation) preferentially when determining the resources it should acquire to meet future load.

Bonneville's response to the new authority in conservation was swift and multifaceted. Within two years after the passage of the Act, utilities were operating five region-wide Bonneville conservation programs, and conservation planning was upgraded to "Office-Level" status. In 1983, Bonneville's Office of Conservation initiated a project to improve its ability to model the effects of its conservation programs and consumer incentive designs for the Pacific Northwest electric power system. A SD model was designed to build from the results of existing models and databases and to facilitate rapid analysis of many scenarios.

THE CONSERVATION POLICY ANALYSIS MODELS

Work on the new model began by adapting relevant structures from a simulation model which had proven useful in studies for a major California utility (Ford & Harris, 1984). The first step was to design a regional model in which conservation programs, system operation, capacity expansion, and electricity pricing were conducted by a single entity. The next step was to construct a subregional model which would distinguish between the loads and resources of the investor-owned utilities (IOUs), the publicly owned utilities, and the federal government. The models were known collectively as CPAM, or the Conservation Policy Analysis Models (Ford & Naill, 1985; Ford & Bull, 1989).

CPAM gave the conservation planner great flexibility in assembling a package of conservation policies. A package might include utility incentives for increased efficiency of electric space heating in older homes and performance standards for improved space heating in new homes. For other end uses, the planner might wish to consider direct utility incentives, performance standards, or even a combination of incentives and standards. The planner may specify a wide variety of incentives for different end uses or for different economic sectors. Programs aimed at the residential and commercial sectors, for example, may strive to acquire low cost conservation savings that will substitute for high cost thermal plants in the region's capacity expansion plan. Programs aimed at the aluminum industry, on the other hand, may strive to improve the industry's profitability and to prevent loss of Bonneville revenues.

Once the model user had specified the package of conservation policies to be examined, CPAM was used to project the likely savings and costs of the package over time. Generating these projections in a self-consistent fashion is the principal objective of the demand sector—the first of five sectors noted in Figure 12.1. This diagram shows the regional model along with support programs which made the model easy to use during the 1980s. For example, the "preprocessor" is a spreadsheet program used to translate Bonneville's information on the cost and savings of thousands of conservation measures into cost-efficiency curves like the one shown for residential space heating in Figure 12.2.

The demand sector determines electric load based on the cost-efficiency curves, the growth in the regional economy, the changes in the price of

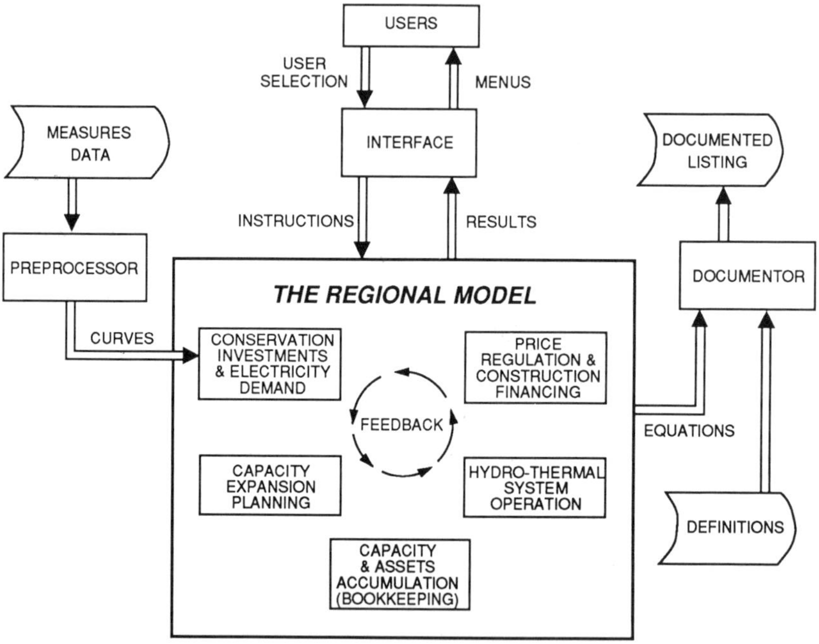

Figure 12.1 The conservation policy model of the early 1980s

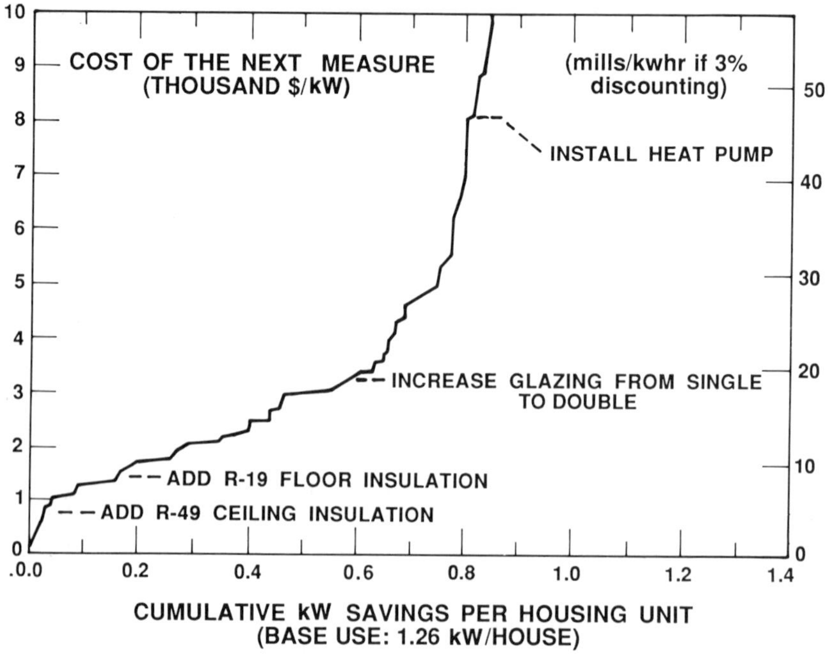

Figure 12.2 Cost-efficiency curve for electric space heating. (1 mill = 0.1 cents)

electricity, and the package of conservation policies to be tested by the model. This sector keeps track of different customer groups' purchases of conservation measures based on their discounting tendencies and the incentives made available by the government and the utility. CPAM finds the different measures that would be purchased and keeps track of how the costs of those measures are split between the customer and the utility. The "end use" style of modeling was combined with simple but plausible behavioral equations to keep track of changes in usage, saturation, and inter-fuel substitution. This combination of "end use" and "behavioral" equations forms the fundamental building block of the CPAM demand sector. The generic equations are repeated for the different end use categories (such as space heating, water heating, and electromotive drive) and for each class of customers. One set of generic equations is reserved for existing buildings to keep track of conservation retrofit investments. A second set is used to keep track of conservation investments in new buildings.

A key feature of the demand calculations is that the same model is used to project conservation investments with and without utility programs. With no programs, the model projects customer conservation investments due to price effects alone. When conservation programs are introduced, the model projects total conservation investment due to the combined effect of electricity prices and utility incentives. A comparison of the two projections shows the "net savings" attributable to the conservation programs. "Net savings" takes into account the possible "redundancy" in conservation programs, i.e. the possibility that a significant portion of the utility incentives may be expended on measures that customers would probably have adopted on their own.

The demand sector was the largest of the five sectors shown in Figure 12.1. This sector was singled out for the greatest detail to permit direct simulation of the wide range of conservation policy options of interest to Bonneville. The remaining sectors complete the representation of an integrated electric system:

- The price of electricity is calculated in the regulatory sector, which sets average retail rates based on allowed expenses, allowed return on investment, and the effects of regulatory lag. This sector also keeps track of construction financing and maintains an accounting of the region's balance sheet and cash flow. The projected rates are passed on to the demand sector to ensure internally consistent projections of customer investment in conservation throughout the simulation.
- The system operation sector obtains an electric energy balance of the system based on the loads and the resources available. Because of the great volatility in hydro generation, CPAM balances the electric system in any one of eight different modes. With "critical year" capacity expansion planning and "average year" system operation, the system is usually balanced by varying the secondary energy sold to utilities in California. Secondary

revenues from California are subtracted from the region's total operating costs to obtain net operating costs for each year of the simulation.

- The capacity and assets accumulation sector performs the bookkeeping functions. This sector keeps track of the utilities' investment in transmission lines, distribution equipment, and six types of generating capacity. Information on utility capacity is passed to the system operation sector to ensure proper dispatching. Information on utility assets is passed to the price regulation sector for proper regulatory treatment.
- The capacity expansion and planning sector prepares an internal forecast of growth in demand in order to plan the initiation of new capacity needed to keep pace with growth in load. The sector usually operates on the assumption of "critical hydro" planning, in which the investment in new power facilities is planned to avoid a deficit under adverse hydro conditions. To the extent that conservation programs reduce the electric load, this sector automatically adjusts the demand forecast downward, and the construction of new power plants is postponed until later in the simulation.

FIGURES OF MERIT

CPAM was designed to generate an internally consistent set of projections and to provide "figures of merit" or yardsticks to gauge relative success of conservation strategies. The three "figures of merit" of most interest in the Pacific Northwest during the 1980s were electric rates, utility revenues, and energy service costs.

Electric Rates

Change in electric rates is one of the most closely studied yardsticks in policy studies, both in the 1980s and especially in the 1990s. CPAM simulations tend to show adverse rate impacts during the first few years after the initiation of an incentive program. Depending on the timing of the program and the utility costs per "net kilowatt" saved, the rate penalties may be quite small (under 1% in the worst year) and may only last for three or four years. Incentives that prove quite costly per "net kilowatt" saved or which generate large savings during the surplus period, on the other hand, show larger and more persistent rate penalties. The short-term adverse rate impacts are primarily attributed to the financing costs of the incentive program and the lower sales of electricity to customers participating in the program. For almost all programs tested, the short-term rate penalties are followed by a period of rate benefits. The benefits usually appear about the time in the simulation when the conservation savings lead to postponements or reductions in power plant construction. The timing of the transition from penalties to benefits is affected by a wide variety of factors, including the cost of the incentive per "net kilowatt" saved, how rapidly customers participate in the program, whether the incentives are capitalized or expensed, the likely duration of the surplus, the rate for selling secondary energy to California, and the cost and lead time of the generating resources being displaced.

Utility Revenues

Total utility revenues is another widely used "figure of merit", and CPAM prepares a discounted sum of annual utility revenues for use in evaluating conservation strategies. This yardstick includes the utilities' traditional costs as well as their costs for conservation incentive programs. In policy studies completed during the 1980s, we found only one or two instances when a utility incentive program did not reduce the utilities' total, discounted revenues.

Energy Service Costs

The energy service costs yardstick takes the consumers' expenditures into account by combining the customers' conservation spending with the utilities' revenues to obtain total system costs in each year of the simulation. The present value of these annual costs is the most comprehensive "figure of merit" for evaluating conservation policies because it combines the customers' electric bill payments with their own investments in conservation measures. Consequently, many companies and agencies adopted energy service costs as the key yardstick for electric resource planning during the 1980s (Sant, Bakke & Naill, 1984; Wellinghoff & Mitchell, 1985). Their view was that planners should strive to minimize the cost of electric energy services, not just the cost of electricity.

In a typical policy evaluation during the 1980s, the energy service costs yardstick was used to judge the overall attractiveness of the conservation strategy, and electric rate impacts were used to gauge the impact on those customers who do not participate in the programs. A typical evaluation also included an examination of key financial and budget variables. These evaluations were conducted for both the region as a whole and for the key utilities within the region.

Utilities within the Northwest System

CPAM distinguishes between the loads, resources, costs, rates and financial situations of the major utilities in the region. The model simulates the operations of an individual IOU, as well as an aggregate group of IOUs that make up the rest of the private portion of the Northwest electric system. The model also simulates the operations of an individual public utility. The focal utility might be a small utility or it may be a collection of public utilities in a particular region (e.g. the Puget Sound). The remaining publicly owned utilities are simulated as an aggregate utility.

The fifth utility in the model is Bonneville, which operates as the federal wholesale supplier in the region. Bonneville sells most of its energy to the public utilities, but it interacts with the private utilities as well. Bonneville also sells a large amount of power to aluminum companies and other direct service industries (DSIs). The DSIs' electric demand is simulated in great detail because of their importance to the region. For example, the model simulates the profitability of smelting operations at each of the ten aluminum smelters in the region. The model incorporates Bonneville's variable rate structure, which allows the price of electricity charged to aluminum companies to vary with changes in the world price of aluminum.

KEY FEATURES FOR POLICY STUDIES

The team members identified three key features which distinguished the SD modeling from alternative approaches. CPAM was valued for its integrated representation of the entire region's electric system, its ability to avoid double counting of conservation, and its automatic representation of information feedback. These features were considered crucial because they permitted Bonneville staff to conduct internally consistent conservation studies with rapid turnaround.

Integrated Representation of the Electric System

The five sectors shown in Figure 12.1 work together to provide an integrated representation of the region's electric system. Information generated in each of the five sectors is available during each time step of the simulation as needed in the remaining sectors. The utility cost from conservation incentive programs, for example, is made available to the price regulation sector, where the expenditures are either capitalized or expensed in the rate-making calculations. Electric rates, in turn, are recalculated after the appropriate regulatory lags, and the rates for the next time step are used in the demand sector.

The CPAM approach to model integration might be contrasted with a more common utility approach, in which several different models (usually constructed in different departments within the same company) are designed to operate together, as shown in Figure 12.3. In this illustration, one begins with a set of electric rates needed as input for an electricity demand model. The output of the demand model takes the form of electric load projected for each of 20 years in the future, and the load projections are used as input for a capacity expansion model. The output of the capacity planning model is a plan for new power plant

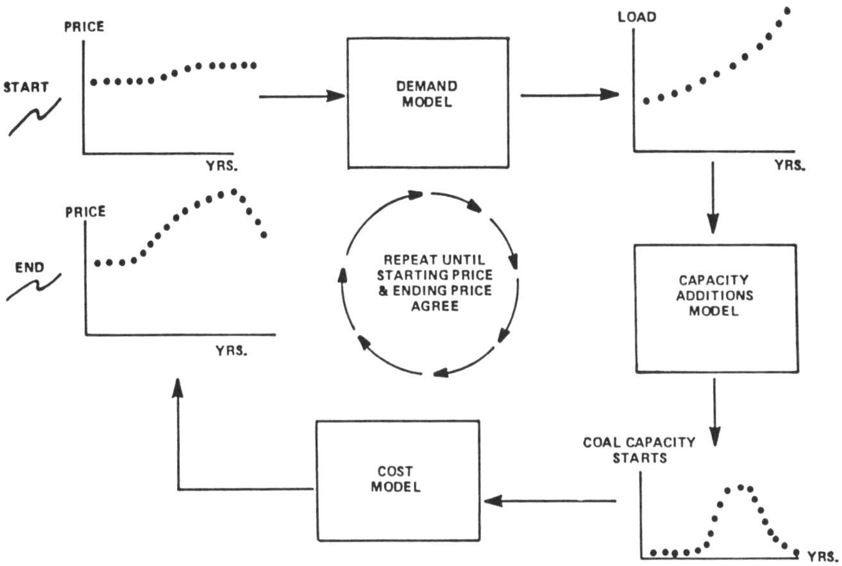

Figure 12.3 The iterative approach often used in utility company analysis of resource strategies

construction during the 20 years, and this plan is used to drive a costing model, which generates a set of electric rates needed to provide adequate revenues. The electric rates emerging from this sequence of model projections are compared with the electric rates used to start the calculations. If the two sets of rates are significantly different, the starting rates are adjusted, and the sequence is repeated. Through artful manipulation of the starting rates, one hopes to obtain a consistent set of projections within a reasonable number of iterations.

With the iterative approach shown in Figure 12.3, the output from an early model is not provided to subsequent models until the early model is finished, with a full 20 years' worth of results. In CPAM, output from the five sectors in Figure 12.1 are available to other sectors as the model proceeds from one time step to another during the course of a 20-year simulation. The main advantage of the iterative approach is the degree of detail that may be permitted in each individual model in the sequence. Separate models may be coded in different computer languages to allow analysts from different departments to find the best fit with their topic area or to take advantage of existing models. With the CPAM approach, all five sectors were designed from the outset to work together automatically over time. Each approach has compelling advantages, and Bonneville used both approaches in the analysis of conservation policies. The idea was to examine a broad range of conservation policies with CPAM before initiating the iterative process with the more detailed models.

Avoid Double Counting of Conservation

A frequent problem with utility analyses of conservation policies is that one approach is used to project electricity demand, and a separate approach is used to estimate conservation savings. The demand forecasting department, for example, might be responsible for demand projections, and this department might employ an econometric approach. The conservation planners, on the other hand, might be located in the customer service department, and they might choose to estimate programmatic savings based on simple engineering calculations (e.g. savings per water heater multiplied by the number of water heaters in a program). Although both approaches make sense given the goals of the separate departments, the company may double-count conservation savings when the separate estimates are combined in a corporate analysis. (Conservation savings from water heater wraps, for example, would be counted in the engineering estimate of programmatic savings even though they might be implicit in the forecasting department's econometric projection of electricity demand without the program.) When using CPAM, Bonneville planners avoid the double counting of conservation investments because the same model is used to represent customer investment in conservation measures with and without an incentive program, and because the model provides an explicit representation of the conservation measures available (see Figure 12.2). Indeed, CPAM projections allow the conservation planner to examine the degree of redundancy associated with different incentive programs and to simulate the effect of alternative incentive designs, which might minimize the amount of utility spending on measures that customers would buy on their own.

Information Feedback

The CPAM models incorporate information feedback directly by allowing information in each of the five sectors in Figure 12.1 to be used in any of the other sectors throughout the simulation period. The flow of information from one sector to another leads automatically to the closure of feedback loops which govern the behavior of the electric system. A well-known example is the so-called "spiral of impossibility" in which higher electric rates lead to lower electric sales; lower sales force the utility to request further rate increases to cover fixed costs; and the new rate increases lead to still further reductions in sales (Ford & Youngblood, 1983). Price feedback is the most common example cited when utility analysts discuss the importance of "closing the loop" in utility modeling of the 1980s (Electric Power Research Institute, 1981; Ford & Mann, 1983). The CPAM representation of price feedback is automatic and tangible. Automatic feedback makes it possible to obtain internally consistent projections without the iterative calculations shown in

Figure 12.3. Tangibility allows the analyst to interpret customer response to simulated price changes in substantive terms, such as increased investment in the measures shown in Figure 12.2.

The simulation of price feedback allows the planner to distinguish between the direct and indirect effects of conservation policies. The direct effects are the reductions in electric load in the customer groups receiving the incentive; the indirect effects are load changes arising from the simulated changes in the price of electricity. In a simulation of incentives for retrofit investments in the residential, commercial, and industrial sectors, for example, the model showed direct savings of around 1000 MW by the end of the 20-year simulation. The conservation savings led to slower growth in regional load, greater sales of secondary power to California, and a postponement of construction of new power plants. The simulated effect of these changes was a slight increase in electric rates for several years followed by a larger, permanent reduction in rates. The lower rates, in turn, led to increased profitability of aluminum smelting and roughly 300 MW of added DSI load by the end of the simulation. Thus, this example shows indirect effects which cause total savings to be 30% smaller than one would expect from the direct effects alone.

Rapid Turnaround

To illustrate the rapid turnaround permitted by the CPAM models, consider how the assessment of a retrofit incentive program might change with a change in the assumed rate of growth of the region's economy. The analysis requires four simulations: two simulations to show the impacts of the incentive program with base case assumptions and two to show the impacts with more rapid growth in the region's economy. Faster economic growth is represented by increasing the rate of housing starts, commercial floor space additions, new acreage under irrigation and new industrial activity. These simulations were performed with a minimum of staff time devoted to repreparation of model inputs—the staff simply increased the rate of housing and business starts, and the model automatically simulated the resulting changes in system operation, capacity expansion, rate making, feedback effects, and the "figures of merit". By allowing planners to conduct these and other sensitivity tests with quick turnaround, CPAM made it possible to examine many different conservation policies under widely varying assumptions about the region.

ILLUSTRATIVE USE IN THE 1980s

CPAM was used under a variety of circumstances, and reports on previous applications are available from Bonneville (Ford & Naill, 1985) and in the modeling literature (Ford, Bull & Naill, 1987; Ford & Bull, 1989; Ford &

Geinzer, 1990; Ford, 1990). I summarize one particular application, which is noteworthy because of the large number of participants in the planning process.

The process began in the spring of 1985 when the Office of Conservation released a "scoping document" to the public (Bonneville Power Administration, 1985). This document posed questions about the appropriate conservation strategy, given a forecast of extended power surpluses, uncertainty about disposition of the region's uncompleted nuclear power plants, and major concerns about the stability of Bonneville's revenues from the region's aluminum smelters. The document explained that Bonneville was asking basic questions about the justification for conservation programs under such circumstances and invited interested parties to contribute to the re-examination. The document outlined possible conservation programs, with particular attention to the size and timing of financial incentives to encourage customer investment in conservation. And, finally, the "scoping document" explained how the Office of Conservation planned to conduct its analysis of alternative strategies. CPAM was presented as the cornerstone of the proposed analysis and as a vehicle for eliciting public comments on Bonneville's planning process.

The structuring of alternative conservation strategies started with two extremes: a "blitz" strategy designed for maximum conservation and a "no programs" strategy, in which existing programs were canceled and no further programs were initiated. Middle-ground alternatives were designed around the continuation of existing programs (primarily residential weatherization) and the initiation of new programs for commercial and industrial customers. A principal component of all programs was an incentive in which the utility would pay a fixed percentage (generally 50–90%) of all conservation measures whose total cost was thought to fall below the levelized cost of a new coal-fired power plant. A variety of strategies were defined by varying the size of the utility incentive and by varying the particular customer classes or end uses to receive the incentives. CPAM was then used to simulate the impacts on electric rates, utility revenues, and energy service costs of each strategy.

The analysis revealed that the commercial and industrial sectors had significant potential for conservation savings. However, Bonneville's information for non-residential conservation was less reliable, and public comments confirmed the popularity of programs targeted at residential customers. Consequently, Bonneville defined several strategies which preserved current residential programs and varied the level of effort in the commercial and industrial sectors. Analysis of a wide variety of combinations led conservation planners to concentrate on five "middle-ground" strategies (Barton & Bull, 1986).

As a general rule, CPAM simulations showed that the more aggressive strategies scored best in terms of reducing the total cost of electric energy services. At the opposite extreme, the "no programs" case scored poorly in all respects (energy service costs, utility revenue requirements, electric rates). The "no programs" strategy scored badly because conservation savings are, in general, the cheapest resource that the region's utilities can acquire to meet long-term load growth, and some resource acquisition was needed over the longer term. When exploring the "middle-ground" strategies, Bonneville staff found that strategies with low incentive levels scored best in terms of electric rate impacts, whereas strategies with high incentive levels did better in reducing energy service costs.

Much of Bonneville's effort in the planning process was devoted to clarifying the required trade-offs between the worthy goals of reducing energy service costs and maintaining low electric rates. Bonneville planners reviewed the model projections and advocated the "medium" and "medium-high" strategies as achieving the best balance between the two goals. These strategies called for a mix of incentives (50–75% of measure costs) to be applied across end uses and economic sectors.

The final, and perhaps the most controversial, question concerned the wisdom of delaying startup of new programs given the projections of an extended period of surplus generating capacity. CPAM simulations were used to compare the impacts of different start dates under a wide variety of assumptions about the region's loads and resources. This analysis showed that delaying aggressive programs until near the end of the surplus period scored only marginally better than strategies in which a steady development of conservation programs was maintained throughout the surplus period. Bonneville planners judged that the marginal benefits of the delay/restart strategy did not justify the administrative difficulties associated with shutting down existing programs only to restart them in the early 1990s. Consequently, Bonneville's 1985 analysis proceeded with conservation strategies that maintained steady programs despite the prospects for an extended surplus.

THE CHANGING ROLE

The Bonneville models have changed over time as the interests and responsibilities of the Bonneville staff have changed. For example, many of the planners from the original Office of Conservation moved into responsibilities of resource planning. They saw a continuing need for simulation modeling which would treat both the demand and supply sides of the electric system, with attention to the individual roles of the public and private utilities. They also saw a continuing need for a model that could close the information feedback loops that tie the pieces of the Northwest electric system together. Accordingly, they commissioned research to undertake major improvements in the model. The most substantial improvement was the addition of a major new sector to account for investment in power generation by non-utility generators.

The new model was called the Non-Utility Generation Model (NUGM; Geinzer, Wood & Ford, 1990).

NUGM simulates the non-utility generating markets in four broad categories:

(1) self-generation
(2) internal co-generation
(3) PURPA co-generation
(4) small power production

Self-generation is comprised of industrial facilities with little or no steam load that generate electricity for their own use (motivated by the retail, industrial price of electricity). Internal co-generators produce both steam and electricity to satisfy internal uses, and their investment is also motivated primarily by the industrial price of electricity. The remaining categories respond to avoided costs. The PURPA co-generators are designed to sell electricity to utilities under the Public Utility Regulatory Policy Act of 1978 regulations. These co-generators typically operate with a much larger electric-to-steam ratio than the internal co-generators. Small power producers use renewable or waste resources to generate electricity under PURPA rules at sizes under 80 MW. NUGM simulates the investments by non-utility generators as a market response to signals that may be issued by any of the utilities in the region. Each utility's signals, in turn, may be represented as a "posting" of avoided costs or by a bidding program which calls for a target MW of generation.

NUGM may be operated in either stand-alone fashion or in concert with the rest of the CPAM models. The combination of CPAM and NUGM gave Bonneville planners broader capability for the analysis of resource issues, so the combined models became known as the Resource Policy Screening Model (RPSM). The "screening" idea was to use RPSM in quick turnaround studies of a wide range of resources issues before turning to Bonneville's more detailed resource models for analysis of a limited set of options. The most ambitious of the screening studies was conducted to support a scenario planning exercise in 1991 (Geinzer & Cranna, 1992). More recent studies with the RPSM have looked at the pivotal role of the aluminum companies (Ford, 1992) and the changes in Bonneville's competitive position (Westman & Neubauer, 1993).

In my opinion, the most important change in recent years has involved the way in which the models are used by Bonneville employees. In the 1980s, the models were typically operated by a small group of employees working alone or with the help of outside consultants. Through their long experience with the model, the small group was well versed in the fundamentals of SD, in the role of information feedback, and in the syntax of the Professional Dynamo software (Pugh Roberts Associates, 1991). They were also comfortable with the fact that various policies could lead to adverse impacts in the short run before delivering benefits to the utility or to its customers in the long run. With this base of knowledge, they were able to perform a variety of screening studies and present their findings to a wider group inside Bonneville. They received generally complimentary feedback on the presentations. But they were left with the troublesome feeling that the key insights from the simulations were not fully appreciated by the wider circle of employees.

To improve the patterns of communication inside Bonneville, the modeling team reviewed the results of research initiatives in learning organizations (Senge, 1990) and the role of models within learning organizations (Morecroft & Sterman, 1994). They were impressed by the People Express "flight simulator" which combines a SD model of an airline company with a Microworlds cockpit to allow students to "take the controls" of the company (Sterman, 1992). They were also impressed by the Microworlds software (Diehl, 1994), both as a means to allow a wider group of Bonneville employees to "take the controls" of a model and to permit the small group of analysts to "look inside the model" via Microworlds' diagnostic lens (Microworlds, 1992). Accordingly, they commissioned the development of the Microworlds gaming version of the model.

The "gaming version" may be used in the traditional manner by a small group of analysts. But the real power of the new approach comes when a wider group of Bonneville analysts and managers is turned loose to operate the model in group sessions. Each participant is free to explore a variety of policies under different assumptions about the future of the system at his or her own pace. With this "learner-directed" approach, the role of the small circle of modeling experts is changed dramatically. They no longer serve as the experts who must teach the wider team about the model insights. Rather, the small group now serves the role of coach. Their new role is to allow members of the wider group to explore the results of a variety of policies and arrive at their own conclusions. In this setting, the key insights emerge as each individual conducts explorations with the model. The insights are then confirmed and strengthened in group discussions. These discussions may roam freely, and considerable time can be devoted to policies that do not necessarily perform well. Some participants may misinterpret simulation results, but their interpretations are often improved through discussion with other participants (from the wider circle) who have had made more progress in their explorations.

Our experience with the "gaming version" of the model is still growing, so we have not completed reports or articles from the group explorations. (An article in the *System Dynamics Review* (Ford, 1995) conveys a general idea of the style of modeling.) But we are encouraged by the initial success in engaging a wider group of Bonneville staff to allow computer simulation to expand the way they view the Northwest electric system. The project team is investing further resources in the "gaming version" of the model. The focus of a recent effort is Bonneville's strategy to compete successfully in an increasing competitive and dynamic electric utility market (Bonneville Power Administration, 1994).

ACKNOWLEDGMENTS

This chapter describes a modeling activity that has spanned over ten years and which has benefited from contributions by many individuals and several organizations. The activity has been supported by research funds from the Bonneville Power Administration and by organizational commitments from the AES Corporation, the University of Southern California and Washington State University. Over a dozen individuals from these organizations have made important contributions to the project. I especially wish to thank Mike Bull and Erik Westman, the project managers, who guided the development and the use of the model at Bonneville. I also wish to thank my two colleagues, who have made major contributions to the development and use of the models—Pat Barton of Portland Energy Strategies and Jay Geinzer of the AES Corporation.

REFERENCES

ACEEE (1986) Proceedings from the ACEEE 1986 Summer Study on Energy Efficiency in Buildings, American Council for an Energy-Efficient Economy, 1001 Connecticut Avenue, NW, Washington, DC 20036, August.

Barton, P. and Bull, M. (1986) Planning conservation programs: decision support with the Conservation Policy Analysis Model, in *Proceedings of the 1986 International Conference of the System Dynamics Society*, Seville, Spain, October.

Bonneville Power Administration (1981) Pacific Northwest Electric Power Planning and Conservation Act with Index, report, DOE/BP-67, Portland, OR, August.

Bonneville Power Administration (1986) Scoping document for the 1986 Long Range Conservation Projection, report DOE/BP-04, Portland, OR.

Bonneville Power Administration (1994) Business plan: draft environmental impact statement, report DOE/EIS-0183, Portland, OR, June.

Commission of the European Communities (1984) Comparison of energy savings programmes of European Community Member States, Brussels, 2 February.

Diehl, E. (1994) Managerial Microworlds as learning support tools, in J. Morecroft and J. Sterman (eds) *Modeling for Learning Organizations*, Productivity Press, Portland, OR.

Electric Power Research Institute (1981) Case study comparison of utility corporate models: utility modeling forum second working group, EPRI report EA-65, October.

Ford, A. (1990) Estimating the impact of efficiency standards on the uncertainty of the Northwest electric system, *Operations Research* July/August.

Ford, A. (1992) Simulating the value of a notice requirement for the closure of aluminum smelters, informal report to Bonneville, June.

Ford, A. (1995) Simulating the controllability of feebates, *System Dynamics Review* **11**(1), 3–29.

Ford, A. and Bull, M. (1989) Using system dynamics for conservation policy analysis in the Pacific Northwest, *System Dynamics Review* **15**(1), 1–16.

Ford, A. and Geinzer, J. (1990) Adding uncertainty to least cost planning, *Energy Policy* **18**(4), 331–339.

Ford, A. and Harris, S. (1984) A simpler method for calculating the cost of conservation subsidies for an electric utility, *Energy Policy* September, 329–341.

Ford, A. and Mann, G. (1983) Summary of the workshop on regulatory-financial

models of the US electric utility industry, report LA-9815-C, Los Alamos National Laboratory, Los Alamos, New Mexico, June.
Ford, A. and Naill, R.F. (1985) Conservation policy in the Pacific Northwest, BPA technical report, DOE/BP-12761-1, Washington, DC, May.
Ford, A. and Youngblood, A. (1983) Simulating the spiral of impossibility in the electric utility industry, *Energy Policy* March, 19–38.
Ford, A., Bull, M. and Naill, R.F. (1987) Bonneville's conservation policy analysis models, *Energy Policy* **15**(2), 109–124.
Geinzer, J., Cranna, M. and Ford, A. (1992) 1991 scenario analysis for the Pacific Northwest electric system, final report to Bonneville, January.
Geinzer, J., Wood, F. and Ford, A. (1990) Dynamic analysis of Pacific Northwest non-utility power generation, final report to Bonneville, September.
Microworlds (1992) S**4: the strategy support simulation system: user's guide and tutorial.
Morecroft, J. and Sterman, J. (eds) (1994) *Modeling for Learning Organizations*, Productivity Press, Portland, OR.
Pugh-Roberts Associates (1991) Professional Dynamo Plus, introductory guide and tutorial, Pugh-Roberts Associates, Cambridge, MA.
Sant, R.W., Bakke, D.W. and Naill, R.F. (1984) *Creating Abundance: America's Least-Cost Energy Strategy*, McGraw-Hill, New York.
Senge, P.M. (1990) *The Fifth Discipline: The Art and Practice of the Learning Organization*, Doubleday/Currency, New York.
Sterman, J. (1992) Teaching takes off: flight simulators for management education, *OR/MS Today* October.
Wellinghoff, J.B. and Mitchell, C.K. (1985) A model for statewide integrated utility resource planning, *Public Utility Fortnightly* 8 August.
Westman, E. and Neubauer, F. (1993) Resource policy screening: BPA competitiveness and non-utility generation, internal Bonneville summary report.

Chapter 13

A Systems Simulation Platform to Support Energy Policy in Colombia

Isaac Dyner and Derek W. Bunn

INTRODUCTION

Energy research in the utilities sector has been dominated in recent years by two different themes. On the one hand, the difficulties in financing new capacity, together with environmental concerns, have motivated both demand-side management (DSM) and integrated energy analysis. This has led to energy resource "planning" based upon hard modelling approaches mainly inspired from engineering and economics disciplines. On the other hand, the more recent worldwide trend towards industry liberalization and privatization moves away from central planning to market-based resource allocation, and its analysis requires the creation and utilization of strategic tools, with much softer specifications, often under the systems thinking tradition.

At one extreme, the US took the lead in the 1980s by approaching the problem in terms of DSM. Special legislation was launched in the early 1990s to attain better results through integrated resource planning (IRP). In fact, the US Energy Policy Act of 1992 establishes that all utilities are required to implement IRP, which implies that a broad set of alternatives have to be considered, including rational energy use, rather than just committing a certain amount of resources for building new generation capacity. The process of

Systems Modelling for Energy Policy. Edited by D.W. Bunn and E.R. Larsen

planning in this context can be seen as the management of links between separate, conventional energy planning models that considers both demand- and supply-side options (e.g. the National Energy Modeling System or NEMS). These, by and large, use nonlinear general equilibrium economic approaches, which intend to provide a picture of long-term trends in energy development. Although myopic, as equilibrium calculations are made on a term-by-term basis, they do not address short-term effects or transitional policy issues. Furthermore, with restructuring and liberalization beginning to gain momentum in the USA, the reconciliation of open competition with existing DSM mandates poses a major problem.

From the other extreme, the alternative perspective was pursued in the UK by way of liberalizing the power supply industry. In this context, it is not so much the liberalization of its energy sector which prompted change of ownership, as the overnight deconstruction of a vertically integrated electricity utility into groups of separate companies competing for generation and distribution. The ability of the newly engineered spot market to deliver fair prices and efficient investment signals, together with a regulatory regime seeking to stimulate competition and control prices, and the absence of demand-side participation, has posed many open questions.

The difficulty in seeking to answer these questions comes from the novelty of the markets being created. Without direct experience or close analogies elsewhere, empirical modelling and economic theory are relatively limited in their practical implications. Because of this, and in the UK case, with the imperfect nature of the competition resulting from too few competitors and high barriers to entry, we have found strategic business simulation modelling to be useful in providing some broad insights into the dynamics of investment, pricing and regulation (Bunn & Larsen, 1992; Bunn, Larsen & Vlahos, 1993). The focus in such an approach is to model some aspect of the system as it is designed to operate over time (e.g. the price setting mechanism), the various important companies in terms of their possible objectives and strategic options, and various regulatory responses, and, through extensive sensitivity analysis, gain some insight into the intrinsic volatility and controllability of the focal variables. Such models are selective in their scope, rather than being comprehensive in detailing the whole sector, yet in terms of focusing upon the behaviour of the strategic players in the system, they can provide an effective way of understanding imperfect competition and regulation in an evolving market. Bunn, Dyner & Larsen (1995, 1996) continue this style of analysis by developing an application to some aspects of the interaction of the gas and electricity markets. These are regulated quite separately in the UK, and our analysis revealed some critical question marks on the viability of continuing without an integrated energy policy and without the joint regulation of these two sectors.

Now, some countries, such as Colombia, are pursuing an intermediate path that makes a synthesis of both approaches. In this case, liberalization and planning have to work hand in hand. Their policy is seeking (i) to promote a gradual evolution of free markets in all activities, when possible, under a regulated framework to protect end users from market "failures", and (ii) to incorporate into planning the major environmental issues involved, appropriate generation capacity requirements to satisfy demand, and essential resources to meet subsidies for low income socioeconomic groups.

However, it is an analytical (as well as philosophical) challenge to manage the system towards these objectives. Hence, the need to develop a modelling framework capable of addressing these diverse issues is crucial. The purpose of this chapter is to demonstrate the feasibility of an approach based upon a broad interpretation and extension of system dynamics (SD).

THE SD APPROACH TO ENERGY ANALYSIS

The history of SD models in energy analysis and planning can be traced back to the early 1970s, when research conducted at MIT was primarily concerned with world dynamics, including factors such as economic and population growth, depletion of resources and pollution. Separate studies soon followed to examine the behaviour of energy sources. This research was the basis for an SD model (COAL2) created under contract to the US government (Naill, 1977). Improvements to the model have led to one of the latest version, FOSSIL2, being used as an important tool to support all US energy plans since 1977 and the National Energy Strategy 1991. During the past fifteen years Ford has applied SD to a number of issues in the electricity sector of the US. Perhaps his most important research is related to policy evaluation in the electricity industry (Ford, 1983); investments and uncertainty (Ford, 1985) and conservation policy analysis (Ford and Bull, 1989). Also in the US, Geraghty and Lyneis (1985) have studied the effect of external agents on utility performance.

Outside the US, Moxness (1990) presents an interesting model on interfuel substitution in OECD–European electricity production. The paper focuses on the fossil fuels oil, gas and coal, in a way which overcomes the inappropriate fuel substitution representation yielded by the constant elasticity demand models. In the UK, SD is being used to explore a diversity of issues related to the privatization of the electricity industry (Bunn & Larsen, 1992; Bunn, Larsen & Vlahos, 1993). Problems related to the reserve margin, market share and plant retirements are examined in these papers. There is still scope for further research in this new area of privatization and regulation.

An application of SD to the electricity sector in Argentina was pursued from the beginning of the 1980s (Rego, 1989), and in Colombia this tool has been used to study energy efficiency penetration and electricity substitution by gas in the residential and industrial sectors (Dyner, Smith & Peña, 1995). However, for the most diverse issues associated with a "planning and liberalization" policy, although an SD based approach has the potential to provide methodological support, it needs a broader functional perspective on its use.

AN OUTLOOK INTO THE COLOMBIAN ENERGY SYSTEM

Briefly, a number of issues arise when making an overall assessment of the Colombian energy sector. On the supply side we can establish that:

- The country possesses abundant energy resources
- The development of resources has had an uneven evolution over the years
- The supply of sources is unbalanced within and across sectors
- Wood continues to be an important source, especially in the rural areas
- Electricity has been very much a monopoly in the residential sector
- The large-scale gas plan is only going through its first stages
- Coal supply to the national market is insignificant
- Energy exports (oil and coal) will continue growing at an important rate during the near future

In particular, some of the problematic features that have an important effect on the power generation sector are:

- Power generation depends heavily on hydroelectricity
- Uncertainty in the hydrology behaviour plays a very important role
- The country is at the verge of another period of electricity rationing
- Thermoelectricity is heavily inefficient
- Planning for new capacity had shown deficiencies (blackouts)
- Heavy non-technical losses are endemic in the system
- The expansion of the transmission network still requires major investments

The energy sector has been driven historically by supply-side management criteria. In practice, this has meant that the central government has played an omniscient role in the specification of required resources. Therefore, interesting characteristics in the Colombian demand-side sector include:

- Energy markets are very underdeveloped
- The domestic sector has not had sufficient energy alternatives
- The variety of fuels supplied to industry has also been limited
- Cross-subsidies exist in the domestic sector for electricity, according to income
- No rational energy use programmes have been implemented yet

The difficulties manifested in the Colombian energy sector are immense and require major intervention from different perspectives. Although economic growth has been an encouraging factor, energy policy has shown weaknesses to support the overall developing climate, from the supply perspective, with respect to alternatives, expansion, security and efficiency. Thus, for example, in most Colombian Departments natural gas is still not available (despite its abundance); electricity is the only alternative in some Departments; power generation is largely hydroelectric, with a small portion of very inefficient thermoelectricity plants; electric non-technical losses exceed 12%; and electricity subsidies are applied extensively.

Against this background, Colombia is adopting a British-style liberalized set-up. The electricity pool has already started operations (in July 1995). Regulators are moving household prices towards those indicated by markets. Independent private investors are building new thermoelectric plants, which could amount to about 10% of the total power generation capacity by early 1997. Also, some government assets in the electricity sector will be sold within the next two years.

In the future, coordination between public policy and private initiatives should have a much more positive widespread effect on the whole system, as a consequence of recent legislation developments, the new governance structure set-up, and initial indications of private investment plans. As a result of these, competition, planning and regulation should work together to correct some major system malformations. For this, the country would probably have to both develop its own strategies and learn from experiences taking place in other developed and developing economies around the world.

ENERGY POLICY ISSUES IN COLOMBIA

Despite the recent Public Services and Electricity laws, the transition towards a more liberalized environment turns out to be a delicate affair, as low prices, low reserve margin and high subsidies appear not to be the most favourable initial conditions. Thus, a number of open questions emerge concerning:

- Evolution of prices
- Subsidies regime
- Supply reliability
- Supply alternatives
- Demand-side management
- Competition in the market of large consumers
- Competition in the retail market

In this sense, there is not an obvious recipe to follow, as a number of crucial interlinked problems arise: the selection of programmes, scheduling and prioritization of projects (seeking possible synergies), incentives for private investors and market regulation, among many others. Therefore, a framework for analysis capable of handling breadth and depth will be required: breadth in terms of being able to address a number of connected policy issues; depth in terms of supporting analysis at a detailed level, as in the case of specific DSM programmes.

Moreover, an incrementalist approach needs to be followed, to analyse the implementation of programmes and projects, as well as for the purpose of policy making. For example, in the 18-month period since this project started, the pool price mechanism was introduced and some DSM programmes were devised. Overall, the modelling requirement is one that provides a platform for simulation, scenario analysis and policy evaluation. This has to focus as much on global strategic issues as on some fairly detailed financial analyses.

In many aspects, with respect to data not yet available in Colombia, the modelling approach needs to make use of experiences from elsewhere. For example, in imposing upon Colombia the UK-style power pool, can we incorporate, as "archetypes" re-customized versions of the SD-type models that we have developed to understand the UK system (Bunn, Dyner & Larsen, 1995, 1996), and thereby investigate whether similar behavioural properties will be exhibited in Colombia? So much of the worldwide trend to restructuring has been promoted by *analogies* that we need a platform to verify if the analogies will transfer to new contexts. Similarly, can the insights from consumer choice models such as in ENPEP (Buehring et al., 1991) be assessed and assimilated into this context?

Thus, one of the primary functions of such a modelling platform is as a workbench for testing analogical reasoning, upon which so much of the advice given to the energy sector is based.

Thus, as a context testing framework, the platform construction process has to evolve through the adaptation of changing circumstances, including new components to address a diverse number of complementary and dynamic issues. Furthermore, transferability of the platform between different groups of modellers requires a modular and transparent structure. End users also need to be trained to use their "own" platform in order to develop it further as new policy issues emerge.

A SYSTEMS SIMULATION PLATFORM FOR COLOMBIA

Figure 13.1 shows the general platform structure, along with its most relevant components and the corresponding interrelationships. This design facilitates the analysis of DSM plans and programmes (the large-scale gas plan and the indicative power generation expansion plan), as will be discussed later. The platform is SD based, with about 350 vector equations (24 of which are differential equations) and about 1000 vector variables. Modelling details of

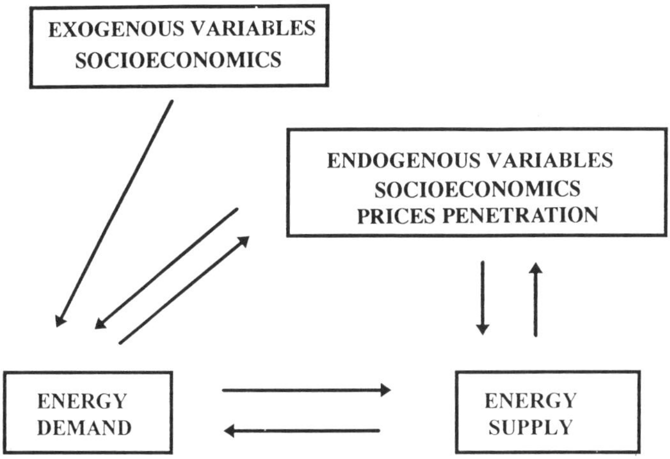

Figure 13.1 General platform structure

this approach are further described in UPME (1995). Work started in 1995 as part of an 18-month development project for the Colombian Ministry of Mines and Energy.

Energy Demand Modules

Figure 13.2 illustrates a specific demand component of the platform design. This represents the end user appliances acquisition selection criteria in the residential sector. This, by and large, corresponds to similar situations taking place in the industrial, commercial and government sectors. Decisions for the acquisition of diverse appliances are based on two fundamental reasons: (i) to fulfil a requirement, or (ii) to substitute an obsolete one.

To determine the consumer's choice between k different appliances, equipment or artefacts, the platform can allow the application of consumer choice models such as the Logit:

$$S_i = \frac{AEC_i^{-\gamma}}{\sum_j AEC_j^{-\gamma}}, \; i = 1, 2, \ldots, k$$

where S_i represents the proportion of appliance i being chosen, out of the total, and AEC_i represents the annual equivalent cost of operating this appliance during one year. The parameter γ (> 0) is a measure of consumers' preference for appliance i with respect to other alternatives. Large values of γ indicate that the decision maker is more readily inclined to choose appliance i when the corresponding AEC is smallest. This parameter needs not be estimated

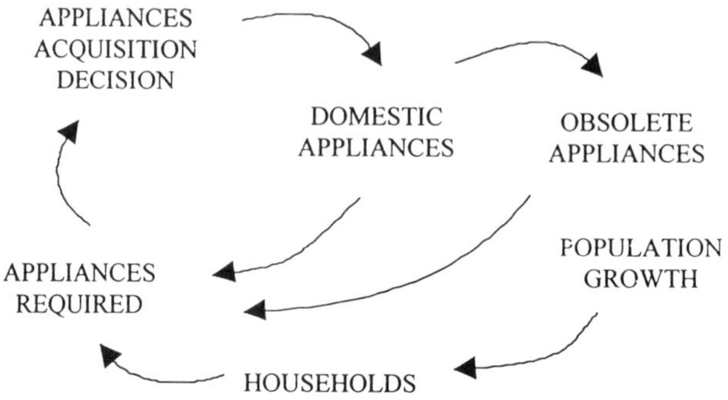

Figure 13.2 Appliances acquisition in the residential sector

statistically, and it is often used for calibration purposes to fit the model as appropriate. A double-check algorithm is implemented here to guarantee that demand never exceeds supply at any point in time.

Energy Supply Modules

The platform incorporates three basic energy supply structures: power supply capacity dynamics, grid expansion to rural areas and the gas coverage plan. The first one offers two alternatives: the government indicative plan and an

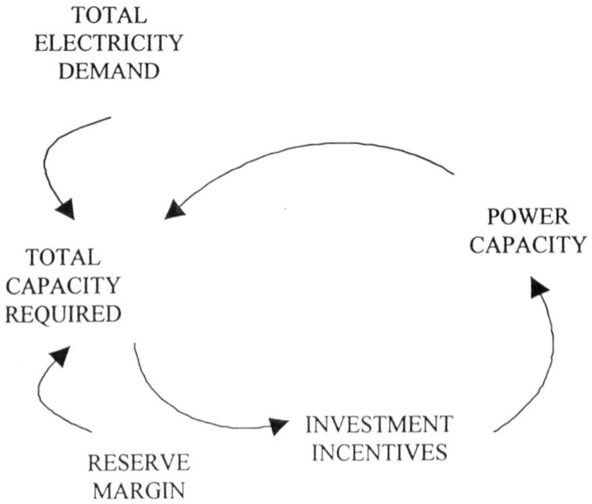

Figure 13.3 Power building capacity dynamics

adjustment to the plan due to postponements of some projected plants, given the success of DSM policies and programmes.

Figure 12.3 shows how the power capacity dynamics are represented in the platform. The reader may observe that building capacity depends on investment incentives, which are created by the electricity requirements to satisfy demand. The capacity requirements are manifested through simple calculations, taking into account systems capacity (existing and being built), short-term demand, system security margin and lead time for plant construction. This is a simplified version of the British model, extensively discussed by Bunn & Larsen (1992), Bunn, Larsen & Vlahos (1993) and Bunn, Dyner & Larsen (1995, 1996), and as such it is taken directly into the platform.

POLICY EVALUATIONS

We explored the implication of pricing policies, when these coincide with programmes and actions on DSM as well as on supply-side management. The intention is to establish synergy effects of several concurrent programmes on the overall energy system. Some examples of these policy analyses are given below.

On the supply side, the gas plan is supposed here to go ahead according to government expectations, but the electricity grid expansion plan is assumed to register some delays. On the demand side, DSM programmes are applied in the areas of lighting, cooking, space cooling and refrigeration. The following specific assumptions are made in this case:

- 100% of the gas plan is reached
- 60% of electricity grid expansion is accomplished
- On average one light bulb is intended to be replaced per household
- Financial arrangements for paying for appliance acquisition are not available
- Sudden price increment of 35% in real terms

A sudden price increment over 35% in real terms may be neither politically viable nor sustainable from the social and political perspectives. By no means can price increments above 5% in real terms, for a period of five years, be considered to have smooth effects. Yet as an alternative to a sudden 35% increase, this scenario seems to be more reasonable as: (i) it offers alternative energy sources to customers, given that the gas plan will only supply satisfactory quantities of this fuel by the end of the 1990s, (ii) total electricity sales will remain sufficiently large, (iii) companies may initiate recovery and be forced to improve their productive efficiency and (iv) inflation effects will be much more under control.

The results shown in Figure 13.4 show a 10% reduction in the average energy bill for households, yet important additional revenues will be available

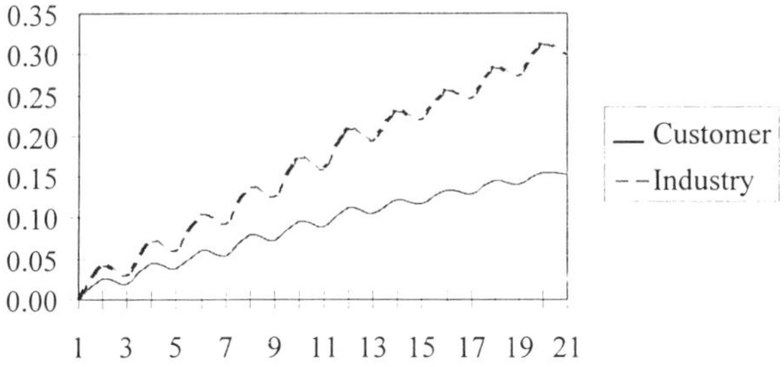

Figure 13.4 Simulation results for case 4. Effects on the average energy bill for each customer and on the additional national revenues. The time scale is quarterly, so the full effect is observed at the twentieth period (fifth year)

to electricity companies across the nation. End users will reduce their energy expenditure by more than 8% during the critical period as a good number of customers will shift to other energy sources and reduce electricity demand (given by the price elasticity effect). In this case there will also be diminished inflationary impacts. For example, inflation due to gradual electricity increments will amount to less than 0.21% (0.07×0.03) in the first year, instead of that originated by a sudden increment which could reach to about 1.05% (0.35×0.03). Note that the average weight of electricity in the consumer price index was about 0.03 in the year 1995.

Furthermore, Figure 13.5 illustrates the effect of this gradual pricing policy on additional sales and on relative savings. The upper line denotes additional sales with respect to the base case scenario; the intermediate one indicates energy savings of this case with respect to a scenario that is exactly the same as the reference scenario, except for the price increment; and, the bottom line signals savings of the gradual price increments with respect to the scenario just mentioned. This means that the gradual price increment scenario accomplishes significant savings in the long term without sacrificing sales in the immediate future. In fact, while this scenario reaches savings close to 1.7% of the total electricity required after one year, the previous scenario brings down demand by over 4.9%.

In a liberalized environment, prices will depend very much on the evolution of what happens in the electricity pool. With a further scenario, based on initial observations of the pool behaviour (during 8 months), when considering the corresponding UK modelling experience, we may infer the average energy bill dynamics by way of simulations. Figure 13.6 actually shows that the average customer may have to be prepared to pay still larger energy bills if tariffs are to follow market prices, assuming that the pool mechanism remains unchanged and that there are no major technology innovations in the years to come.

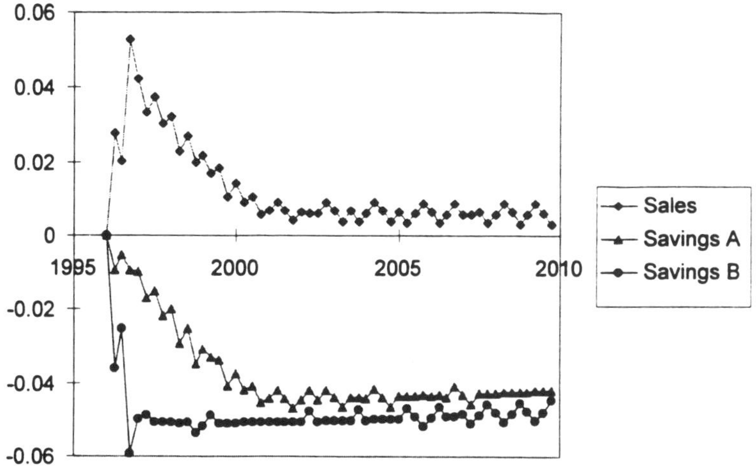

Figure 13.5 Electricity sales and saving accomplished

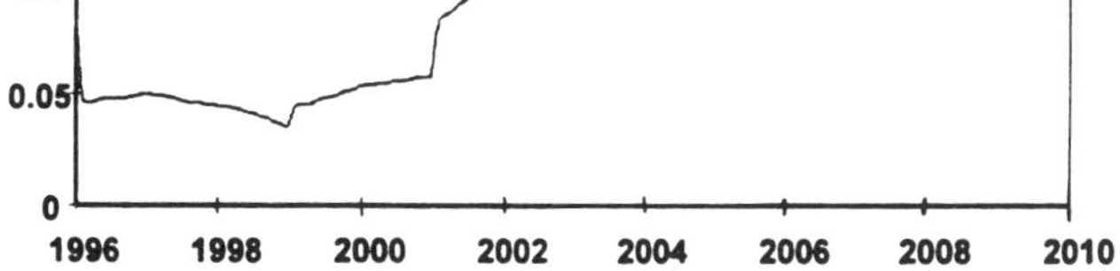

Figure 13.6 Average energy bill increments

CONCLUSION

The modelling protocol of the platform exhibited in this chapter is based upon the need to implement generic, modular, adaptable and transportable structures, capable of supporting the process of system analysis for intervention. It is evident that with Colombia seeking to balance market liberalization and demand-side planning, the platform has had to provide the basis for examining various policies, which might be difficult to reconcile, making use of analogies from elsewhere upon how various submodules may operate (e.g. power pool, consumer choice).

The platform developed here therefore provides a context-rich basis for investigating a number of different policy issues. Although a similar framework has been undertaken in the US continuously since 1976 (Naill, 1992), the open literature gives limited information with respect to the modelling protocol that needs to be followed if such a platform is to provide a sustained level of policy support through changing circumstances.

As has been shown in this chapter, an SD based platform has the potential to provide such evolutionary policy support under complex policy environments, although it then needs to have a multifaceted perspective on its use. In particular, the application of SD in these circumstances goes beyond the single focus of providing insight into a preselected policy issue. Yet, if the traditional approaches of optimization and econometrics are losing some of their relevance in the new era of market liberalization, and SD is well placed to fill the new modelling requirements, it seems that it must offer governmental energy advisors some of the same features of longevity, modularity, adaptability and detail that the other approaches have provided. The features then constitute requirements for a clearer modelling protocol to be developed upon a systems simulation platform.

ACKNOWLEDGEMENTS

We wish to acknowledge the support of UPME (Colombian Ministry of Mines and Energy), the London Business School (University of London), Universidad Nacional de Colombia and COLCIENCIAS during this research.

REFERENCES

Buehring W.A., Hamilton, B.C., Guziel, K.A. and Cirillo, R.R. (1991) ENPEP: an integrated approach for modelling energy systems, Environmental Assessment and Information Sciences Division, Argonne National Laboratory, USA.

Bunn, D.W. and Larsen, E. (1992) Sensitivity reserve margin to factors influencing investment behaviour in the electricity market of England and Wales, *Energy Policy* **29**, 420–429.

Bunn, D.W., Larsen, E. and Vlahos, K. (1993) Complementary modelling approaches for analysing several effects of privatization on electricity investment, *Journal of Operational Research*, **44**(10), 957–971.

Bunn, D.W., Dyner, I. and Larsen, E. (1995) Strategic opportunities across the new UK electricity and gas markets, in *Proceedings of the 18th IAEE Conference, 1995*, International Association for Energy Economics, Cleveland, Ohio, pp. 477–484.

Bunn, D.W., Dyner, I. and Larsen, E. (1996) Modelling strategic behaviour across energy markets, research report, Decision Technology Centre, London Business School.

Dyner, I., Smith, R. and Peña, G. (1995) System dynamics modelling for energy

efficiency analysis and management, *Journal of Operational Research* **46**(10), 1163–1173.
Ford, A. (1983) Using simulation for policy evaluation in the electric utility industry, *Simulation*, 85–92.
Ford, A. (1985) Short lead time technologies as a defence against demand uncertainty, in J. Plummer, E. Oatman and P. Gupta (eds) *Strategic Management and Planning for Electric Utilities*, Prentice Hall, Englewood Cliffs, NJ.
Ford, A. and Bull, M. (1989) Using system dynamics for conservation policy analysis in the Pacific Northwest, *System Dynamics Review* **15**(1), 1–16.
Geraghty, D.M. and Lyneis, J. (1985) Feedback loops: the effect of external agents on utility performance, in J. Plummer, E. Oatman and P. Gupta (eds) *Strategic Management and Planning for Electric Utilities*, Prentice Hall, Englewood Cliffs, NJ.
Moxness, E. (1990) Interfuel substitution in OECD European electricity production, *System Dynamics Review* **6**, 44–65.
Naill, R. (1977) *Managing the Energy Transition*, Ballinger, Cambridge, MA.
Naill, R. (1992) A system dynamics model for national energy policy and planning, *System Dynamics Review* **8**, 1–19.
Rego, J. (1989) Schedule delays and new financing for the Argentine electricity sector growth, in System Dynamics Conference, Stuttgart, Germany in P. Milling and E. Zahn (eds) *Computer-Based Management of Complex Systems*, Springer-Verlag, Berlin, 1989.
UPME. (1995) SD modelling for integrated energy analysis, Colombian Ministry of Mines and Energy, Bogota, Colombia, report no. 1, December (in Spanish).

Chapter 14

Preparing for a Competitive Environment: Developing Strategies for America's Electric Utilities

James M. Lyneis

INTRODUCTION

America's electric utilities for the first time face the prospect of direct competition. This has already occurred at the generation level (those utilities that need new capacity are seeking competitive bids, rather than simply building the capacity themselves as in prior times), and at the wholesale level (a number of municipalities have put contracts out to competitive bid, and traditional suppliers have in many cases lost those bids). In addition, by the late-1990s, the existing utilities' retail service monopoly for large customers, if not for all customers, will likely be rescinded. As a result, other suppliers will be able to compete to provide power to the utilities' customers and "wheel" that power over the utilities' transmission and distribution system for a price. Eventually, utilities are likely to be deregulated and forced to set competitive prices.

Generation competition threatens the traditional earnings model of the electric utility. Under that model, guaranteed returns on investments to meet demand growth generally assured reasonable growth in earnings and dividends. But without those investments, where is earnings growth to come from?

Systems Modelling for Energy Policy. Edited by D.W. Bunn and E.R. Larsen

Wholesale and retail competition threaten more than earnings growth. This threat arises from the typical starting condition of the regulated monopoly:

(1) The historical requirement to serve all customers on demand (the "universal service obligation"), in exchange for protection from competition and assurance of an adequate rate of return on investment.
(2) Cross-subsidization of customer classes (usually benefiting residential customers).
(3) Uncompetitive cost structures, service, and customer relations, nurtured by years of monopolistic protection.
(4) Payment of high levels of dividends to shareholders, who have come to expect bond-like returns.
(5) Significant excess capacity, which contributes to high costs and prices and raises the possibility of a destructive price war.

Couple this starting condition with the emergence of new, cheaper generating technologies, and competitors can easily come in and "cream skim" the utility's customers, thereby causing the former monopoly to lose sizable, and generally high-profit, customers. This leaves the utility with a significant dilemma: if it raises prices to the remaining customers (to make up the lost revenues), it risks losing even more customers to competitors; but if it holds the line or even cuts prices, profits will suffer unless costs are also reduced.

Fortunately, electric utilities have several transition years in which to prepare for competition and mitigate or avoid this dilemma. The key strategic need of these utilities is determining the right mix and order of the many possible actions open to them, including: downsizing, investing to improve service, mergers and/or acquisitions, becoming an aggressive competitor in another utility's service territory, and investing in new technologies, products, and markets "on the other side of the meter".

This chapter discusses the use of a system dynamics model to develop competitive strategies for an American electric utility. The analyses presented herein are illustrative, but reflect the spirit of the results from several actual applications of the model.

THE MODEL AND ITS DEVELOPMENT

The Strategy Modeling System—Utility (SMS—*Utility*, Pugh-Roberts Associates), was originally developed in the early 1980s and applied to investment, financing, and regulatory strategy issues (Geraghty, 1981; Lyneis & Geraghty, 1982; Lyneis, 1984). It was recently refined and extended to incorporate competition at the generation, wholesale, and retail levels.

Model Architecture

SMS-*Utility* is a generic, integrated business model that represents, at a strategic level, the interactions among:

- Patterns and growth of energy demand and peak load
- Competitive position and market share
- Construction of new generating plants and T&D
- Investments in demand-side management
- Dispatching of existing generation
- Cost structure (by business unit)
- Financial structure (by business unit)
- Response of the capital markets to financial performance
- The regulatory process

As illustrated in Figure 14.1 the model is organized into nine sectors:

(1) Demand sector: computes the utility's energy and peak load based on number of customers, kWh per customer, effect of price elasticity, and retail competition.
(2) Supply planning sector: forecasts energy and peak load and determines the mix of purchased power (from "traditional" suppliers and/or new generator competition), new baseload generation, new peaking generation, and demand-side management to meet that load; also plans for T&D capacity.
(3) Power generation sector: dispatches the available generation to meet load, thereby determining fuel mix and emissions.
(4) T&D sector: determines system reliability based on requirements relative to capacity, and on the technical "newness" of the capacity.
(5) Accounting sector: computes profit and loss, balance sheet assets and liabilities, and cash flow based on inputs from the other sectors.
(6) Financial planning sector: meets the demand for funds (if possible) via a mix of debt and equity; determines dividend payout.
(7) Capital markets sector: determines the utility's interest rate and stock price based on financial performance and economic conditions.
(8) Regulation sector: sets allowed rates based on costs, rate base, perceived customer satisfaction, and economic conditions.
(9) Competitor sector: represents the cost, capacity, and price characteristics of various categories of competitors.

The important interactions among the sectors are also illustrated in the figure. For example, changes in demand from the demand sector, when matched against the available capacity from the supply sector, alter the plant and fuel mix dispatched in the power generation sector. These changes in power delivered are used in the accounting sector to determine costs. Cost information

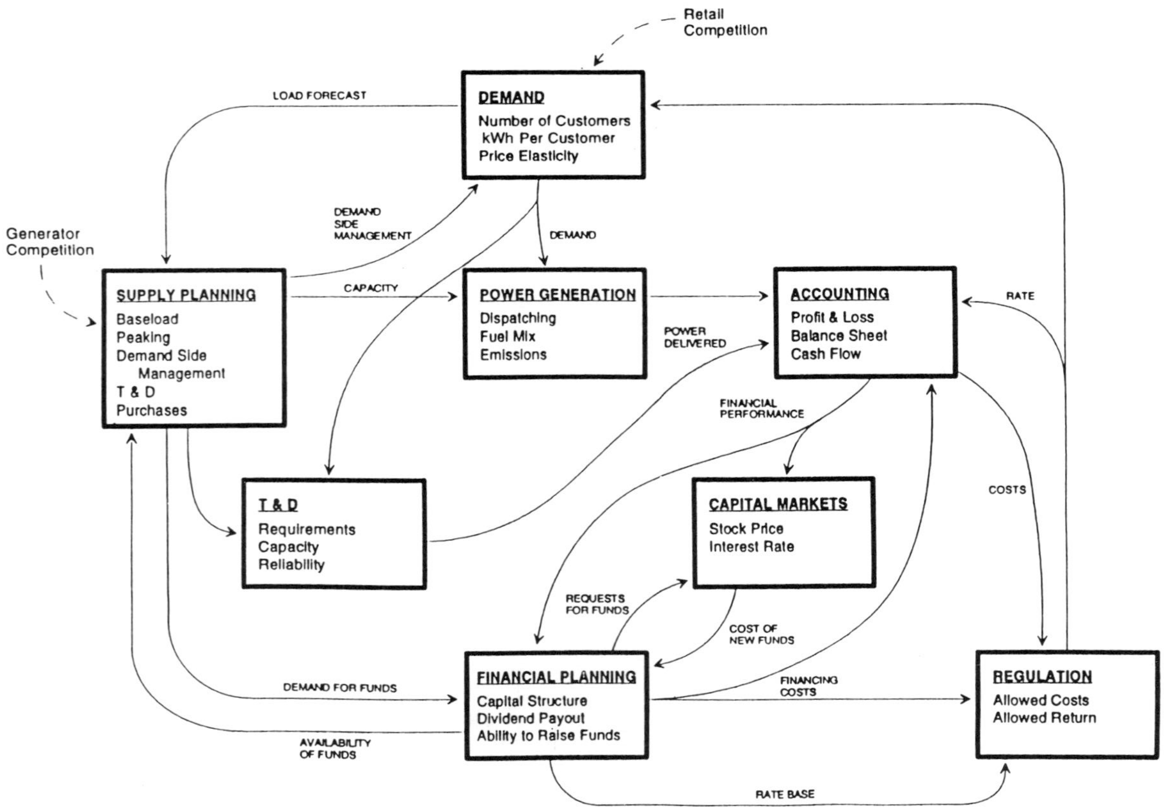

Figure 14.1 Model architecture

is then provided to the regulation sector where, after appropriate delays, rates (prices) are changed. Price changes then "feed back" to the demand sector to potentially cause further changes in demand. Figure 14.1 shows other such feedback relationships, particularly among the accounting, financial planning, capital markets, and regulation sectors.

In taking a strategic view of the electric utility business, SMS-*Utility* focuses on the key cause–effect relationships which drive short-, mid- and long-term performance. For example, as described above, the model explicitly links changes in demand to capacity utilization (e.g. reserve margin), fuel mix, and hence costs and profits in the short term, and to capital investment requirements, financial structure, and earnings growth in the mid-term. The model includes the many factors that cause generation and distribution margins, fuel mix, and staffing levels to vary over time. It reflects the way in which capital investment requirements and earnings growth determine financing requirements and combine with the utility's dividend and capital structure policies to affect interest costs and stock price. It also "closes the loop" by tying operating and financing costs to rates and demand for the utility's power, in the short-term through price elasticity and in the longer term through ability to compete with alternative fuels, with self- and co-generation and ultimately with other retail or wholesale competitors.

While taking a strategic view, SMS-*Utility* represents important details:

(1) Customer classes: residential, industrial, commercial and major wholesale customers.
(2) Capacity: baseload capacity by fuel type, peaking capacity, and purchased are all separately represented, as are transmission and distribution capacity.
(3) Business units: financial performance is computed for five distinct business units (distribution, existing generation, new generation, demand-side management and diversified businesses).
(4) Competitors: co- and self-generation, non-utility generators (NUGs), and other traditional utilities.

Methodology

At the heart of SMS-*Utility* is a dynamic simulation model. This unique methodology explicitly represents:

(1) Cause–effect relationships, as in price changes cause changes in electricity usage:

Price changes → Electricity usage

(2) The time delays between an action and its ultimate effect, for example, between price changes and changes in usage (Figure 14.2). There can also

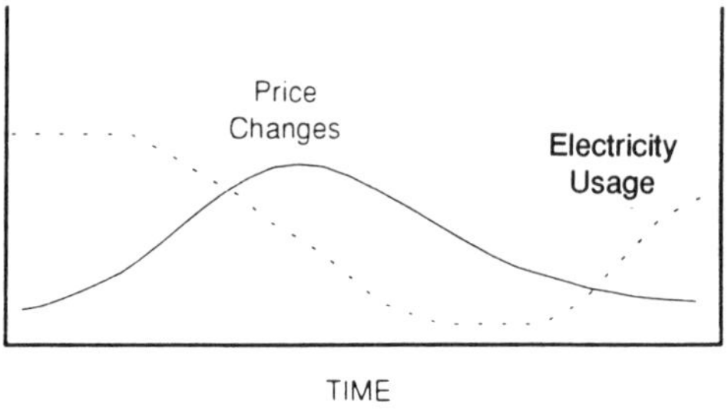

Figure 14.2 Time delay

be time delays in management responses: for example, in obtaining information and then acting upon it.

(3) Nonlinear responses to actions: for example, small price decreases may not increase usage much, but then past a point further decreases lead to significant increases in usage; finally, usage may "saturate" and further price changes yield little, if any, additional usage (Figure 14.3).

(4) The decision-rules which produce actions: for example how prices are based on costs, competitive conditions, and regulatory constraints.

This simulation methodology then takes these cause–effect relationships, time delays, nonlinearities, and decision-rules and combines them in feedback loops that describe how the system works. Some feedback loops are "positive" or self-reinforcing feedback loops; for example, price reductions stimulate increased usage, thereby lowering unit costs and allowing further price reductions, other things remaining equal (Figure 14.4).

Other feedback loops are "negative", or controlling feedback loops. For example, as increased usage approaches capacity, service reliability deteriorates. Poor service reliability slows or even reverses the growth in usage, particularly as businesses expand elsewhere, move from the area, or even generate their own power. By varying service reliability, this negative feedback

Figure 14.3 Nonlinear response

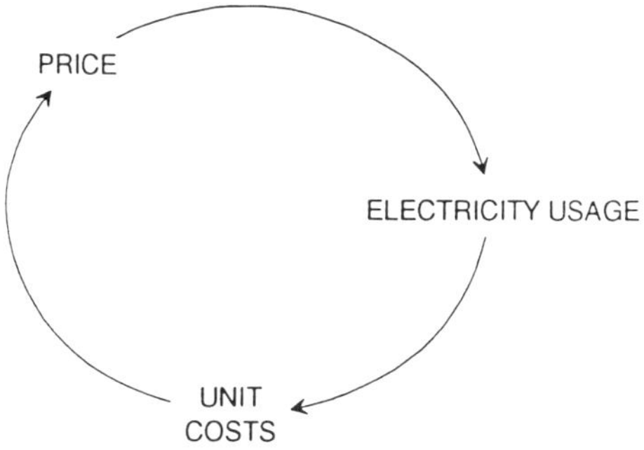

Figure 14.4 Feedback loop

loop attempts to control electricity usage to the level which the available capacity can support (Figure 14.5).

One can begin to imagine adding other feedback loops to represent management responses. For example, capacity might be expanded as usage increases in order to avoid or minimize the reduction in service reliability. But as capacity increases, unit costs increase, thereby slowing or reversing the reduction in prices.

As more and more cause–effect relationships are added, the complexity of the "mental simulations" necessary to estimate how alternative actions would affect behavior become impossible. We therefore use the computer to simulate the company's response. First, we recreate a period of past history to establish confidence in the model (Figure 14.6).

Then, we simulate forward a "base case" assuming business as usual (i.e. the continuation of current business strategies and management policies). For example, in Figure 14.6, we might see electricity usage increasing (as prices

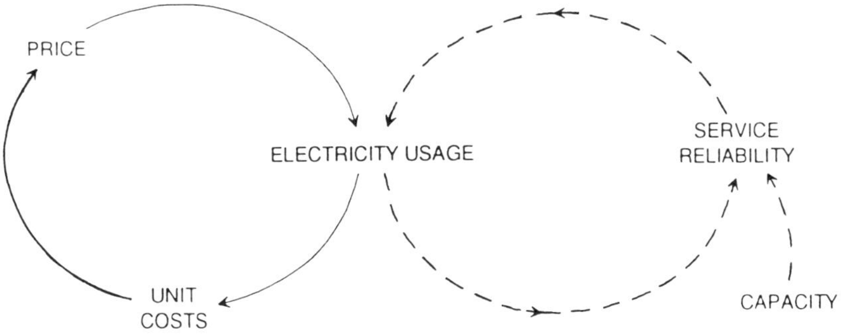

Figure 14.5 Negative feedback loop

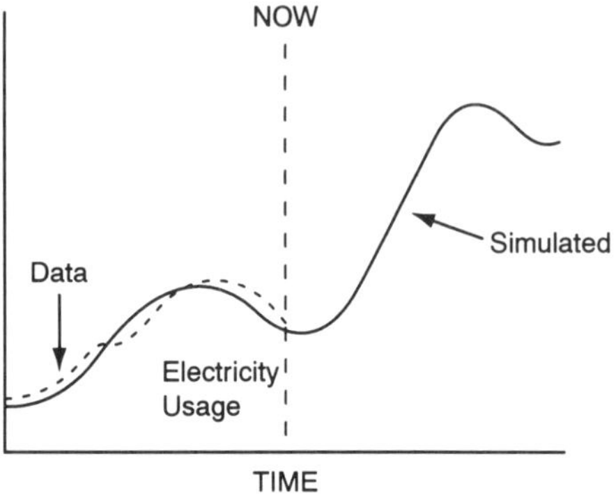

Figure 14.6 Period of past history

steadily decline), but then reversing because the company was late in adding capacity (resulting in a reduction in service reliability). Once additional capacity comes on line, reliability improves and volumes can increase again (assuming costs and prices do not increase too much), until the additional capacity is overloaded. In this case, the simulation recreates past performance problems and shows that they will continue into the future unless an alternative strategy is adopted.

Simulation is used to perform a wide range of "what if?" tests. These departures from the base case explore alternative strategies, under a range of alternative economic and competitive scenarios (Figure 14.7). For example, should capacity be expanded in advance of growth in usage, or only after we are sure the volume is there? Should we expand capacity, or invest in demand side management instead? If we do decide to expand capacity, what size of

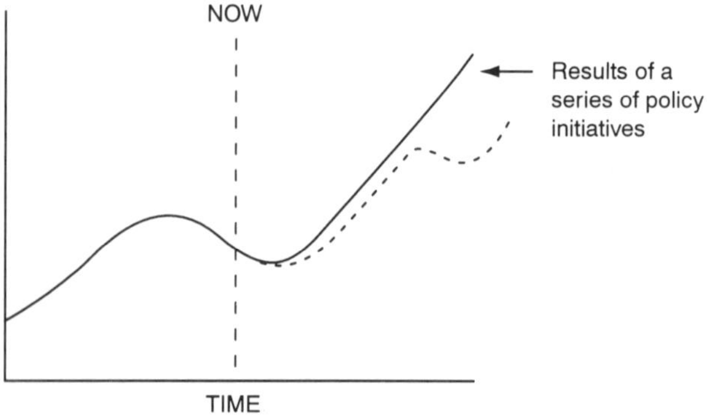

Figure 14.7 Base case

plant and fuel mix? What is the impact on costs and profits if the expected usage does not materialize? How are potential competitors likely to affect our usage?

Model Dynamics

SMS-*Utility* consists of numerous cause–effect relationships of the type described above. These relationships specify the internal workings and management decision processes of the utility, customers, regulatory bodies, capital markets, and competitors.

Figure 14.8 provides a simplified overview of these cause–effect relationships. It focuses on the utility, and the utility's interactions with the external environment. In Figure 14.8 we have highlighted the feedback loop corresponding to the positive, self-reinforcing loop described in the previous section. Starting with demand (electricity energy usage), an increase in demand, which

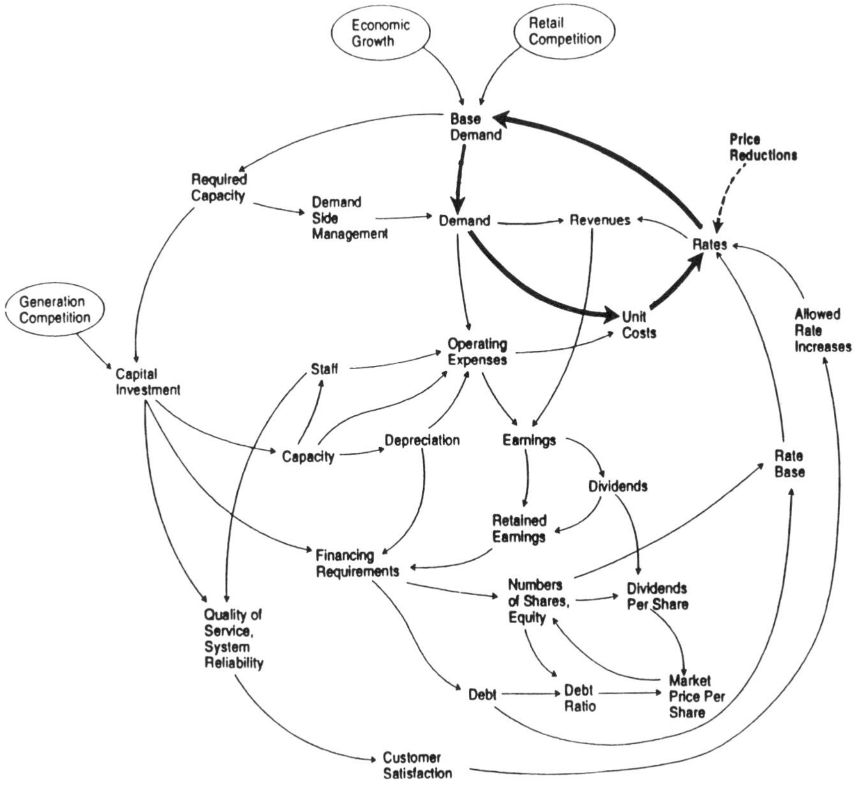

Figure 14.8 Key feedbacks: self-reinforcing loop through rates

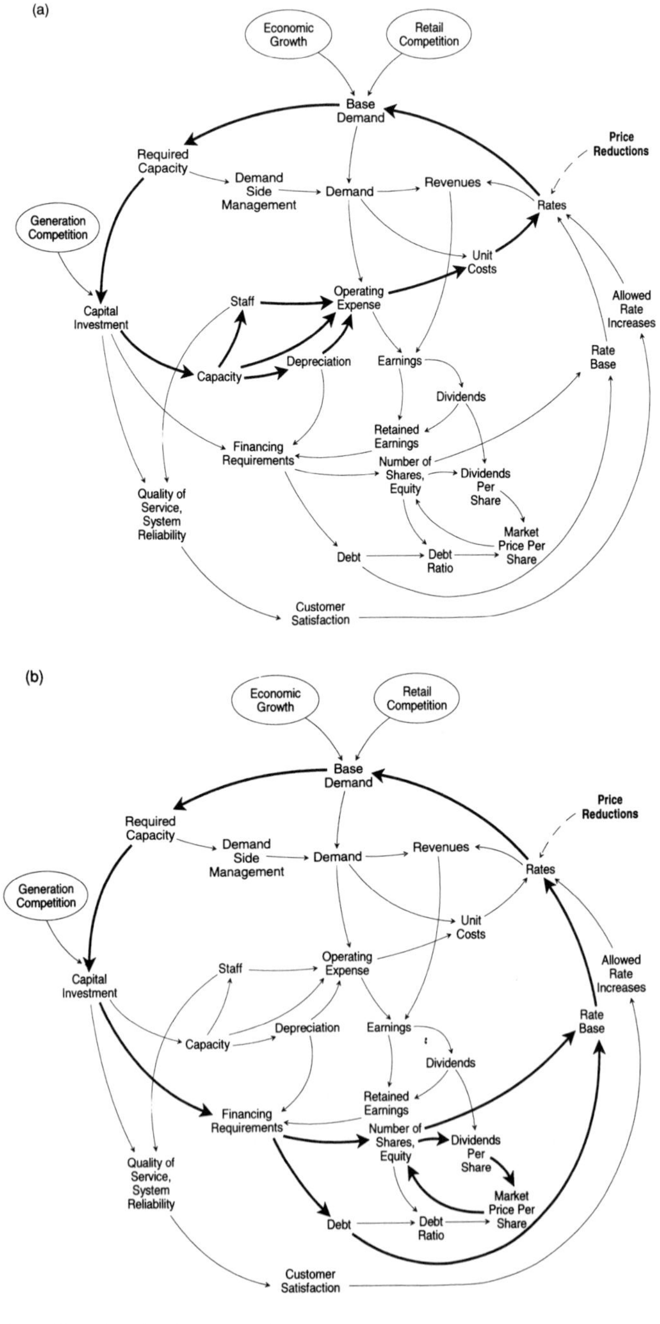

Figure 14.9 Key feedbacks: negative loops suppressing growth

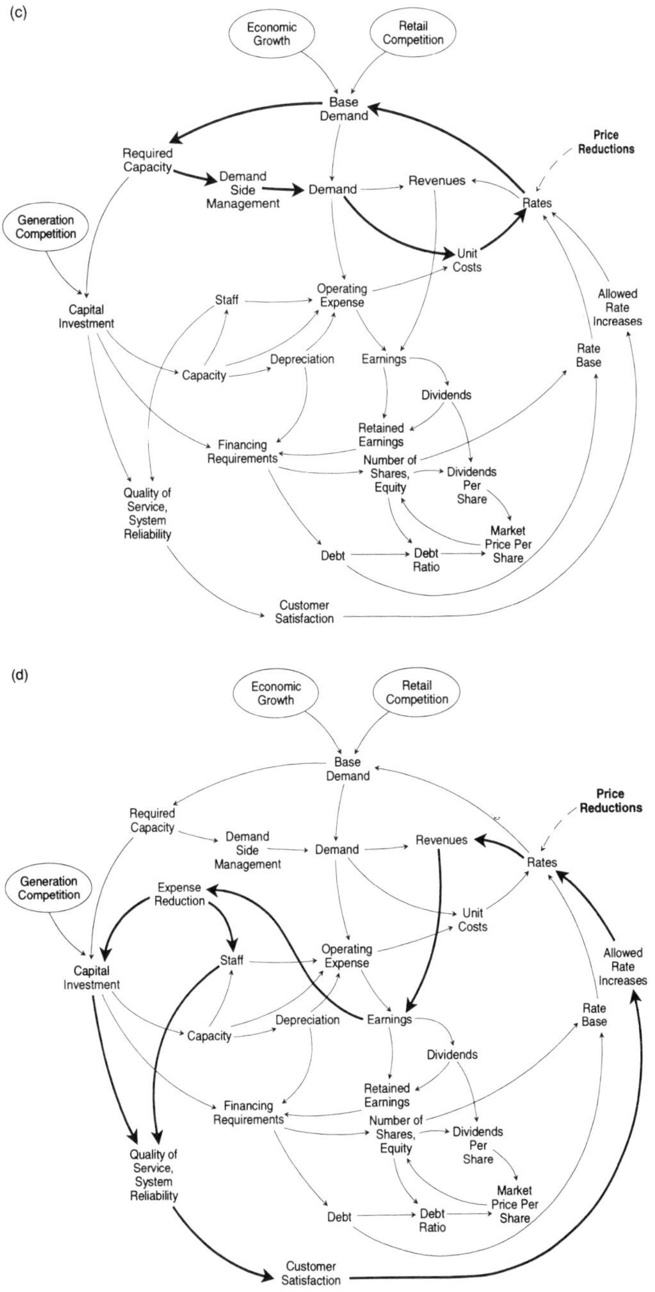

Figure 14.9 *Continued*

might be caused by improved economic conditions, reduces unit costs, thereby allowing the utility to reduce rates (while maintaining the same profitability). A reduction in rates further increases demand in a self-reinforcing manner (under present conditions, by encouraging greater usage per customer; under potential future conditions, by improving the utility's competitive position, thereby increasing market share).

Figure 14.9 highlights several negative feedback loops which act to suppress demand growth. The first works through increases in capacity and operating expenses (see Figure 14.9a): an increase in "base demand" increases required capacity; the utility invests to increase capacity; as this capacity comes on line, operating expenses increase, thereby increasing unit costs and requiring an increase in rates, which slows or reverses the initial increase in demand.

The second feedback loop works through financing (Figure 14.9b): increases in capacity investments increase financing requirements; as debt and equity are added, the utility's rate base increases, thereby requiring an increase in rates to provide an adequate return on that investment. Capital market feedbacks can amplify these effects. For example, as shares are issued, earnings and dividends per share decrease, and market price falls (relative to what it otherwise would have been). As a result, even more shares must be issued to raise the required equity.

A third feedback loop works as an alternative to investments in traditional capacity (Figure 14.9c): investments in demand-side management reduce the demand on the utility's generating capacity; because demand does not grow, unit costs are not reduced and the self-reinforcing loop through rates is inactive; further, if the utility is allowed to earn a return on these investments and on the lost revenue, rates will increase and cause demand to fall further.

A fourth set of feedbacks works through customer satisfaction (Figure 14.9d): if, for any number of reasons, the utility ends up short of capacity (e.g. underestimating demand growth, financial constraints on capital investment, reductions in staff beyond any "fat"), capacity availability, system reliability, and service adequacy deteriorate; this causes a reduction in customer satisfaction, and in response the regulator is likely to be less favorable in granting rate increases. The resultant reduction in profitability is likely to cause the utility to further reduce needed spending, just at the time when constraints on rate increases might be stimulating demand above expected levels!

Both the self-reinforcing and the constraining dynamics work within the model, and real life, simultaneously. Which will dominate, and during what periods of time, depends upon the strengths of the causal relationships, the time delays, and management's responses. The purpose of a simulation model is to test how alternative management responses affect the utility's ability to achieve the best performance, under a range of alternative economic and competitive scenarios.

Figure 14.10 shows the major policy levers through which management can try to achieve corporate objectives in response to a changing economic and competitive environment:

- Desired generation and distribution reserve margins (affect capacity availability and system reliability in the face of uncertain demand)
- Investments in demand-side management
- Size and fuel mix of any new generation, or, alternatively, the purchase of power from traditional sources or from alternative generating companies
- Cutbacks in capital investments, staffing levels, and/or non-staff operating costs (a certain fraction of which may be "fat")
- Investments in new types of equipment, in "re-engineering", and/or in training to improve the productivity of current staff
- Desired capital structure

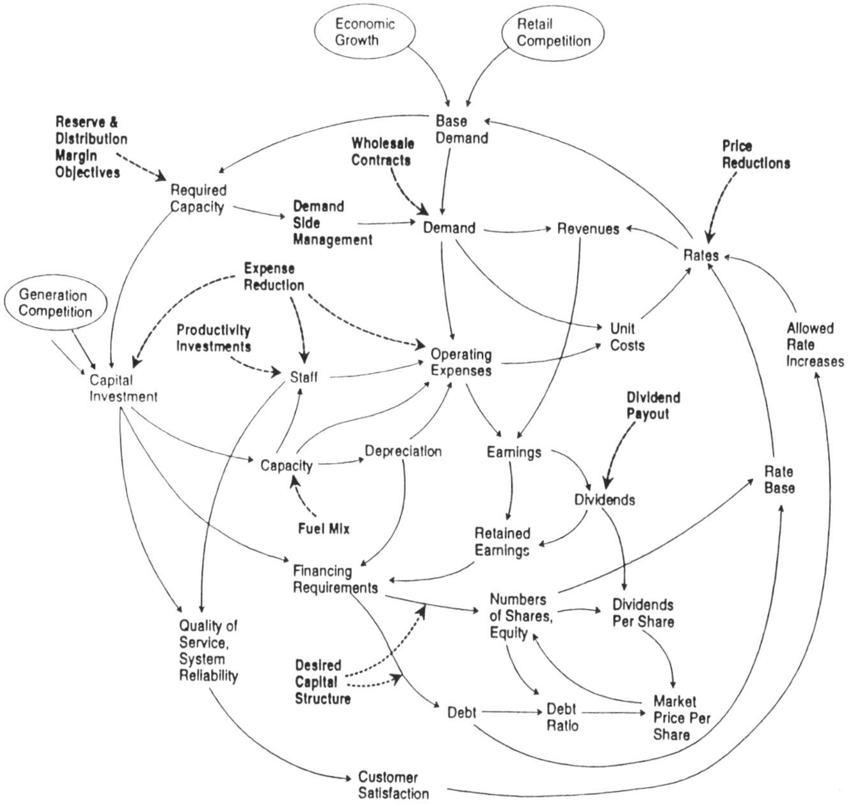

Figure 14.10 Key feedbacks and policy levers

- Dividend payout
- Volume and terms of wholesale contracts
- Regulatory policy

SMS-*Utility* represents our general theory of how competitive utility markets and electric utility companies work. For example, the theory specifies the key cause and effect relationships operating to determine usage of electricity and market shares; the drivers of productivity, operating expenses, and service reliability at a utility; and management's responses to performance problems. This general theory has evolved over many years from work with several utilities, and from work with dozens of clients in other similar industries (e.g. telecommunications, transportation services, financial services).

Setting Up and Calibrating SMS-*Utility* to a Specific Company

The general theory identifies cause–effect relationships, with "industry-average" parameters used to determine the relative importance of relationships. When SMS-*Utility* is set up to represent a particular company and market, changes are made in the parameters to specify how this general theory pertains to the particular characteristics of that company and market (e.g. relative importance of price versus service quality in determining market shares). This is a multi-step, iterative process:

(1) Assemble "hard" data on number of customers, kWh, peak load, prices, number of employees, financial performance, and so on.
(2) Review qualitative (and quantitative) studies of the utility.
(3) Conduct interviews with utility managers regarding the strategic issues, performance concerns, and decision processes at the utility.
(4) Starting from generic industry-average parameters, set up the SMS-*Utility* model:
   (a) Change parameters to reflect the size of the market, size of utility and competitors, cost structure, and so on.
   (b) Modify "decision-rules" to reflect known characteristics of utility's pricing, capacity expansion, staffing, financial, and other decision processes.
   (c) Modify parameters that reflect the characteristics of the market (e.g. the effect of price on electricity usage), the regulators (e.g. allowed rates of return, caps on rate of change in prices), and competitors.
(5) Simulate the SMS-*Utility* model and compare simulated behavior to historical data and to managers' recollections of how "soft" factors changed over time (e.g. customer satisfaction).
(6) Adjust parameters and decisions outlined in step (4) to correct discrepancies between the data and simulated behavior.

(7) Iterate through steps (4) through (6) until a satisfactory calibration is achieved.

The iterative calibration process is done in consultation with managers at the utility, and under a rigorous set of guidelines, which ensure that reasonable adjustments are made to parameters and decision rules.

PREPARING FOR COMPETITION: USING THE MODEL TO ANALYZE ALTERNATIVE STRATEGIES

The "Business-as-Usual" Earnings Model

The historical compact between the monopoly utility and regulators created an "earnings model" driven by investment. Under that model, illustrated in Figure 14.11, the utility added capacity to meet projected demand growth. These investments increased the utility's rate base, and an allowed return on that rate base was added into the price charged for electricity. Even when the increase in prices caused demand growth to fall below expectations, any potential shortfall in the coverage of operating costs and return on rate base was generally made up by raising unit prices on remaining usage. Therefore, investment to meet projected demand growth drove earnings and dividend growth. Were the traditional earnings model to continue into the future a typical utility could expect "business-as-usual" growth in earnings and dividends.

Generation Competition Threatens the Traditional Earnings Model

The traditional earnings model is threatened by competition. As illustrated in Figure 14.12, generation competition breaks one of the key links in the earnings model chain. Generation competition requires a utility to put its future supply needs out to competitive bid, and purchase power from the lowest bidder. While the utility can bid for these generation contracts through a subsidiary company as in the "business-as-usual" case, most utilities have not been cost-competitive with alternative sources of supply. Without utility investment in future generation, rate base will not increase and earnings will not grow. As old assets are retired and replaced by purchased power, earnings could even decline. While the utility will still need to grow its distribution and transmission assets, this is not likely to offset the decline in generation.

Retail Competition Threatens Shareholder Value

As illustrated in Figure 14.13 retail competition adds some new links to the earnings model chain. A utility's high unit price, relative to that of competitors, will now cause a loss of market share. The amount of share loss may be

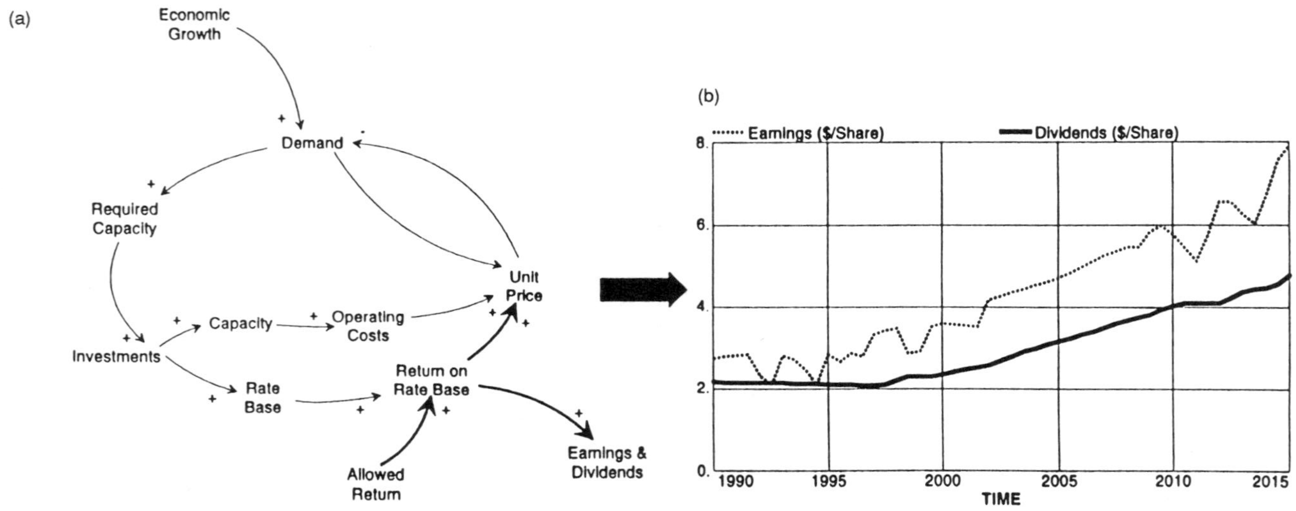

Note: Arrows indicate causal connections. A plus (+) sign at the end of the arrow means change in the same direction, e.g. an increase in demand increases required capacity; a negative (-) sign means change in opposite directions, e.g. an increase in unit price reduces demand.

Figure 14.11 (a) The traditional earnings model produces (b) continued earnings and dividend growth

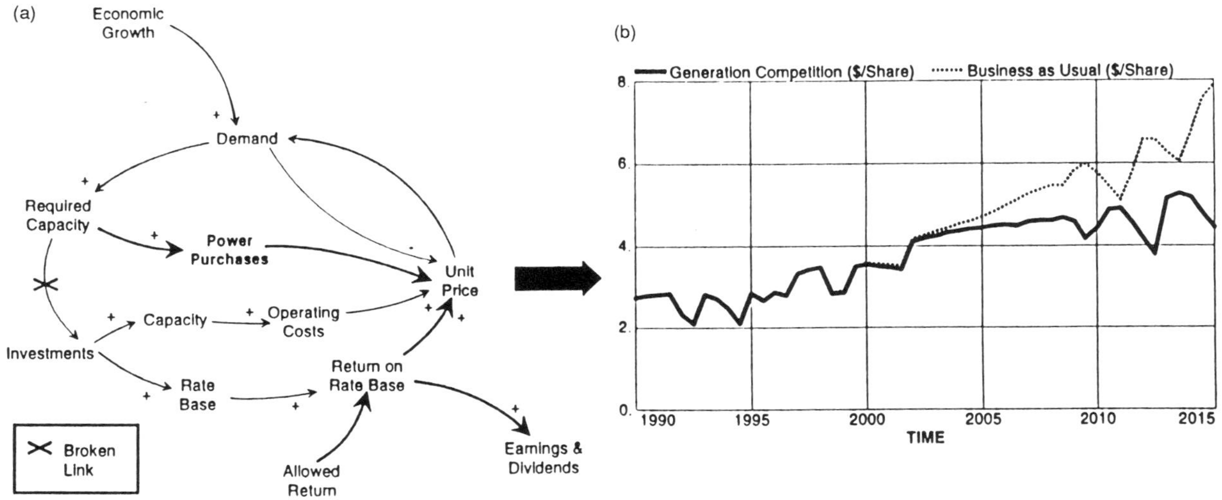

Figure 14.12 (a) Generation competition breaks one key link in the earnings model chain, causing (b) earnings growth to stop

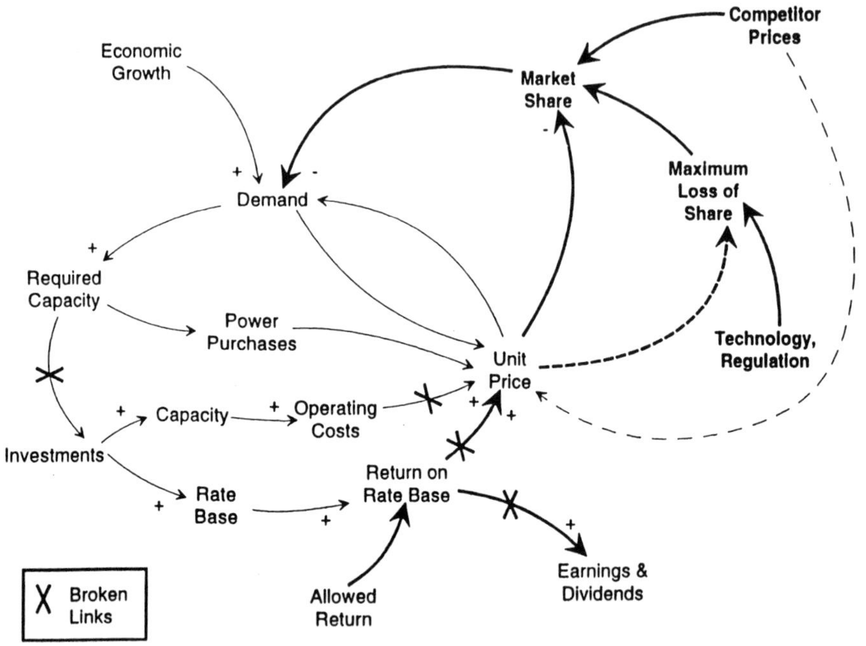

Figure 14.13 Retail competition adds new links and breaks others

constrained by technological and regulatory constraints. The loss of market share reduces demand, raises unit costs and prices, and potentially leads to a downward spiral. That downward spiral can be averted if either the regulators or the utility break two additional links in the traditional earnings model chain: (i) the link between operating costs and prices, and (ii) the link between return on rate base and prices. Constraints placed by regulators on the maximum allowed rate of price increase can break these links, as can actions by the utility to respond to competitor prices. But with these links broken, a reduction in demand means that revenues no longer increase to cover costs and return on rate base, and earnings and dividends decline.

How might retail competition affect the utility? For the purposes of illustration, we created a "retail wheeling" scenario in which the retail market for electricity is opened to competition in 1997, and competitors are assumed to discount the former monopoly utility's prices by 10%. We further assume that: (i) the utility does not make any pre-emptive move or special response to competitors' lower prices; and (ii) regulators limit the rate of price increase on remaining utility customers to the rate of inflation. The utility's performance under this retail wheeling scenario (which includes generation competition as well) deteriorates substantially (see Figure 14.14). The utility's share of

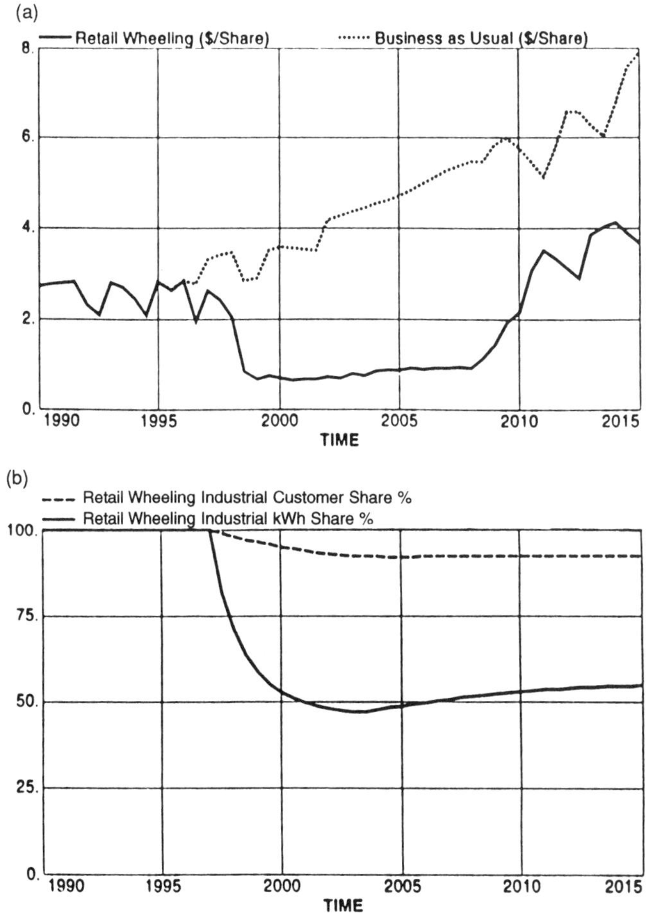

Figure 14.14 Without pre-emptive moves or responses, retail competition (a) devastates earnings per share, because (b) utility loses significant market share

industrial customers within its current service territory declines by about 10%, with a loss of nearly half of industrial energy sales because the most intensive users will likely be those targeted by competitors (the corresponding loss of commercial customers is approximately 1%, and 25% of energy). This produces a substantial increase in reserve margin, with a corresponding increase in unit costs.

Because regulators are assumed to allow prices to increase only at the rate of inflation, some of the utility's investment is "stranded" in the sense that returns are less than costs. As a result, the utility's profitability and earnings drop dramatically, and only begin to recover after 2007 as reserve margins improve and unit costs fall. However, earnings then bump up against the limit imposed by the requirement to buy power rather than build. Significant external financing is required to cover operations and remaining capital expenditures.

Retail competition adds the threat of near-term bankruptcy to the long-term earnings stagnation of generation competition. This leaves the utility with a significant dilemma: if it raises prices to the remaining customers (to make up the lost revenues), it risks losing even more customers to competitors; but if it holds the line or even cuts prices, profits will suffer unless costs are also reduced.

These simulation results clearly show the need for decisive action on the part of the utility. What are the options? They fall into three basic categories: (i) regulatory bail-out; (ii) downsizing and/or consolidating via merger or acquisition; and (iii) investing for the future. We will examine the consequences of each of these in turn.

Regulatory bail-out

Regulatory bail-out is the easiest extension from the current regulatory model, and the one that many utilities are likely to hope for. It assumes that there will always be a class of customers that are too small to be served by competitors, and in exchange for serving them the utility should be allowed to recover all of its reasonable costs and earn a fair return on all of its assets. Therefore, regulators should allow the utility to raise its prices at whatever rate is required to achieve this. In its extreme, this policy assumes full cost-of-service rate recovery, which includes stranded investment.

But this hope in a regulatory bail-out ignores the key feedbacks added by retail competition—the higher unit prices are, the more market share will be lost, the higher will be prices to remaining users, and so on. If prices get high enough, residential customers will begin to band together to extract better deals from competitors. Even if regulators are at first sympathetic to the utility's plight, at some point, and at very plausible losses of market share, the price increases for remaining users will become so high as to be politically unacceptable (see Figure 14.15). Further recovering stranded investment from the remaining customers will not be allowed, and the utility's earnings and dividends will decline.

Arguments can also be made that lost customers (via competitors) should be required to make up some if not all of the investment left stranded by their departure. However, this may be difficult to justify for self- or co-generators, and experience in other industries and in other countries indicates that regulators often give competitors a "grace period" to become established before they are required to pay such charges.

Therefore, while arguments of fairness will likely lead to some relaxation of any price cap, and also to some recovery of stranded investment (e.g. via

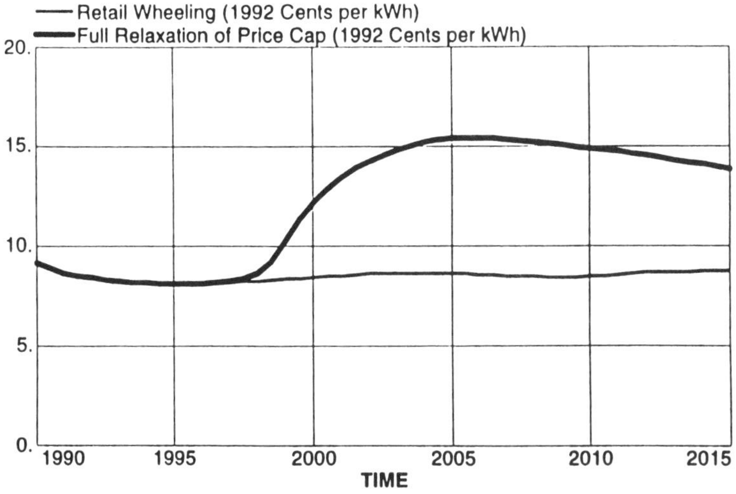

Figure 14.15 Price increases required to make up lost revenue would be politically unacceptable

backup charges as well as direct fees), the utility's shareholders will suffer some of the revenue and profit loss from retail competition if the utility fails to take appropriate actions. This can have a significant effect on shareholders. Retail wheeling can devastate stock price. Utilities are no longer the safe, stable investment for assured dividend growth!

Downsizing and/or consolidating via merger or acquisition

To become competitive, the utility must reduce its cost base. How and when this is done will determine the success of the utility in preparing for competition. Following the lead of many other industries, the obvious strategy may be to downsize by cutting staff and non-staff expenses. In most cases, however, costs exist because the utility's current technologies, processes and/or organization require them—without changes in these, the staff and expenditures reductions are not likely to be "fat". In addition, downsizing generally produces experience dilution and morale problems that adversely affect productivity and work quality. Therefore, insufficient staff and low productivity cause customer service to deteriorate and backlogs of "deferred" maintenance to build up. These backlogs eventually:

- Increase needed expenditures, often on an emergency basis
- Adversely affect plant availability system reliability, and quality of service.
- Create longer term "costs" which more than offset the short-term savings.

As a result the short-term improvements cannot be maintained in the longer term. Expenses increase, customer service deteriorates, the utility loses market share, and unhappy regulators squeeze the utility's earnings. Downsizing produces short-term improvement, but fails to solve the long-term problem (see Figure 14.16).

Effectively reducing the cost base requires a mix of actions and a long-term outlook. Three actions are required:

(1) Investments in productivity-improving technologies (e.g. automated meter reading, "squirrel-safe" transformers) and processes (consolidated handling of service, billing and other inquiries), which allow reductions in staff without adverse side effects. Productivity investments cost money and take time, and therefore are less appealing in the short term, but they are essential to achieving reductions in the cost base without long-term problems.
(2) The phasing in of staff reductions over a sufficient period of time to minimize adverse morale and experience dilution problems (growing the business in related areas can help here, as discussed in the next section).
(3) The deferral of energy conservation measures until they are needed as an alternative to new generation. Investing in energy conservation when the utility has excess capacity only raises costs and prices, and thereby makes the utility more vulnerable to competition (peak load shifting is acceptable because it is revenue neutral).

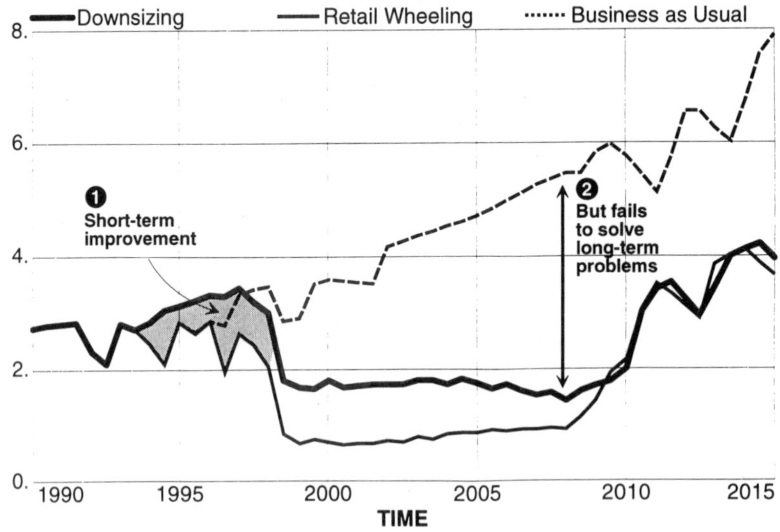

Figure 14.16 Downsizing produces different short- and long-term impacts on earnings per share

Actions like those listed above, taken now, can successfully deal with the threat of competition. As an example, Figure 14.17 shows that while earnings are lower in the short term as a result of the up-front investments (which are assumed to come out of the shareholders' pockets), earnings improve in the longer term as the investments reduce costs, allow lower prices, and prevent a loss of market share. But while these investments work, it is critical that they be taken now. Waiting four years, until competition is a certainty, undermines the success of the strategy—while the utility is making the investments to

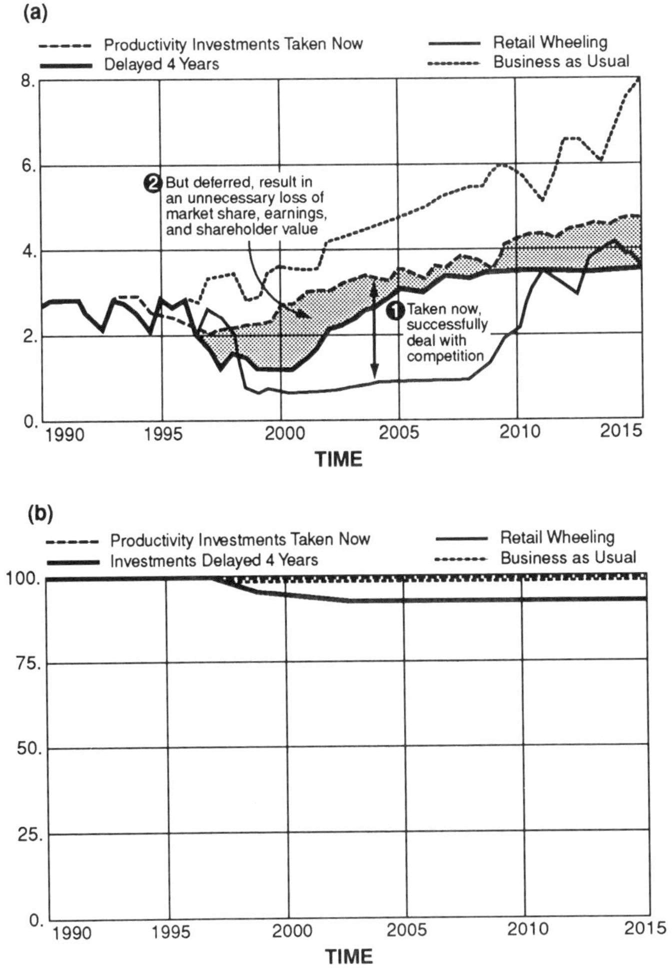

Figure 14.17 The impact of productivity investments and deferred conservation depends on timing

reduce costs, competitors gain share and the utility suffers a further reduction in earnings (see Figure 14.17).

As a final note, mergers and acquisitions are often discussed as possible solutions to the problem of becoming more competitive. Potential benefits of a merger include (i) economies of scale, (ii) a more robust fuel mix, (iii) diversity of loads and non-coincident peaks, and (iv) more political "clout" accompanying the larger size. However, mergers and acquisitions provide real economies of scale only if the combined entity can serve more customers and provide the required power with fewer employees and with no degradation in customer service or system operation. Typically, a merger/acquisition does not save any money in the short term, and requires investments to facilitate the economies (especially in T&D, generation, and customer service systems). Therefore, while consolidation in the industry is likely to be beneficial in the long term, it will still require short-term investments to achieve the benefits.

Investing for the future

The actions above address the short-term problem of retail competition. The long-term difficulty of slow earnings growth can only be dealt with by creating a new earnings model. Borrowing from other industries, electric utilities need

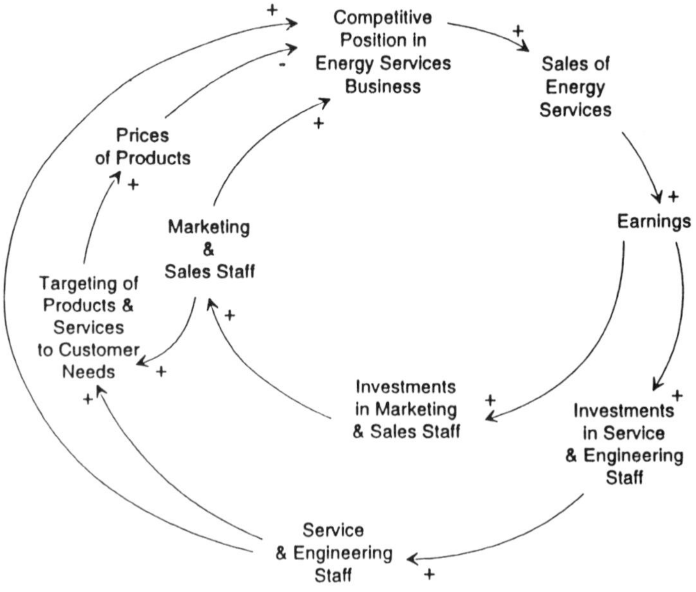

Figure 14.18 Creating new links for a future earnings model

to make investments in order to develop a competitive position in new business areas, most naturally as providers of energy services and products. As illustrated in Figure 14.18, the utility needs to invest in marketing, sales, service, and engineering staff for the new businesses. It will take time for these investments to bear fruit. Staff must gain experience, learn customer needs, and develop (or procure) products and services targeted to those needs. Sales effort and service effort directly affect the utility's competitive position in this industry. Indirectly, they determine the effectiveness of the utility's targeting of products and services, and the prices that can be charged. Competitive position drives utility sales in the energy services business. These sales will eventually create growth in earnings, a portion of which can be reinvested in the new business.

As an example, suppose that the hypothetical utility aggressively invests in energy services, which might include conservation, peak load shifting, electric vehicles, heat pumps, and so on, in addition to the productivity investments and conservation deferral noted above. As the new earnings model takes hold, an increasing share of the utility's income will come from these new businesses. Part of this can come from substituting its own energy conserving investments for purchases of power from other investors; the remainder from new products and services sold to existing and new customers. As illustrated in Figure 14.19, this strategy will allow the utility to begin growing earnings again.

There are other business options for the utility, for example (i) investments in new generation via a subsidiary company, (ii) investments to gain market

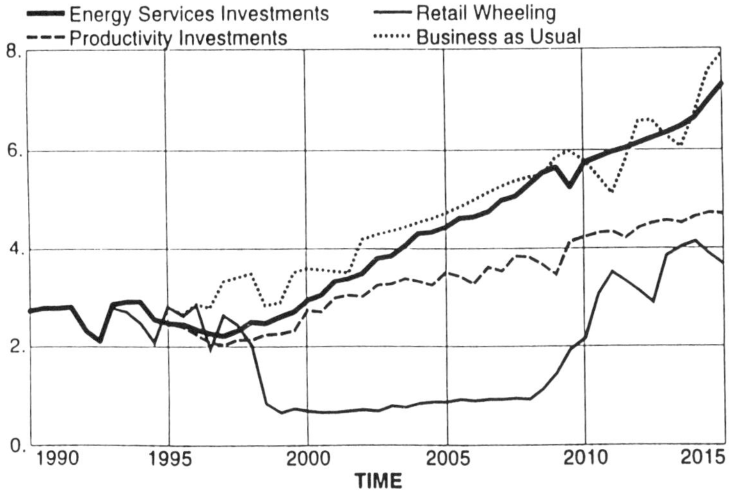

Figure 14.19 A new earnings model will restore growth in earnings per share ($/share)

share in traditional businesses from neighboring utilities, and (iii) investments in non-related businesses ("diversification"). Attempts by companies in a given industry to diversify into unrelated fields have rarely met with success. Investments in new generation via a subsidiary, or gaining share from competing utilities, may become more feasible with significant reductions in utility costs.

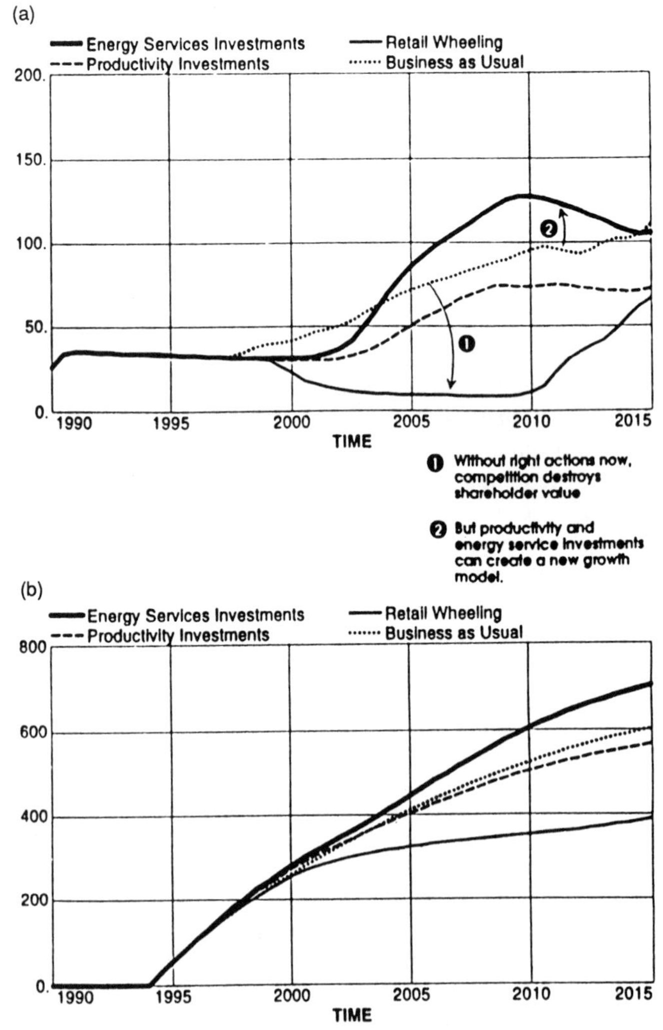

Figure 14.20 The bottom line: maintaining shareholder value, (a) stock price ($/share), (b) present value of dividends @ cost of capital ($ million)

The Bottom Line: Maintaining Shareholder Value

Competition will destroy the traditional earnings model for electric utilities, and with it shareholder value, unless the right preventive steps are taken immediately (see Figure 14.20a). Some of the steps involve investments and actions to improve the cost structure of the utility's traditional investment. In addition, the utility needs to create a new earnings model to replace the old. Taking the right actions will return the utility to continued growth in earnings and stock price, and provide a higher stream of dividends to shareholders (see Figure 14.20b).

CONCLUSIONS

The analyses in this chapter clearly demonstrate that:

(1) Relying on the regulators for a bail-out is not a viable approach. Regulators are not likely to allow a utility to recover all lost revenue through higher rates. Therefore, lost customers means lower returns. Moreover, increasing prices to recoup even some of the lost revenues merely drives more customers to competitors. And with larger price increases, the utility risks having smaller customers band together into buying groups to get lower prices from competitors.

(2) Significantly reducing the utility's cost base is an essential part of any long-term strategy. Some of this reduction can come from cutting "fat", but the utility must be very careful to avoid cutting beyond this point (without enabling investments in productivity, such as new technologies, training, and process re-engineering). Otherwise, the utility risks:
   (a) Adversely affecting plant availability, system reliability, and quality of service.
   (b) Increasing needed expenditures in the longer-term, often on an emergency basis.
   (c) Losing goodwill with customers and regulators, with inevitable adverse consequences.

(3) For aggressive and lower-cost utilities, adopting an "offensive" strategy to acquire new customers can improve performance, but also requires a careful balancing of the short-term benefits against the additional capacity required in the longer term.

(4) Whatever the strategy, the utility must be proactive in preparing for competition *in order to avoid losing any market share*:
   (a) The "hit" from a loss of market share significantly reduces earnings, dividends, and stock price, and requires a large infusion of debt to cover the short-term losses.

   (b) Longer-term recovery is then hindered by the high debt ratio and low stock price.
(5) Investing for the future in productivity enhancing technologies and processes, and in demand-side management, offers the possibility of both lower costs to consumers and reasonable and growing profits to the utility.
(6) The "bottom line" difference between successful and unsuccessful strategies is significant. The difference in present value to shareholders between the successful "energy services" strategy and the unsuccessful "do nothing" strategy, in the face of full competition, can be as much as $150 million (20% improvement) over a 10-year period and $300 million (75% improvement) over a 20-year period.

Whatever the future brings, the monopolistic utility/regulatory structure is going to change. Utilities, unlike other deregulated industries, have an opportunity to seize the moment and take advantage of the time before deregulation becomes a reality. Whether or not deregulation occurs in the next few years, utilities will be well served by aggressively reducing costs, improving service and building better relationships with their customers. Those that do not will be acquisition targets or worse—stripped of large, profitable customers.

REFLECTIONS

Over the years, versions of the SMS-*Utility* model have been used to address issues of demand-side management, cancellation of plants under construction, construction strategy, and financial strategy and dividend policy for the Electric Power Research Institute as well as for individual utilities (Geraghty, 1981; Lyneis & Geraghty, 1984). Most recently, the model has been used to examine competitive strategy for individual utilities (Lyneis, Bespolka & Tucker, 1994).

The model's primary successes are in the areas of consciousness-raising, stimulating thinking, and consensus-building within a management team. These successes stem from the model's ability to handle the whole of the business, competition, and intangibles, and to quickly assess the consequences of alternative actions. It allows managers as a group to visualize their future. Often, this has been done in a "gaming" mode, where teams manage the computer-based utility against a hypothetical competitor.

The "negatives" of this approach stem from the model's lack of detail in specific sectors (e.g. finance and power generation) compared to the in-house models of a utility. The model is therefore less well accepted by the entrenched functional areas within the utility.

REFERENCES

Geraghty, D. (1981) Utility corporate models applied to the "end use versus supply" decision, *Public Utilities Fortnightly* 10 September.

Lyneis, J.M. (1984) Modeling and valuation of utility investments, EPRI report EA-3531, Electric Power Research Institute, Palo Alto, CA, May.

Lyneis, J.M. and Geraghty, D. (1982) Feedback loops: the effect of external agents on utility performance, *Public Utilities Fortnightly*, 4 September.

Lyneis, J.M. and Geraghty, D. (1984) Electric utility strategies for the eighties: taking advantage of a "window of opportunity", *Public Utilities Fortnightly* 24 May.

Lyneis, J.M., Bespolka, C.G. and Tucker, B. (1994) Preparing for a competitive environment—the prospects for America's electric utilities, in E. Wolstenholme and C. Monaghan (eds) *Proceedings of the 1994 International System Dynamics Conference*, University of Stirling.

Chapter 15

Complementary Modelling Approaches for Analysing Several Effects of Privatization on Electricity Investment

Derek W. Bunn, Erik R. Larsen and Kiriakos Vlahos

BACKGROUND

The UK electricity supply industry was substantially transferred from the public to the private sector in March 1991. Prior to 1990, a single, integrated public utility, the Central Electricity Generating Board (CEGB) had operated the whole generation and transmission business, together with twelve regional distribution companies. These twelve distribution companies were also sold to the private sector in 1990. Figure 15.1 summarizes the key points in the new industry structure. The two new generating companies, National Power and PowerGen, inherited 50% and 30% of the capacity, respectively, with the remaining 20% (the nuclear plants) staying in the public sector. A National Grid Company has taken over the transmission business and is responsible for operating the daily "power pool", which is the market-place for the distribution and generating companies to compete in buying and selling electricity. All of this has been motivated by the UK Government's belief in "popular capitalism"

Systems Modelling for Energy Policy. Edited by D.W. Bunn and E.R. Larsen

Generation Business

- Two new private companies created, National Power & PowerGen
- All Nuclear Power Stations transferred to a new Public Company, Nuclear Electric
- Independent Power Producers encouraged
- Deregulated competitive environment

Transmission Business

- Transmission system controlled by National Grid Company (NGC), a private company
- NGC jointly owned by the distribution companies
- NGC responsible for the operation of the electricity pool: the daily "market" for power
- NGC tightly regulated

Distribution Business

- The 12 regional distribution companies privatized
- Separate distribution / supply business
- Distribution business is effectively a monopoly, heavily regulated.
- Supply business will gradually become competitive in the commercial and industrial sectors

Figure 15.1 The new electricity supply industry in England and Wales

and its faith in a deregulated generation business attracting independent producers and thereby promoting greater industrial efficiency.

This chapter has, primarily, a methodological focus. We describe the use of two modelling approaches, optimization and system dynamics, to address separate classes of effects attributable to the privatization process. The insights gained from this joint exercise were found to be quite complementary. Thus, in thinking about the issues and consequences of privatization for *capacity investment*, it is useful to distinguish between those effects due to (i) the transfer of ownership from the public to the private sector and (ii) the competitive structure of the market. For example, these include:

(1) Transfer of ownership effects:
   (a) Rate of return implications

   (b) Capital structure (debt) implications
   (c) Tax implications
(2) Competitive structure of the market
   (a) Market incentives to invest
   (b) Regulatory effects
   (c) Uncertainty and risk
   (d) Competitive strategies

Thus, we found that an optimization framework broadly supported analysis of (1), whereas system dynamics became more useful in developing insights on (2).

The work that is reported here was undertaken during the transitional period to privatization, 1989–1991, and is an extension and elaboration of some early policy studies that were undertaken during that time (Bunn & Vlahos, 1989, 1992; Bunn & Larsen, 1992). The aim in this chapter, however, is not to provide an empirical commentary on the estimated effects of privatization so far (reports from OFFER (1991, 1992) and the UK Government (House of Commons Energy Committee, 1992) document this extensively), but to identify several distinct effects of the privatization process and describe the application of models to facilitate long-term scenarios for their potential impacts. The implications of the modelling are intended to be more general than a particular set of forecasts for the UK electricity industry as seen at that time.

Hence, the reference point for this analysis is the pre-privatization outlook of the industry seen from the perspective of CEGB, a public monopoly, at the end of 1989. Through collaboration with the CEGB on capacity planning models at that time, the modelling framework was well calibrated to providing a continuing baseline situation (Bunn & Vlahos, 1989). This baseline is contrasted with the prospects of a fragmented, competitive industry in private ownership assessed soon after the privatization. Thus, in terms of modelling, we used the previous practice of the CEGB as a starting point, which was based upon a marginal analysis of generating options, supported by a large-scale, long-term, cost optimization model. The results indicated that even before accounting for market inefficiencies, the impact of the transfer of ownership on the pattern of investment decisions is very large.

In trying to explore the implications of the competitive structure of the market, we soon hit the limits of this modelling framework. We realized that these issues required new ways of thinking and different modelling tools. A variety of speculative inputs concerning the behaviour of the market and the strategies of the competing companies became as important as the usual range of assumptions on energy costs, efficiencies and demand. Therefore, in addressing the second set of issues, we adopted the "softer" approach of a system dynamics modelling framework. We found the insights from this approach to be quite complementary to the optimization model. Whereas the optimization model facilitated some understanding of capacity mix, the systems model indicated how cycles in the reserve margin might evolve. While the deterministic optimization model was effective in giving relative generation costs, the system dynamics model demonstrated various ways in which uncertainty in the market behaviour could evolve and hence influence the decision making in the industry.

We will discuss the two sets of issues, transfer of ownership and competitive behaviour, in turn, and then reflect in the final section upon general methodological aspects.

MODELLING THE TRANSFER OF OWNERSHIP EFFECTS

The first set of issues, concerned with the transfer of ownership, can be analysed as an extension of traditional electric utility planning models. Capacity planning for investment generation has generally been undertaken within the framework of large-scale mathematical programming models (Noonan & Giglio, 1977; Bloom, 1983; Vlahos & Bunn, 1988). The decision variables are when and in what type of technology to invest, the objective being to minimize the total discounted capital and operating costs for the whole generation business over a long time horizon (commonly about 40 years, to reflect normal plant lifetimes) and subject to the constraints of meeting demand and various operational considerations. The major uncertainties of fuel costs, demand growth and load shape are investigated by sensitivity analysis, leading to a number of different scenarios from which to assess the robustness of various investment plans (Jenkin, 1988a). It is also usual to include the extra details of average seasonal load shapes and the probabilistic availability of individual plant to fulfil the load, as required. Some utilities may, in addition, need to include transmission and/or hydrological constraints in the planning model (Pereira et al., 1985), although these did not used to be significant in the CEGB context. The CEGB made extensive use of such a large-scale linear programming model for long-term capacity-mix planning and we would expect the new companies to maintain the use of a similar approach in order to take a view on baseline, cost minimizing plans.

Our series of experiments on transfer of ownership issues was therefore undertaken using a linear programming model of the full system in England and Wales, plant by plant, and assuming that the cost minimization investment policies are still maintained (the competitive issues will be considered in the next section). A software program, ECAP-PC, was used to provide a fast, PC based capacity expansion planning aid using Benders decomposition for computational efficiency. This had previously been validated against the

MIXLP program used by the CEGB for capacity planning (Vlahos & Bunn, 1989) and was therefore very suitable for an extensive series of experiments under many different planning assumptions. A summary of the model and the background planning assumptions for the experiments are given in the appendix.

Rate of Return Implications

In the years prior to privatization, as a public utility, investment proposals were discounted at a 5% real rate of return to assess viability. To satisfy the private market, real rates of at least 8%, and possibly closer to 12%, were suggested (Dimson, 1989). It could be argued that a change of ownership by itself would not cause such a large step in the required rates of return, but it was a reflection of the divergence between public and private sector rates. More recently, for example, public sector rates have increased to around 8%, closer to those in the private sector. Nevertheless, to the extent that the privatized company faces more business and financial risks, its market rate of return has to incorporate a larger risk premium effect.

The first impact of this change in the discount rate for new investment is to make the high capital, low operating cost plants less attractive. Capital costs occur earlier than operating costs and are therefore penalized more by higher discount rates. By running our planning model under varying discount rates, this switch in the capacity mix from capital intensive (nuclear and coal) to higher operating cost plant (combined and open cycle gas turbines; CCGT and OCGT) due to higher discount rates is shown in Figure 15.2. OCGT plant are typically the small, gas turbines that have been used to provide peak load capacity. CCGT make additional use of the waste heat from gas turbines to provide a further thermal source of power; they are thus more efficient than OCGTs, but have higher capital costs. Thus, we can observe in Figure 15.2 the major switch from coal to CCGT for quite modest real rates of return, and understand the economic pressure behind the so-called "dash for gas" which characterized the capacity building intentions of all generators during the early years of privatization. According to the National Grid Company (1992) in 1992, 24 GW of new CCGT plant was being planned for commissioning by 1998, representing a massive reserve margin of 60%, *if* it had all gone ahead without further retirement of current plant. The political consequences of this became evident as early as autumn 1992, when a major pit closure programme was announced by British Coal. More recently, we have seen a slowing down, with several proposed new power plants being deferred or cancelled. Nevertheless, the uncertainty in the prospect of overcapacity actually adds to the business risk, thereby increasing the cost of capital even further.

Figure 15.3 shows the impact of this technology switch on total annual operating costs. At 12%, the annual operating cost is well over twice that of the

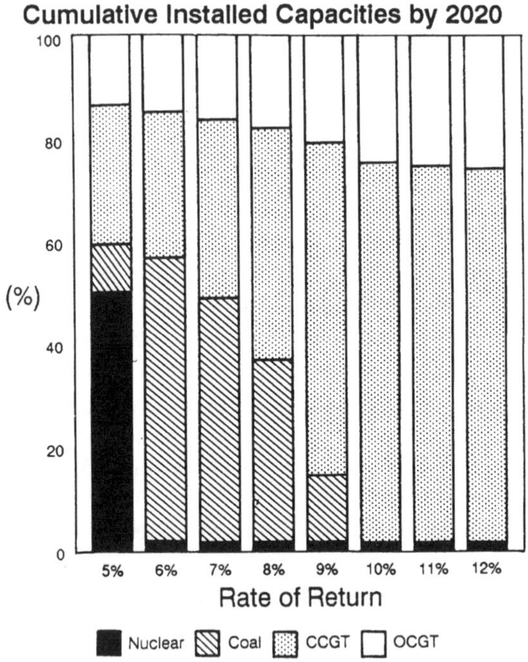

Figure 15.2 The impact of rate of return on capacity mix

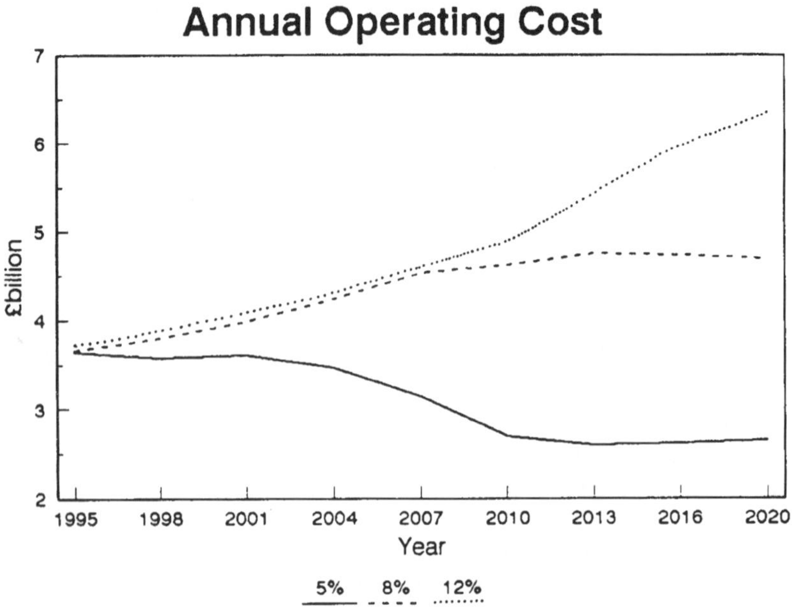

Figure 15.3 The impact of rate of return on operating costs

(unprivatized) 5% case, although the effect takes time to evolve because the new investment in more capital intensive plant is gradual.

Although we are not seeking to model prices in this study, it is worth noting that the higher rate of return has a second effect through increasing the profit margin required for the private sector. The impact on prices of a switch to higher operating costs and the higher required margin has been analysed more extensively in an earlier paper (Bunn & Vlahos, 1989). These two effects are somewhat offset by the reduced capital costs, but the net effect over the long term is a considerable increase in prices through rate of return requirements alone. An assessment made on this basis in 1989 (Bunn & Vlahos, 1989) of an overall price increase of around 30% is close to more recent estimates of the actual outcome over the first three years of privatization (Yarrow, 1992), during which it was largely the initial transfer of ownership effects that became evident. These results are an empirical contrast to an often expressed view that transfer of ownership by itself should not exert a significant influence on prices (e.g. Yarrow, 1992). Privatization without an increase in required rates of return would seem to be very rare. Thus, it is interesting in this context to observe that in order to reduce upward pressure on prices, the Government sold the industry at the heavily discounted average price of £75/kW (compare a new CCGT with capital costs of about £375/kW at the time). Had the Government chosen a stricter regulatory policy to restrict rates of return on new investments and existing capital assets (as with the water privatization), thereby reducing the pressure for higher prices, then presumably they would have had to discount the sales proceeds even further.

Tax Implications

In the public sector context, corporation tax was not explicitly considered in the capacity planning process (Jenkin, 1988b). Prior to 1984, there was no liability to tax, and even afterwards, when profits were taxed, it was not seen as a significant issue: the taxation of a public utility appeared to be a rather circular process.

However, in a private sector context, assuming that the company is profitable and pays tax, then corporation tax has an asymmetric influence on capital and operating costs. In general terms, operating costs are completely tax-deductible whereas capital costs are only recovered through the tax benefit of capital allowances (i.e. depreciation). The year-by-year tax situation for an electric utility is, of course, very complicated and we are only looking for very broad, longer term insights from our experiments. For a more precise analysis of how different corporation tax policies can affect investment see Mole (1987, 1988) and London Economics (1990).

Figure 15.4 demonstrates an aspect of the way in which accounting for tax in the planning model favours the less capital intensive technologies. The

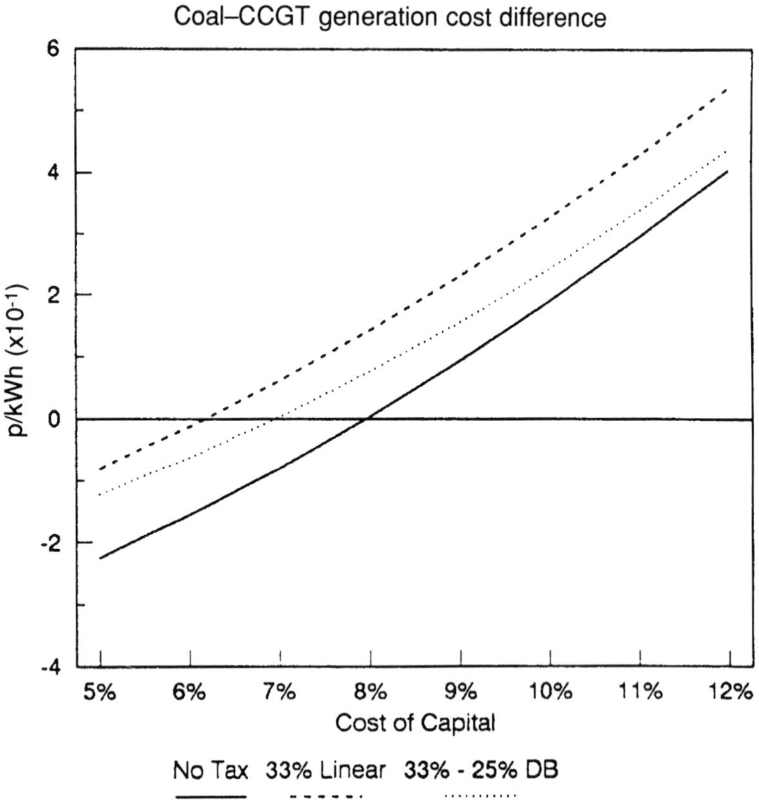

Figure 15.4 The impact of taxation on investment decisions

continuous line shows the average generation cost difference between coal and CCGT. At just less than 8%, they are equivalent. The dotted line shows that the usual 25% declining balance depreciation process causes the equivalence point to move down to just less than 7%. With a linear schedule replacing declining balance, the cost difference moves further in favour of CCGT. This is because the depreciation is spread more to the later years and is therefore adversely affected by the discounting process. In both cases corporation tax at 33% was assumed. Mole (1987, 1988) observed that inflation can further complicate this picture through tax liabilities being paid a year later. Our analyses showed this to be a second order effect in this problem.

Debt Implications

Pursuing the theme of corporate taxation further, it is clear that if there is a substantial amount of debt in the capital structure (as indeed is the case with the new companies, which have debt of about 25% of total assets), because interest payments on debt are tax deductible, but dividends are not, then the effective cost of capital is reduced as the fraction of debt capital goes up. Figure 15.5 shows this in the context of our model. We assume a declining balance depreciation schedule and, in order to isolate the impact of the interaction of taxation and debt, also assume that the costs of debt and equity capital are the same. It is apparent that debt financing therefore favours capital intensive technologies by reducing the effective cost of capital. For example, 50% debt counteracts that coal–CCGT difference attributable to the basic taxation effect which we saw in Figure 15.4.

To the extent that both National Power and PowerGen have negotiated substantial financing arrangements and with the independent power consortia also looking to leverage their incipient power projects with large borrowings, the above results are encouraging for capital investment and lower operating

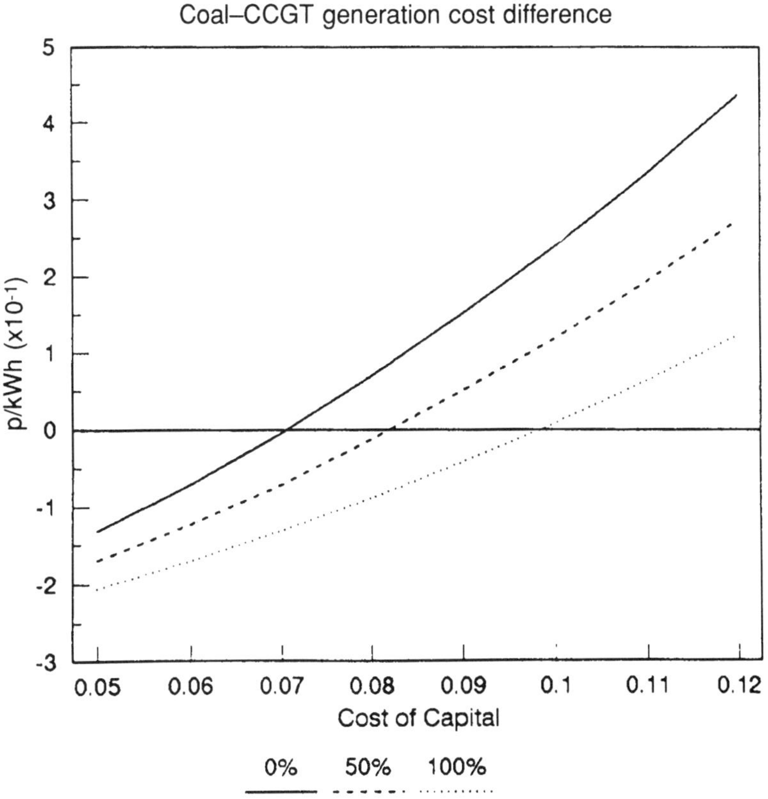

Figure 15.5 The impact of capital structure on investment decisions

costs. However, in practice the cost of debt and equity capital are not the same, and higher debt situations do induce increased risk aversion in the investment criteria. Clearly, various differential cost-of-capital scenarios could be run in our context; we just wish to make a basic point about the impact of debt on after-tax discount rates. We will analyse the risk situation more in the next section, where we look at the effects attributable to market competition.

MODELLING THE COMPETITIVE EFFECTS

In order to speculate on the competitive factors influencing investment behaviour by the new private companies, it is first necessary to look at the way in which the market has been structured and how it will be regulated. Figure 15.6 shows a schematic representation of the central role played by the power pool in seeking to produce half-hourly prices depending upon the demand/supply balance.

The power pool, administered by the National Grid Company (NGC), every day invites bids from generators (National Power, PowerGen, Nuclear Electric, Electricité de France, Scottish Power and the independent power producers) to supply power over the next 24 hours. These are matched against the requirements from the twelve regional distribution companies to produce a least cost, half-hourly generation schedule. The price per half-hour is fixed at that of the marginal unit scheduled for that period, i.e. the system marginal price (SMP). There are several other payments to generators for "ancillary services", such as spinning reserve, frequency and voltage control, but these will not be considered here as being significant in driving investment decisions. Clearly, any unit that has consistently cheaper average costs than the marginal

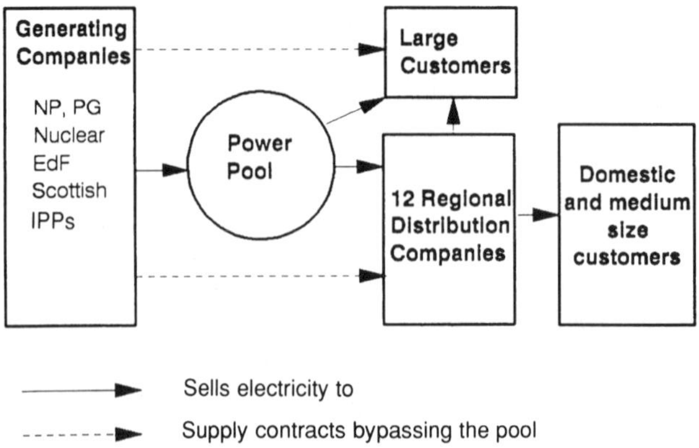

Figure 15.6 Competitive structure of the market

plant will make money for the company in that period. However, in the longer term, if investment in new capacity is converging on the same type of technology (we have seen in the previous section how attractive CCGT has become through the transfer of ownership effect), the scope for undercutting SMP will become more limited, resulting only from possible economies of scale and cost of capital advantages. The taxation and debt leverage effects, discussed previously, could become important instruments in reducing average costs.

Realizing that the potential to undercut SMP may not be sufficient to stimulate new capacity investment, the industry regulator has made the price received by generators to consist of SMP plus an extra "capacity payment", which is determined as:

$$\text{LOLP (VLL} - \text{SMP)}$$

where VLL is the value of loss of load (initially set at £2/kWh) and LOLP is the loss of load probability per half-hour. LOLP is computed 24 hours in advance, taking account of demand uncertainty and the probabilistic reliability of individual plant in meeting the load as planned. Thus, in periods where there is a lot of excess capacity, LOLP will be small and there will be little incentive to invest, whereas when capacity is short, LOLP will be high and thereby encourage investment. Although it has been the case that the distribution companies have hedged against pool price by the extensive use of contracts with the generators, the belief of the regulator was that new capacity will ultimately be signalled by pool price expectations and the capacity payments in particular (OFFER, 1991). Furthermore, by adjusting the VLL, the regulator can, in principle, influence the reserve margin for the industry as a whole.

Any model seeking to represent the structure of the market and the investment behaviour of the companies in it must therefore capture the feedback effect of LOLP on capacity and its effect on the incentive to invest. While it is the case, as indicated in Figure 15.6, that direct contracts between generators and both the distribution companies and large industrial customers can bypass the pool, the pool price will still be the reference point for such hedging (OFFER, 1991). Thus, a system dynamics framework for modelling the competitive effects was considered more appropriate than further extensions of the traditional linear programming approach, as used in the previous section. Furthermore, to the extent that we wish to consider a variety of strategies for the companies and understand their effects on investment, the methodological emphasis switches from optimization to simulation. Geraghty & Lyneis (1985) developed a large system dynamics model of the US electric utility industry in order to simulate the behaviour of pricing and investment, and Ford, Bull & Naill (1989) used a system dynamics approach to model conservation policies for the Bonneville electric utility. In our context, a simulation model using system dynamics has been developed to help understand

(i) the effects of LOLP in determining capacity and (ii) the competitive behaviour of the newly privatized companies. The structure and principal results of this model are summarized here. For further details, the model is described more fully elsewhere (Bunn & Larsen, 1992), although the results reported below represent updated extensions of those previously published studies.

Figure 15.7 shows the feedback structure in this system dynamics model. The model was originally developed in STELLA (Dynamo and iTHINK versions have also since been implemented, and a more general object-orientated model has been programmed in SmallTalk (Ninios, Vlahos & Bunn, 1995). The feedback effect of the capacity payments (through LOLP) is the central feature of this model and the experiments focused not on capacity mix, as in the previous section, but on overall reserve margin, where this is defined as the difference between installed capacity and demand, and expressed as a fraction of demand.

One type of technology (CCGT) is taken to be the preferred choice, and investment will occur when the projected capacity element in the pool price

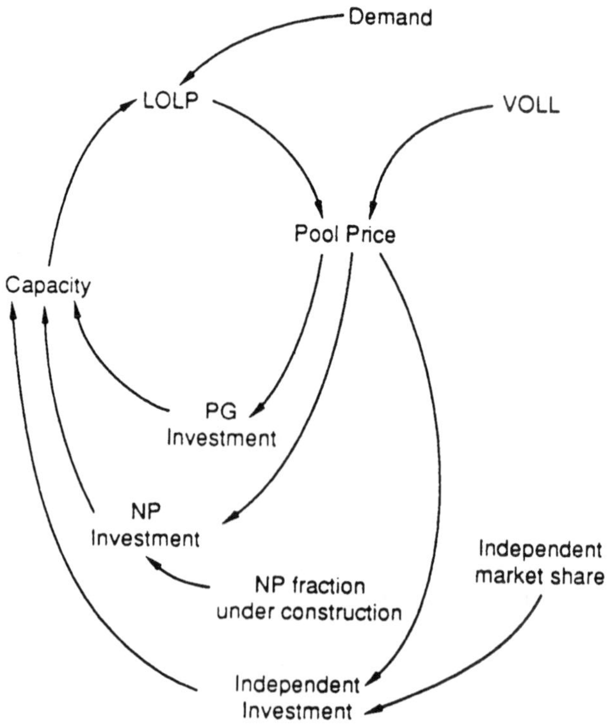

Figure 15.7 A system dynamics model of capacity investment

exceeds the annuitized capital cost of a new CCGT plant. Clearly, as before, the cost of capital to the investing companies becomes a key parameter in this decision rule. This is a simple basis for initial modelling, but the design of the overall simulation model is being explicitly structured to accommodate more elaborate investment decision-rules involving SMP as well as multiple technologies. The functional relationship of LOLP to reserve margin, shown as base LOLP in Figure 15.8, is taken from a previous study by the Electricity Council (1985), although several sensitivity analyses were done to account for changing characteristics of plant mix. One of them, convex LOLP, is also shown in Figure 15.8 and represents an (arbitrarily drawn) function intended to be more representative of the switch to a larger number of smaller plants (CCGTs) on the system. A baseline discount rate of 13% (pre-tax) was used and the system was modelled net of nuclear, Scottish and French power in order to concentrate on competition between the new private companies. As before, sensitivity analysis on this cost-of-capital baseline was extensively undertaken.

As Figure 15.7 indicates, in our initial model, the decision rules for investment by National Power, PowerGen and the independent power producers are slightly different, and this was to introduce some strategic differentiation into the investment simulations. With National Power being the larger, and initially the more aggressive, company, we assume that it may pursue a more market-share orientated strategy compared to PowerGen, which may concentrate on maintaining an attractive rate of return (Thomas, 1990). Thus, to reflect competitive differentiation, we assumed that:

- PowerGen seeks an "economic" baseline discount rate for investment.
- National Power will take lower effective discount rate for investment, to

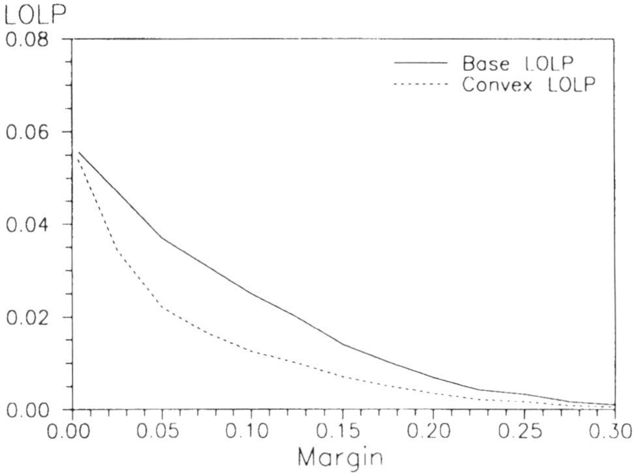

Figure 15.8 Average LOLP as a function of reserve margin

reflect its primary objective of market share maximization. However, we have increased its actual cost of capital according to how much capacity is under construction at the time (to reflect increased risk).

- The independent power producers will be looking for a higher return (we have assumed a differential of 3% above the PowerGen baseline, consistent with a similar analysis by UBS Phillips and Drew (1990), because of higher costs of project finance for a smaller utility). However, if their market share is below 20%, then for strategic reasons, the distribution companies (who are allowed to generate up to 20% of their power) will encourage the formation of consortia with the independents, and this would have the effect of introducing lower required investment returns.

The baseline scenario and economic background assumptions for the UK system are comparable to those used in the previous section, and are given in detail in the appendix. We looked at a number of competitive and structural features that would affect investment behaviour:

- The degree of *foresight* that each company applies to the investment decision is varied between 0 (myopic) and the full four years. Thus, in the former case, the company is just responding to the immediate signal, whereas in the latter case the investment calculation is based upon a projection of the demand and plant retirement trends for four years into the future in order to infer the LOLP, and hence the capacity element.
- *Uncertainty* in the forecast demand over four years is introduced, by random simulation of errors in expected demand, within *ranges* of 0%, 5%, 10% and 20%. According to the values given by UBS Phillips and Drew (1990), even 20% does not seem unreasonably high.
- Also coming into the investment calculation are scenarios for *information exchange* between the competing companies. Information on the building of new plant, retirement or life-extensions of existing plant by competitors can be complete or absent (until it actually occurs). We consider factors of 1.22, 1.00, 0.66 and 0.33 as estimates of competing companies' plans over the following three years to be combined with *foresight* to estimate LOLP.

We have used this model to consider some effects of the transition to a new competitive environment with respect to (i) market structure, (ii) risk and uncertainty and (iii) strategic competition.

Effects of Market Structure

One of the radical consequences of privatization has been the formation of an industry without any obligation to supply. Through VLL, the regulator may seek to influence the reserve margin, but in this case the margin is clearly market determined, and will depend upon the generators' propensity to invest.

Figure 15.9 shows a simulated profile in the reserve margin which would occur if the LOLP were to drive capacity investment in the simple case of investors reacting directly to the year-by-year capacity element in the pool price. Without any foresight or information exchange, the nonlinearity of the pricing signal and the lag in construction, as expected, induce cycles of overcapacity and undercapacity. While clearly being myopic in its assumption of investment behaviour, it nevertheless shows the potential problems that this market mechanism, by itself, can incur. More realistically, if we assume a 33% knowledge of competitive activity, reliance on the market signal still produces noticeable cycles, but of much lower amplitude, and with an apparently lower average value.

A more optimistic reference scenario is with four years' foresight, full information exchange and 5% uncertainty. Under these circumstances Figure 15.10 shows VLL at a third less and a third greater than the reference case of £2/kWh. This baseline simulation (the fine dotted line) seems to be varying about an average reserve margin of around 24%, which is consistent with the presumed intentions of a VLL of £2/kWh (Mole, 1988), and is compatible with the margin assumed in the optimization model used for capacity mix in the previous section. A reduction or an increase in VLL by 33% seems to be associated with an asymmetric change in the reserve margin, with an increase in VLL being much less effective than we would have thought. It is important to realize that in such system dynamic studies, a simulation is shown as an indicative scenario, not a forecast. They are useful mainly for analysing the

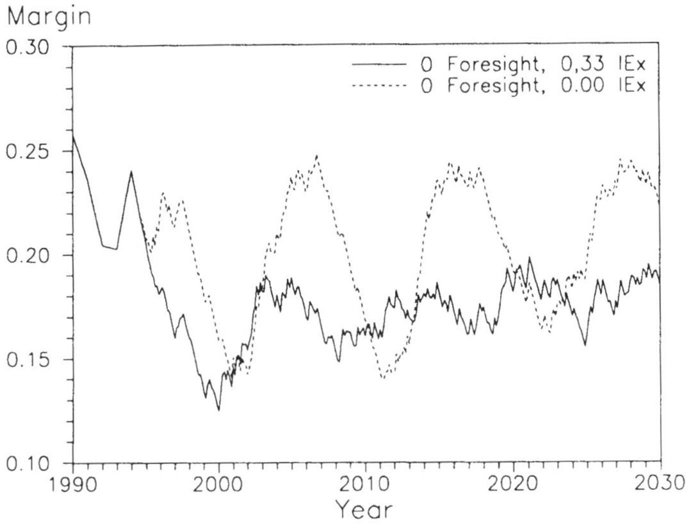

Figure 15.9 Reserve margin as a function of foresight and information exchange

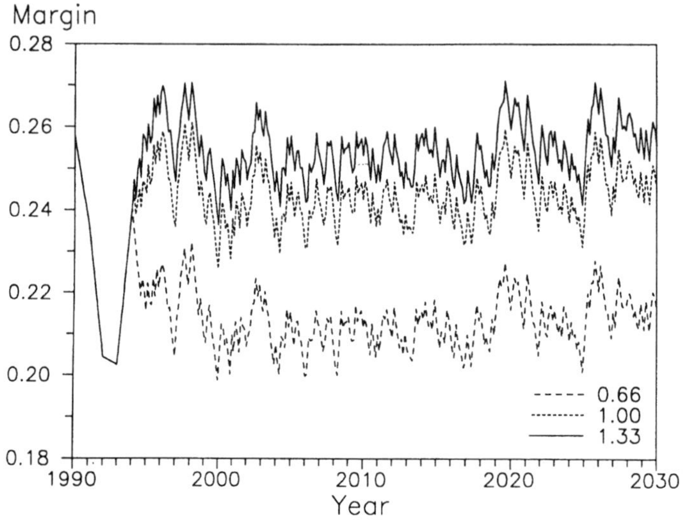

Figure 15.10 Reserve margin sensitivity to VLL, plus and minus 33% on £2/kWh

comparative sensitivities of different factors (although further simulations revealed that our sample here is a reasonably typical realization).

Figure 15.11 shows the average reserve margin, calculated over the final thirty years of our simulations (to avoid the "initial effects" in the 1990s) for the baseline industry model, as a function of the basic "economic" discount rate. Recall that in the baseline model, PowerGen uses an "economic" 13%, with National Power and the independents being differentiated for strategic reasons. The simulations in Figure 15.11 vary this "economic" rate (i.e. PowerGen), while maintaining the strategic differentials. The base LOLP is the baseline function relating LOLP to reserve margin, as produced by the Electricity Council (1985). Recall from Figure 15.8 that the convex LOLP is a similar, but more convex, version estimated to reflect the new type of capacity mix now coming into operation (i.e. more smaller CCGT units than the larger coal and nuclear plants envisioned in 1985). It is apparent that capacity mix, as discussed in the earlier part of this chapter, will have important implications for the reserve margin and investment behaviour through its influence on the shape of the LOLP function.

Effects of Risk and Uncertainty

Uncertainty on demand, fuel prices, competition and the workings of the power pool all induce risk averse investment behaviour, resulting implicitly in higher

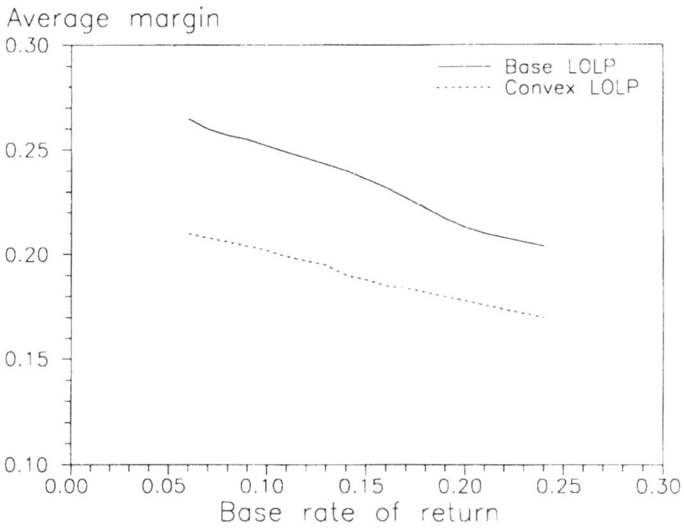

Figure 15.11 The impact of rate of return on average reserve margins

discount rates for capital investment. This favours both less capital intensive options (e.g. CCGT) and also plants with flexible technology (e.g. mixed fuel capabilities). Furthermore, an understandable response to risk aversion on the part of the distribution companies, in the face of pool price uncertainty, is to take action to hedge against price exposure by extensive contracting and to increase their equity stakes in new generating capacity. The cost to the country of this market uncertainty is therefore higher prices and higher reserve margins, as became evident by the end of 1992.

As far as reserve margin is considered, the results reported by Bunn & Larsen (1992) suggest that:

(1) Increased demand forecasting errors induce higher amplitudes in the cycles of capacity and slightly higher average reserve margins.
(2) Providing a reasonable amount of foresight and information exchange could be maintained in the industry, the cyclical behaviour of the reserve margin could be largely eliminated.

Both of these observations suggest a role for the regulator in providing an information management and forecasting service to the industry in terms of timely notifications of planned capacity additions and retirements, together with state-of-the-art forecasts for peak demand and LOLPs. The more that uncertainty can be reduced in the pool, the less risk averse will be the behaviour of the competing companies.

Strategic Competition

Figure 15.12 gives a market-share projection based on the competitive assumptions for the three private companies, outlined previously. Despite National Power being modelled as seeking to maintain market share at the expense of rate of return, our analysis shows that this 1.5 percentage point differential against PowerGen (recall the basic discount factors of 11.5% for National Power, against 13% for PowerGen) was not sufficient. Some sensitivity analysis showed that this differential would have to be doubled, with National Power having a basic discount rate of 10% in order to maintain its share of the market. In this latter case, it would be PowerGen who would lose out to the independents. Such a large difference in cost of capital would appear impossible to maintain over the long term, even with some of the creative taxation and debt leverage possibilities indicated in the previous section, particularly in view of the higher business risks that a high capacity building programme involves.

Figure 15.12 is not meant to be a forecast of the business future for PowerGen and National Power, only a projection of what would be implied by two different investment strategies and the extent to which different cost-of-capital assumptions can be modelled into the competitive (i.e. market share) dimension. In the previous section we related such assumptions to capacity mix.

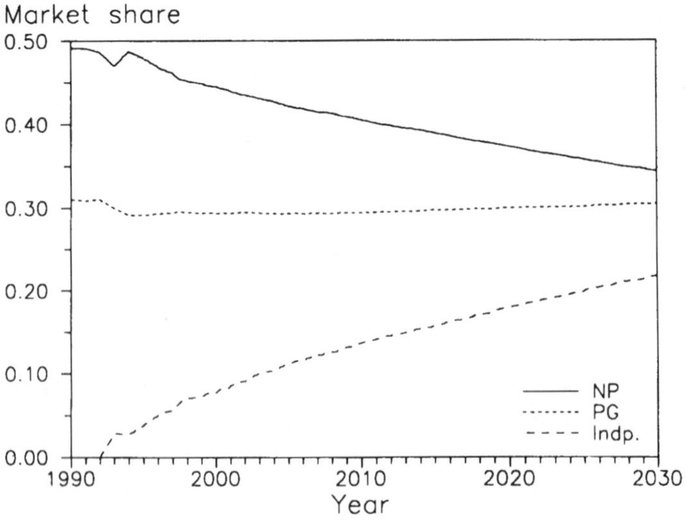

Figure 15.12 Market shares under three simple strategies

SUMMARY AND DISCUSSION

In looking at the effect of privatization on capacity mix, we observed that the higher required rate of return in the private sector implied a switch to less capital intensive plant. A large-scale, long-term optimization model facilitated an analysis of this transfer of ownership effect. For rates above 9%, CCGT becomes the preferred technology. The impact of corporation tax encourages this trend, but debt leverage may offset it slightly. Taking account of inflation in the capital budgeting evaluation had only a minor effect. The competitive structure of the industry increases the business risks to the generation companies, and this in turn may encourage investment in less capital intensive technologies (e.g. CCGT). By 1992, a major trend towards accelerated retirement of older coal plant and a CCGT building programme had indeed become evident (National Grid Company, 1992).

Apart from the capacity mix changes, we also looked at the effect of the new market structure on reserve margin. Here a system dynamics model proved to be effective in generating insights on potential cycles in capacity, the degree of uncertainty in the pool price and the sensitivity of these to a number of features. The margin is potentially quite unstable and will require the regulator to provide (i) information exchange on capacity intentions (new plant and retirement) by generators and (ii) state-of-the-art forecasts on demand and load shapes in order to rescue this susceptibility. The market price signal, by itself, will not provide a stable signal for investment. Furthermore, such uncertainty in the operation of the market, as perceived by the distribution companies in 1990–1992, promotes risk averse behaviour and is a major factor in the extent of contracting to hedge the pool price. It is also a motivation behind the desire for equity stake-holding in new independent power producer generation projects. The premiums paid for hedging and the potential overcapacity in the industry are essentially the costs to the consumer of uncertainty in market behaviour.

In addition, following much of the earlier analysis of the rate of return issue in driving investment and its consequent effects on capacity mix and operating costs, we obtained some indication of its competitive impact. Again, the system dynamics model was useful here. To the extent that different companies may have cost-of-capital advantages, or may effectively use different rates for strategic reasons, then this has, over the long term, very significant effects on market share trends.

As well as focusing on different aspects of the problem (capacity mix versus reserve margin), the complementary use of the optimization and systems models provided useful inputs for each other. Reserve margin, the main output of our systems model, is exogenous to the optimization model. Although not implemented so far in our model, LOLP and the retirement schedules, which are exogenous to the systems model, can be outputs of the optimization model.

Moreover, the uncertainty displayed by the system simulations can give insight into the business risks and hence the cost of capital required in both models. Our overall approach was really a synthesis of "hard" and "soft" OR modelling.

Finally, some more general comments on methodology are, perhaps, appropriate. Electric utility planning has been one of the major application areas for operational research methods. This has been because of the well-specified, but continually challenging, nature of the problems, with major payoffs in terms of cost minimization for improved operational procedures, a technically orientated managerial stock and extensive data resources. As the companies move into the more competitive world, the methodology also needs to change. A broader range of objectives, strategic as well as financial, has to be faced. Risk and uncertainty become major concerns, caused now more by competitive elements, in addition to the previous concerns of demand, fuel price and regulatory factors. The possibility of business failure now exists as it never did before. This will make the models less well specified, require analyses more in the form of multiple scenarios and, most significantly, will involve managers and decision makers much more closely in the modelling process. The engineering culture of the old organizations, where the results of modelling efforts are summarized for management, should be replaced by an entrepreneurial culture, where interactive decision support models aid organizational learning and the understanding of strategic investment decisions. Commenting upon the difficulty of applying system dynamics models in the electric utility industry in the US, Ford, Bull & Naill (1989) relate this to the needs of managers coming from an engineering background for considerable amounts of detail in the models in order to have confidence in their validity. This is because their concerns have traditionally been more operational than strategic. Developing a useful mixture of optimization and simulation models to support planning in the privatized companies will therefore challenge not only the modelling process, but also the managerial attitude towards using models. In most implementations of operational research, we sense a desire for a more technically proficient managerial culture. In this case, for strategic modelling of investment, we may be encouraging managers to work with models of less technical detail than they would normally feel comfortable with.

APPENDIX: PLANNING ASSUMPTIONS

The linear programming model for the capacity planning experiments in the second section was developed on data available in 1988, and validated against that used by the CEGB at the time (Vlahos & Bunn, 1988). The planning horizon is split up into twelve planning periods from 1994, the first eight of which are of 3 years, the last four being of 5 years. The planning horizon is thus 1994–2037. Each planning period within this has eight different seasonal components to take account both of seasonal variations in the shape of the demand profile and the plant unavailability at different times of the year for maintenance and other reasons.

Plants are represented individually in the model with their own operating costs, planned lives and reliabilities. The usual procedure of annuitizing capital costs over the plant life, at a predetermined discount rate, is followed. The data on existing plant (operating costs, plant life, reliability) and the seasonal demand profiles are broadly consistent with those that became available in published reports by the Electricity Council, the CEGB and others at the time. The number of existing plants comprises the full CEGB system, and four plant expansion options were considered, namely nuclear, coal, CCGT and OCGT. A planning (i.e. reserve) margin of 24% was assumed. In addition, the following assumptions were made:

(1) Peak demand growth from 51 GW in 1995 at 1% per annum.
(2) World price of imported coal in 1995 delivered to UK of \$50 per tonne, in real terms constant thereafter. This converts, at an exchange rate of \$1.7/£, to fuel costs for new coal plants of £11/MWh. Construction cost for a standard new station with two 840 MW generating units is taken as £770/kW, with additional interest during construction computed over 3 years at a specified discount rate. Annual fixed operating costs of £23/kW, and initial average availability of 60% for the first 5 years, with 77% for the remainder of a 45-year life for new coal plant were also assumed.
(3) For nuclear, basic construction cost of £1260/kW for an 1175 MW station, with additional interest during construction computed over 4 years, fixed annual operating costs of £19.6/kW and fuel costs of £4.5/MWh were assumed. Initial average availability is 62% for the first 5 years, and thereafter 72% for a 40-year life.
(4) Gas turbine assumptions were: basic construction cost of £240/kW for OCGT and £400/kW for CCGT, interest during construction for 1.5 years, fixed operating cost of £6/kW/year, fuel and variable costs of £45/MWh for OCGT and £17/MWh for CCGT, with 80% availability for a life of 40 years.
(5) The installation of renewable energy sources such as wind and tidal is not treated as a decision variable, but account is taken of some progressive introduction. Specifically, the percentages of energy delivered by renewable sources are taken to be 0.3% in 2000 and 3.3% in 2004, rising to 6.0% in 2035. Imported power of 1.5 GW is taken account of in the merit order, and pumped storage is treated as contributing to the operational reserve margin.

The system dynamics model developed in the third section is consistent with the above data. Its initial baseline discount rate for investment was 13%. For National Power, the market share seeking strategy, the effective rate was 1.5% lower but increased according to the amount of capacity under construction. Thus, with 12% capacity under construction, the discount rate was the same as the baseline, but with 18% capacity under construction, it was assumed to be 3.5% higher. Other values were interpolated. Similarly, independent power producers had a discount rate of 3% higher than the baseline, but dropping to 0.25% below according to how far below a 20% market share they had established.

The net demand, which has to be met from the activities of National Power, PowerGen and the independents, was assumed to grow from the base in 1990 of 36 000 MW (which is the peak demand of 48 000 MW less 12 000 MW available from nuclear, hydro and imports). The capacity situation (MW) in 1990 was taken as 29 664 MW for National Power, with 600 under construction for 1993 and a retirement rate of 740 per annum; 18 712 MW for PowerGen, with 900 under construction for

1993 and a retirement rate of 460 per annum. Thus we were assuming that there will be no increases in the nuclear, hydro and imported power sources. In fact, our uniform assumptions for retirement were an underestimate for 1991/92, as National Power and PowerGen phased out their stock of older coal plant sooner than had been anticipated pre-privatization. However, retirement rate did not affect our main conclusions materially. We assumed the same capital cost of £400/kWh for CCGT, with a "planned life" of 40 years but an "economic life" (i.e. for NPV calculation purposes) of 25 years. These plants can be built in tranches of 2000 MW, 1500 MW, 1000 MW and 500 MW, with a company always selecting the largest economic size. Construction time is four years.

ACKNOWLEDGEMENTS

This chapter is based on a paper previously published by the Operational Research Society.

REFERENCES

Bloom, J.A. (1983) Solving an electricity generating capacity expansion planning problem by generalized Benders decomposition, *Operations Research* **31**, 84–100.

Bunn, D.W. and Larsen, E. (1992) Sensitivity of reserve margin to factors influencing investment in the electricity market of England and Wales, *Energy Policy* **20**, 420–429.

Bunn, D.W. and Vlahos, K. (1989) Evaluation of the long-term effects on UK electricity prices following privatisation, *Fiscal Studies* **10**, 104–116.

Bunn, D.W. and Vlahos, K. (1992) A model-referenced procedure to support adversarial decision processes: application to electricity planning, *Energy Economics* **14**, 242–247.

Bunn, D.W., Larsen, E. and Vlahos, K. (1993) Complementary modelling approaches for analysing several effects of privatization, *Journal of Operational Research Society*, **44**, (10), 957–971.

Dimson, E. (1989) The discount rate of a power station, *Energy Economics* **11**.

Electricity Council (1985) Report on the generation standard, Electricity Council, London.

Ford, A., Bull, M. and Naill, R. (1989) Bonneville's conservation policy analysis models, *Energy Policy* **15**.

Geraghty, G.M. and Lyneis, J.M. (1985) Feedback loops: the effect of external agents on utility performance, in J. Plummer, E. Oatman and P. Gupta (eds) *Strategic Management and Planning for Electric Utilities*, Prentice Hall, Englewood Cliffs, NJ.

House of Commons Energy Committee (1992) *Consequences of Electricity Privatisation*, HMSO, London.

Jenkin, F.P. (1988a) Hinkley Point C power station, CEGB proof of evidence, Central Electricity Generating Board, London.

Jenkin, F.P. (1988b) Hinkley Point C Power Station: memorandum on the economics of nuclear and fossil fuelled plant, CEGB proof of evidence, Central Electricity Generating Board, London.

London Economics (1990) Corporation tax and investment in UK manufacturing, London Economics, 91 New Cavendish St, London.
Mole, R.H. (1987) An analysis of the burden of corporation tax upon investment of retained profits in plant and machinery, *Accounting and Business Research* **17**, 311–318.
Mole, R.H. (1988) Corporation tax on investment in plant and machinery: two extreme cases. *Omega* **16**, 175–180.
National Grid Company (1992) Seven year statement, NGC, Sumner St, London.
Ninios, P., Vlahos, K. and Bunn, D.W. (1995) Industry simulation: systems modelling with an object oriented/DEVS technology, *European Journal of Operational Research* **81**, 521–534.
Noonan, F. and Giglio, R.J. (1977) Planning electric power generation: a nonlinear mixed integer model employing Benders' decomposition, *Management Science* **23**, 946–956.
OFFER (1991) Report on pool price enquiry, Office of Electricity Regulation, Hagley House, Birmingham B16 8QG, UK.
OFFER (1992) Review of pool prices, Office of Electricity Regulation, Hagley House, Birmingham B16 8QG, UK.
Pereira, M.V.P., Pinto, L.M.V.G., Cunha, S.H.F. and Oliveira, G.C. (1985) A decomposition approach to automated generation/transmission expansion planning, *IEEE Transactions on Power Apparatus and Systems* **104**, 3074–3081.
Thomas, D. (1990) Competition jolts the electricity industry, *The Financial Times*.
UBS Phillips and Drew (1990) The electricity industry in England and Wales, London.
Vlahos, K. and Bunn, D.W. (1988) Large scale evaluation of Benders' decomposition for capacity expansion planning in the electricity supply industry, London Business School research report, electricity planning project, London Business School, Regent's Park, London NW1 4SA, UK.
Yarrow, G. (1992) British electricity prices since privatisation, research report, Regulatory Policy Research Centre, Hertford College, Oxford.

Index

Note: Page references in *italics* refer to Figures; those in bold refer to Tables

Annual Energy Outlook 10, 17
(1994) 12, 13, 16, 18, 27
(1995) 13, 20
(1996) 13, 18, 19, 27–8
Australian Energy Planning System Optimization Model (AEPSOM) 80, 81–7
areas for further research 85
changes in Australian energy policies 86
industrial development strategies 84–5
Parametric Programming Search (PPS) and 83–4
sustainable industrial developments and 86
usefulness of 85–6
usefulness of MLO approach 85–6
validation tests and policy implications 84
autonomous energy improvement (AEEI) 53, 54, 153

blitz strategy 252
bottom-up models 90–2, 141

C++ 18
carbon dioxide
resource pricing with an environmental constraint 145–7
specificity as a resource problem 144–5
specificity as an environmental problem 143–4
Clean Air Act Amendments (1990) (US) 19, 23, 25
Climate Challenge 41
Climate Change Action Plan (CCAP) 32
IDEAS model and 39–45
Climate-Wise 41
COAL2 261
Colombian energy system 262–70
energy policy issues 263–4
policy evaluations 267–8
systems simulation platform 264–8, *265*
energy demand modules 265–6
energy supply modules 266–7
Computable General Equilibrium (CGE) 3
conservation policy analysis models (CPAM) 243–6
changing role 253–5
energy service costs 247
figures of merit 246–8
electric rates 246–7
gaming version 255
illustrative use in the 1980s 251–3
key features for policy studies 248–51
avoiding double counting of conservation 250
information feedback 250–1
integrated representation of the electric system 248–9
rapid turnaround 251
utilities within the Northwest System 247–8
utility revenues 247
Corporate Average Fuel Economy (CAFE) standards 40
cost–benefit analysis 141, *142*

cost curves method 92–5

Data Resources, Inc (DRI) US Quarterly Model 19–20, 36
demand-side management 259
Downstream Challenge 207

Earth Summit (1992) 31
ECAP 2
ECAP-PC 306
EFOM (Energy Flow Optimization Model) 72, 90, 91
  computer resources and organizational requirements 97–8
  cost curves method 92–5
    assumptions for Danish study 93–4
    discounted energy systems costs 92, *93*
    scenario assumptions 93
  development 96–7
  dissemination of software 98–9
  econometric models and markets with few agents 107–9
    future market prices in Northern Europe 107–9
  electricity market with several agents 105–7
  explanation of results and model limitations 110–15
  international cooperation on modelling 115–16
  model quantification of market conditions 109–10
  modular structure *91*
  presentation graphs for several scenarios 110
  sectoral model for electricity 99–105, *99*
    Danish utilities in a competitive market 100
    sensitivity to economic parameters 102–5
    short-term marginal costs for electricity generating technologies 101–2
  software for 97
EFOM-ENV 96, 122, 127
EGEAS 2
electric utilities, American 273–300
electricity supply industry privatization, UK, modelling effects of 303–24
  modelling the competitive effects 312–20
    effects of market structure 316–18
    effects of risk and uncertainty 318–19
    strategic competition 320
  modelling the transfer of ownership effects 306–12
    debt implications 310–12
    rate of return implications 307–9
    tax implications 309–10, *310*
  new structure 303–4, *304*
  planning assumptions 322–4
energy–economy–environmental interactions, Turkish model for 119–37
energy–economy interactions 118
Energy Information Administration (EIA) (US) 9–10, 12
Energy Policy Act (1992) (US) 19, 259
ENPEP 264
ERB model 150, 152–3
  market shares of energy sources 156–7
ETA-MACRO 119
EURIO 97, 115
exhaustible resources, theory of 150

Federal Energy Administration 10
Fishbanks 198
FOSSIL1 model 33
FOSSIL2 model 33, 261
Framework Convention on Climate Change 31
free lunch option 141, 142

game theory approach 71, 100
GAMS (General Algebraic Modelling System) 97
GEMINI 36
general equilibrium model 70
GHG emission models
  comparison **148–9**, 153
  comparison of energy price paths 157–9
  comparison of main results 159–61
  market shares of energy sources 154–7
  strategic behaviour and incomplete competition on fossil fuel markets 161–2

Global 2100 150–1
  market shares of energy
    sources 154–5
GREEN 150, 151–2
  market shares of energy
    sources 155–6
Green Lights Program (EPA) 40, 41, 42
GRI model 36
gross refiners margin (GRM) 210

IEA-MARKAL model 69
IKARUS project 47
IKARUS-LP 48, 55, 57
Integrated Dynamic Energy Analysis Simulation (IDEAS) model *35*
  Climate Change Action Plan (CCAP) and 39–45
  history 32–4
  market-clearing in 36–7, *37*
  previous use for global warming policy analysis 38
  structure 34–8
integrated resource planning (IRP) 259
Intermediate Future Forecasting System (IFFS) 10

JOULE Energy Research Programme (EU) 96, 115
JOULE Framework Programme 96, 97

learning-by-doing 10, 24
least cost planning approach 93
LINPROG 98
Logit 265

MACRO 2
macroeconometric modelling 2
management flight simulators 198
MARKAL 2, 69, 71–5, 90
  Extended 72
MARKAL-MACRO 2, 70
Message 72
MICRO-MARKAL 70
Microworld's diagnostic lens 255
Microworld Creator 191
Midterm Energy Forecasting System 10
MINOS 125
MIS (Macroeconomic Information System) 47–65, *49*
  Autonomous Energy Efficiency Improvement (AEEI) factor 49, 54
  carbon tax calculations for Germany 54–9
    business-as-usual 55, 59–60, *60*
    carbon dioxide factors 58, **59**
    CES parameter 57
    cost/carbon tax 57
    energy prices 57, **58**
    key assumptions 55–6, *56*
    reduction case 55, 60–2, *61*
    sectoral AEEI 56–7, *56*
  calculated macroeconomic effects 59–62
  Constant Elasticity of Substitution (CES) production function 49, 53, *54*
  elasticity of substitution (ESUB) 57
  formulation 63–4
  IO approach 50–2, *51*, 63
    employment 52, 64
    energy 53
    housing 52
    investment 50, 63–4
    population 52
    price model 64
    transport 52–3
  overview 48–9
MIXLP program 307
Motor Challenge Program (DOE) 40, 42
MSG-EE 70
multi-level optimization (MLO) 79–87
  AEPSOM 80, 81–7
  areas for further research 85
  industrial development strategies 84–5
  justification and theoretical formulation 80–1
  Parametric Programming Search (PPS) and 83–4
  usefulness of 85–6
  validation tests and policy implications 84
multiperiod energy-technology models 118

National Energy Act (1978) (US) 33
National Energy Modeling System (NEMS) *14*, 260
  development of 11–18

National Energy Modeling System (NEMS) (*cont.*)
- alternative platforms and outreach 18
- content 12–13
- decisions on 12–18
- design reports and the independence expert review program 15–16
- documentation 16–17
- fast turnaround capability 17–18
- modularity 13–15
- National Academy of Sciences recommendations 11–12
- regionality 15

modules 18–27
- coal market 22
- commercial demand 26
- conversion 23–5
- demand 25–7
- electricity market 23–4
- geothermal–electric 23
- industrial demand 26–7
- international energy 19
- macroeconomic activity 19–20
- natural gas transmission and distribution 21–2
- oil and gas supply 20–1
- petroleum market 24–5
- renewable fuels model 22
- residential demand 25
- solar–electric 23
- supply 20–3
- transportation demand 27
- wind–electric 22

uses 27–8

National Energy Plan (US) 32, 33
National Energy Policy Act (1992) (US) 33
National Energy Strategy (US) 10, 33
Natural Gas Policy Act (1978) (US) 11
no programs strategy 252
no regret measures 94, 141
Non-Utility Generation Model (NUGM) 254
Nordisk energimarkedsmodell 70
Northern Europe electricity trade 67–76

oil market model
- diary of model building process 209–30
  - Ref_Proto_V0.0 213–15, *214*
  - Ref_Proto_V0.7 217–18, *217*
  - Ref_Proto_V0.8 218
  - Ref_Proto_V0.911 218
  - Ref_Proto_V1.1 219
  - Ref_Proto_V1.20 220, *220*, 222
  - Ref_Proto_V1.25 222
  - Ref_Proto_V2.00 223, *223*, *224–5*
  - Ref_Proto_V2.02 *228–9*, 227, 230
  - Ref_Proto_V2.01 226
- planning as learning initiative 205–8
  - specific project background 206–8
- reflections on the study 230–7
  - cybernetics thread 231
  - information processing elements 233–4
  - intuition creation 235–6
  - longer term consequences of the work 230–1
  - mental model access 235
  - non-conformist economics 232
  - servomechanisms thread 231
  - social negotiation 236–7
  - social process elements 234–5
  - soft systems methodology 231
  - technical elements 231–3
  - viable system model 231

oil producers' model
- development as a gaming simulator for training 188–91
  - design of pre-game briefing 188–91
  - design of user interface 191–5
  - Oil Producers' Microworld 188–99
- generating scenarios using 180–8
  - published scenarios 186–8
  - quota busting in a green world 184–6
  - ten-year supply squeeze followed by supply glut 180–4
- independents 172–4, *173*
- mapping the decision making logic 170
- model development process 168–70
- opportunists 175–6, *176*
- price and demand 177–8, *177*
- quota setting 174, *175*
- structural map of swing producer 178–80, *179*
- swing producer 170–2, *171*
- value of 195–9

Oil Vulnerability Study 33
optimization models 2
  modular structure 95–6, *96*
Organization of Petroleum Exporting Countries (OPEC) 13

Pacific Northwest Electric Planning and Conservation Act (1980) (US) 242
Pacific Northwest electric system (US) 241–55
Parametric Programming Search (PPS) 83–4
partial equilibrium model 70
People Express flight simulator 198, 255
Petroleum Administration for Defense Districts (PADDs) 15, 19
Project Independence Blueprint 10
Project Independence Evaluation System (PIES) 10
Public Utilities Regulatory Policy Act (PURPA) (1978) 33, 254

quasi dynamics 89

Reduced Instruction Set Computer (RISC) system 18
reduced-form modelling 17
reference energy system (RES) 68–9
  generic description *69*
Regional Macroeconomic Model (RMM) 20
Resource Policy Screening Model 254

Samkjøringsmodellen 70
scenario models 168
scenario planning 167, 168
SODA 237
spiral of impossibility 250
STELLA model 191, 213
STELLA/Ithink simulation language 178
strategic assumption surfacing and testing 236
Strategic Choice approach 236
Strategy Modeling System—Utility 274
  business-as-usual earnings model 287
  feedback loops 283–6, *285*
  maintaining shareholder value 299
  methodology 277–81
    base case 279, *280*
    cause–effect relationships 277
    feedback loop 278, 279
    negative feedback loop 278, *279*
    nonlinear response 278, *278*
    past history 279, *280*
    time delays 277–8, *278*
  model architecture 275–7, *276*
  model dynamics 281–6, *281–3*
  retail competition threatens shareholder value 287–98, *290–1*
    downsizing and/or consolidating via merger or acquisition 293–6, *294*
    investing for the future 296–8
    regulatory bail-out 292–3
  setting up and calibrating 286–7
  traditional earnings model 287
Synfuels Task Force 33
systems dynamics (SD) simulation models 205–38, 242
  approach to energy analysis 261–2

technological optimism 10
top-down models 90–2, 141
traditional earnings model 287, *288*

Upstream Model project 206
video game syndrome 188

World Energy Outlook 28

Index compiled by Annette Musker